Augustus George Vernon-Harcourt, Henry George Madan

Exercises in Practical Chemistry

Vol. 1: Elementary Exercises

Augustus George Vernon-Harcourt, Henry George Madan

Exercises in Practical Chemistry
Vol. 1: Elementary Exercises

ISBN/EAN: 9783337275846

Printed in Europe, USA, Canada, Australia, Japan

Cover: Foto ©berggeist007 / pixelio.de

More available books at **www.hansebooks.com**

Clarendon Press Series

EXERCISES

IN

PRACTICAL CHEMISTRY

VOL. I

Elementary Exercises

BY

A. G. VERNON HARCOURT, M.A., F.R.S., Sec. C.S.

Senior Student of Christ Church and Lee's Reader in Chemistry

AND

H. G. MADAN, M.A., F.C.S.

Fellow of Queen's College, Oxford

LONGUM ITER EST PER PRAECEPTA, BREVE ET EFFICAX PER EXEMPLA.
Seneca, Lib. I. Ep. VI.

Third Edition

REVISED BY

H. G. MADAN, M.A., F.C.S.

Oxford

AT THE CLARENDON PRESS

M DCCC LXXX

PREFACE.

THE object of this volume is to furnish a systematic course of study to those who are beginning to learn Chemistry practically. The course commences with a series of exercises upon the preparation and properties of some of the most familiar substances, in which precise directions are given as to the apparatus and materials to be used, and as to the manner in which each experiment is to be made. In the earlier exercises, especially, the directions given extend to minute details; the aim of the authors being to provide, as far as possible, all the guidance that a beginner, working by himself or with only occasional supervision, may require. A few exercises have been introduced on parts of manipulation which, from their importance, it seemed desirable to treat separately; but, as a rule, each operation is described where the occasion for its use first arises: and the subjects have been so chosen as to furnish examples of all the usual chemical operations. The attempt has been made to arrange each exercise so as to provide the student with continuous occupation, and economise his time by giving two or three operations which can be carried on simultaneously and to which he should direct his attention in turn. The time required for the completion of different exercises cannot but vary considerably, since some operations, too important not to be included, do not admit of being hastened; but in most cases about two hours will be found sufficient.

The list of apparatus given at the outset shows what it is desirable that a student should have for the performance of the exercises; but those who have not access to a chemical laboratory, and who are unwilling to incur the expense of providing themselves with a complete set of apparatus from a chemical dealer, will find in the Appendix a number of suggestions for the construction of pieces of apparatus which can be made very cheaply.

The authors have endeavoured, by repeating nearly every experiment in exact conformity with their written description, to insure that the student who carefully follows their directions shall command success; but as they cannot but fear that many errors and omissions may, nevertheless, have escaped them, they will be grateful for any suggestions from those engaged in teaching who may make use of their book. It is only by a large experience of the errors into which beginners are liable to fall, that the many ways of going wrong can be discovered and stopped.

The question of chemical nomenclature is at present in such a condition that every lecturer or writer must choose for himself the names he considers least objectionable. The names assigned to substances in this volume are, with a few exceptions (made with a view to consistency), the same as are used in Roscoe's 'Lessons in Chemistry,' and Watts' edition of 'Fownes' Manual.'

The illustrations in the text have, with very few exceptions, been drawn directly on the wood from photographs, taken by one of the authors, of the apparatus actually used.

PREFACE TO THE THIRD EDITION.

In preparing the present edition, the Editor has, in the first place, to regret the loss of the co-operation of Mr. Harcourt, whose care, accuracy, and discrimination gave to the former editions a value which the present one cannot justly claim to possess.

Some more or less important alterations will be noticed in the present issue : chiefly of the following kinds,—

1. The succession of the Exercises has been to a certain extent re-arranged, so as to make them follow more nearly in the order in which a beginner would probably study the radicles in the course of his reading.

2. A few additional Exercises, such as those on Weighing and Measuring, and on Chemical Action, and a rather large number of additional experiments have been introduced; such experiments alone being selected as can be made without much risk, and with the simplest apparatus [1].

3. Short headings have been placed before most of the experiments, in order that the student may appreciate more clearly what the experiment is intended to illustrate, before he performs it.

Practical Chemistry seems in danger of being made far too much a study of a few reactions of salts, got up for the purpose of detecting them in the course of an analysis. This is, of course, due to the requirements of examiners, to satisfy which

---

[1] Suggestions for an advanced course of experimental work, which requires experience and more elaborate apparatus, will find a more appropriate place in the second volume.

nearly all the very moderate time available for practical instruction in schools must at the present day be spent. Moreover analytical work (in the narrow, technical sense) entails, like Latin verses, less trouble to the teacher and less risk to the pupil than other kinds of practical work; while it undoubtedly affords, when intelligently pursued, a very excellent training in the application of logical methods.

But it may well be doubted whether a more real and valuable advance in a scientific education is not made by the careful preparation and examination of the properties of such a substance as oxygen, or by an exact study of a few examples of oxidation and reduction, than by simply observing, for instance, that chlorides give a white precipitate with silver nitrate which is soluble in ammonia.

It is hoped that this book may do something towards tracing, for those at any rate who have time and enthusiasm for the work, an outline of a rather wider range of study.

H. G. MADAN.

Eton, *September*, 1880.

TABLE OF CONTENTS.

PART I.

Experiments on the Preparation and Properties of Substances.

SECTION I.

Preliminary Exercises.

SECTION II.

Preparation and Examination of Non-metallic Radicles and their Compounds.

SECTION III.

Preparation and Examination of Metallic Radicles and their Compounds.

Group I.

Group II.

Group III.

PART II.

Qualitative Analysis of Single Substances.

SECTION I.

SECTION II.

SECTION III.

SECTION IV.

SECTION V.

SECTION VI.

CORRIGENDA.

Page 201, line 3, *for* 80 *read* 160.
  ,, 262, ,, 8, *for* nitrate *read* chloride.

Memoranda.

1. BE orderly and neat in manipulation. Cleanliness stands at the head of the chemist's scale of virtues. All messes must be cleared away with the zeal of a sanitary inspector.

Never go to work, or continue to work, with the table covered with a litter of bottles, flasks, basins, and test-tubes; but replace each bottle on its shelf as soon as you have done with it, and have a basin at hand in which to put dirty test-tubes, &c.

Always wash your test-tubes twice : once, before they are put away; and again, with distilled water, immediately before they are used. Probably more puzzling reactions occur from the use of dirty test-tubes than from any other cause.

2. Do not begin work in a hurry. What is expended in time is very often gained in power, in grasp of a subject. Yet, on the other hand, learn to be economical of time. Several filtrations and evaporations, for instance, may be going on at once. The chemist may sometimes, in spite of the proverb, do more than one thing at a time, by allowing things to do themselves.

3. Be economical of materials. In analysing a substance, do not (without the strongest reasons) use up at once the whole quantity at your disposal. Reserve at least one-fourth of it in a corked tube or covered watch-glass, in case unforeseen accidents should occur and the other portion should be lost. In making a gas, the residue left in the generating vessel will often be of use, at any rate interesting as a specimen. It should not,.as a rule, be thrown away, but purified by recrystallisation or otherwise.

4. Never begin an experiment until you have looked over all the preparations for it, to make sure that you have everything that is necessary within reach. You will not then have the mortification of seeing the half-performed experiment fail for want of some requisite which cannot be procured at the moment.

5. Do not think merely of what will *do,* but what is *best,* of the means at your disposal.

6. Never add one chemical substance to another without considering for what purpose you add it, and what various effects may be produced.

7. Be exact and methodical. Let nothing pass unnoticed, although you may not see its significance at the moment. Make written notes of everything that you do, analyses of lectures, sketches of apparatus. Whatever is worth doing is worth recording.

8. Do not attempt to devise a modification of an experiment until you have tried it in exact accordance with the directions given. Then, and then only, if you fail, you will find it possible to blame the book and not yourself.

9. Do not expose yourself needlessly to vapours which you know to be injurious, e.g. chlorine, hydrogen sulphide, hydrogen arsenide. Remember that the bad effects may not be perceptible immediately.

10. Finally, do not look upon Chemistry as a mere amusement, as a means of getting up a few explosions, creating a few unsavoury smells, producing a few striking changes of colour. Chemistry is worthy of better treatment; it is no longer a 'black art,' but a refined science, and should be thoughtfully and reverently studied.

Nor, again, give up hopes of making discoveries in the science because the land appears to be already highly farmed, and you have not all the refined apparatus which the optician and operative chemist can supply. Records of close and accurate observations of some of the (apparently) simplest phenomena of Chemistry are much needed; and such it is in the power of every student to contribute.

Ad Chemeiam.

Narem illaudatis vehementer odoribus angis,
  Aurem terrifico percutis usque sono.
Quippe oculum, Chemeia, magistra haud alma, vel ipsum
  Corripis, et digitos igne petisque caput.
Quid manet? una fides manet inviolata tuorum
  Semper, et ut crescunt vulnera crescit amor.

H. G. M.

LIST OF APPARATUS

*Required for the Course of Practical Work contained in
Parts I and II of this Book* [1].

[The abbreviation cm. stands for centimetre.
„ „ mm. „ millimetre.
„ „ c.c. „ cubic centimetre.
„ „ grm. „ gramme.
For tables of weights and measures see Appendix E.]

1. A Common Metre (or **Half-metre**) **Rule**; the first decimetre divided into millimetres, the rest into centimetres.

2. A Cylindrical Graduated Glass measure; to contain 200 c.c.; graduated into spaces of 5 c.c.

3. A Pair of Scales, with beam about 20 or 25 cm. long, sensitive to a weight of 1 centigramme. The ordinary grocers' scales, if well made, are sufficiently good.

4. A Set of Weights, from 100 grms. to 1 decigramme. These need not be of the highest accuracy: such sets, including 100, 50, 20, 10, 10, 5, 2, 2, 1, ·5, ·2, ·2, ·1 grm., mounted on a wooden stand, are sold in France for a price equivalent to less than half-a-crown, and can be procured in England at a reasonable cost [2].

[1] In making this List it has been thought right to err rather on the side of completeness than of deficiency. Some pieces of apparatus, e. g. a Bunsen's holder, are not absolutely necessary; and in several cases the student will be able, by the aid of a few tools and a little ingenuity, to make substitutes for himself. Some suggestions for the construction of economical apparatus will be found at the end of the book, but no direct reference to such substitutes is made in the text, since the student, who is ingenious enough to make a piece of apparatus for himself, will be at no loss to bring it in where it is wanted.

[2] Thus, in Paris, they can be obtained from M. Carette, 12 Rue du Château d'Eau; and in England from Mr. Solomon, Red Lion Square.

5. A Pneumatic Trough, about 36 cm. long, 24 cm. wide, and 16 cm. deep, Fig. 1. This is an apparatus for collecting and experimenting upon gases, and consists of a cistern for holding water or any other fluid, furnished with a

Fig. 1.

movable shelf placed across it about 5 or 6 cm. below the top. In the centre of the shelf is a hole, which should not be less than 2 cm. in diameter, and immediately underneath this hole is soldered a broad, very shallow funnel, the mouth of which is of the same width as the shelf : the accompanying figure represents a section of the shelf at this point [1]. This funnel serves to catch bubbles of gas, in case the delivery tube is not exactly under the centre of the hole, and to direct them upwards through the hole into a jar placed above it. In addition to these essentials, the trough represented in the figure has a ledge running along the whole length of one side on which jars may stand when filled with gas, and

---

[1] A long narrow opening, extending nearly the whole length of that part of the shelf which is unsupported, is more convenient than a round hole. Its width need not exceed 1 cm., and on each side of it should be a strip of metal, soldered to the under surface of the shelf, serving to strengthen the shelf and to guide bubbles of gas to the opening. The figure above represents a section of this part of the shelf.

also an overflow-pipe *a* at one corner, the mouth of which is about 3 cm. above the level of the shelf, to convey any excess of water into the supplementary trough *b*. These troughs are made of japanned tin, and it is a decided advantage to have the inside japanned white, as a much better

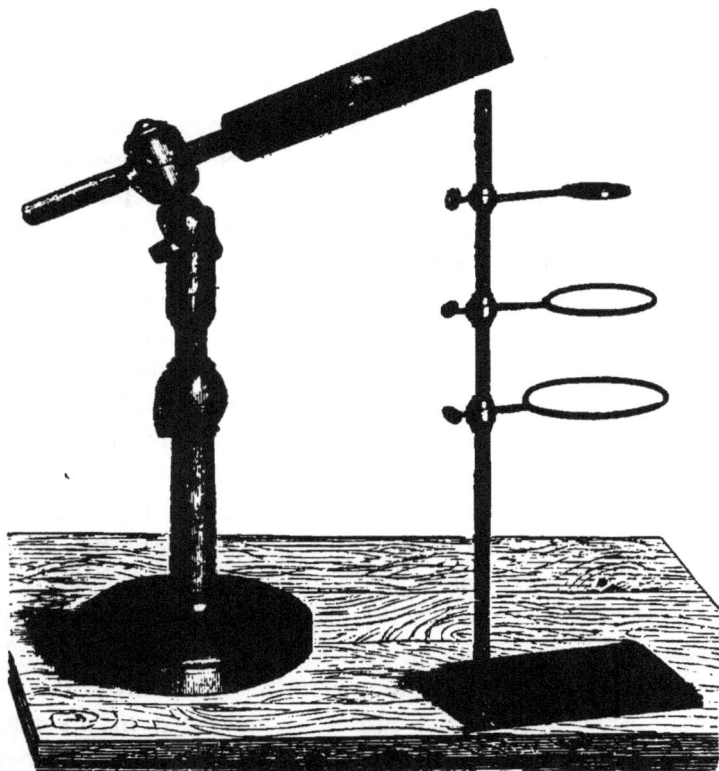

Fig. 3.   Fig. 2.

view of the position of tubes, jars, &c. when immersed in the water, is thus obtained. For the method of using the Pneumatic Trough, see p. 79.

6. A Retort Stand, Fig. 2, with rectángular iron foot, iron rod about 36 cm. in height, and three brass rings.

7. A Bunsen's Universal Holder, Fig. 3, of stained beechwood, the smaller of the two sizes which are sold. This, although rather expensive (about 5*s*.), will be found a most useful piece of apparatus, both for chemical and physical experiments. Its construction will be sufficiently evident from the engraving. Care must be taken not to use undue force in tightening the screws, or the threads may be torn away. If the screws work stiffly, a little black-lead, in preference to tallow, should be applied to them.

8. A Bunsen's Burner, Fig. 4, small size, with rose top.

This burner consists of a brass tube, about 1 cm. diameter, having several holes close to its lower end, which, in the best form of burner, can be closed when desired by a revolving cap. Gas is introduced through a small flattened central jet, the orifice of which is just above the holes. Thus the gas is allowed to mix freely with air which enters through the holes ; the supply of air being increased by the upward rush of the gas, in the same way as the draught in the chimney of a locomotive is increased by the blast of escaping steam.

This mixture of gas and air burns, when a light is held at the top of the tube, with a blue, non-luminous flame, very different in appearance to the usual bright flame of coal-gas.

The theory of the burner is this : — Coal-gas consists mainly of hydrogen and compounds of hydrogen with carbon. Both these elements unite with oxygen, producing heat in doing so : but oxygen has more affinity for hydrogen than it has for carbon, and hence, when insufficient air is supplied, the oxygen in it combines with all the hydrogen, but with only part of the carbon. The rest of the carbon is separated in the flame, and its particles heated to whiteness give the brilliancy to the ordinary gas-flame. The presence of these particles of unconsumed carbon in the flame may be proved by holding a white plate in it for a moment ; the particles are thus cooled down and deposited as soot.

If more air is supplied, so as to afford sufficient oxygen to combine with all the· carbon as well as the hydrogen, none of the former element is separated, and the flame loses its brightness but gains in temperature, since more chemical combination takes place. Thus we have the ordinary bluish flame of the Bunsen's burner, which deposits no soot upon anything held in it, while its temperature is sufficiently high to melt readily a piece of copper wire held in it [1].

One or two precautions should be observed in using this burner.

1. The supply of air must not be too great.

If much more than half the total volume of air required to burn the gas completely, is admitted through the holes, the mixture becomes too explosive, and the flame passes

Fig. 4. Fig. 5.

down the tube, continuing to burn at the jet below, with a small smoky flame, the heat of which is mainly expended on the tube itself. In such a case, the gas must be entirely turned off, and kindled afresh. If the flame is still unsteady and shews a tendency to recede down the tube, owing to a low pressure of gas in the main, the air-holes should be partially closed by turning the cap (or if there is no cap, by plugging up one or more of them with bits of cork).

In fact, it is always best to cut off the supply of air

---

[1] The actual temperature of the flame is estimated at more than 2000°, but unless it is enclosed in a furnace, a large amount is lost by radiation.

altogether before lighting the burner, and then to admit just sufficient air to destroy the luminosity of the flame. In this way even a small flame, 2 or 3 cm. high, can be safely obtained.

2. The supply of gas must not be too great, and it must be thoroughly mixed with the air before reaching the top of the tube. This latter condition is effected by the peculiar shape of the jet.

If the flame is white even though the full supply of air is admitted through the holes, the gas should be partly turned off. If the flame still shows luminosity, there is reason to suspect that the jet is out of order, or that the air-holes are clogged by dust or some fused substance which has dropped down the tube. The main tube should be unscrewed and the jet and air-holes examined and cleaned.

3. The hottest part of the flame is a little below the top and near the border. This can be proved by holding a piece of platinum foil in various parts of the flame and observing where it glows most brightly.

4. The interior of the flame has a strong 'reducing' action (for an explanation of this term see the exercise on the Use of the Blowpipe). This should be borne in mind when dealing with glass containing lead, which will be blackened, owing to the separation of lead, if held in the middle of the flame.

The cast-iron cap shown in the figure, when placed upon the burner, divides the single flame into a ring of small jets which distribute the heat over a large surface, and are well adapted for heating an evaporating basin or sand-bath.

9. A Cast-iron Stand, Fig. 5, with screw, on which may be fitted the blowpipe-jet *a*, the Argand burner *b*, and the fish-tail burner *c*. The blowpipe-jet is described in the exercise on the Use of the Blowpipe. The Argand burner, although not absolutely necessary, is extremely convenient for applying a gentle heat to a flask (e. g. in making oxygen gas), since it is more under control than the Bunsen's burner, and may be regulated to give the least possible flame. The fish-tail burner is chiefly intended for bending tubes, p. 31.

10. A Set of Five Wooden Blocks, about 12 cm. square, and respectively 2, 5, 7, 10, 12 cm. in thickness.

11. A Test-tube Stand, of the usual form, for supporting test-tubes while in use. It should have twelve holes in one row, and a strip of slate should be fitted in front of the holes, on which may be written the contents of each test-tube when it is placed in the stand.

12. A small Dish, of tinned iron or copper, about 12 cm. in diameter, for use as a sand-bath.

13. Two pieces of fine Iron-wire Gauze, about 12 cm. square.

14. A Mouth Blow-pipe, Fig. 6, about 20 cm. in length.

Fig. 6.

15. A small Ladle, with bowl about 7 or 8 cm. in diameter, and an **Iron Spoon,** or **Capsule,** about 4 or 5 cm. in diameter, for ignitions.

16. A pair of Crucible Tongs, Fig. 7, about 20 cm. in length.

Fig. 7.

17. A Set of four Cork-borers, from 3 mm. in diameter upwards. These are short pieces of thin brass tube, sharpened at one end, and having a thick collar soldered to the other end, to afford a better hold. For the method of using them, see p. 37 [1].

---

[1] When the cutting edge becomes blunt or bent, it should be sharpened on a hone, or by a very fine file, the borer being constantly rotated while the hone or file is passing over it.

18. A piece of Platinum Foil, about 2 cm. × 5 cm. It should weigh about 0.4 grm. This is used chiefly as a support for substances on which we wish to try the effect of a high temperature, in order to test their fusibility, volatility, &c. If the edges of the foil are turned up round a spherical mould, such as the end of a pestle, the foil being laid on the palm of the hand, and the pestle pressed forcibly upon it, we obtain a very convenient capsule for fusions on a small scale, e. g. for the decomposition of barium sulphate by sodium carbonate.

It should, however, be a rule—1st. Never to use a platinum vessel when a piece of porcelain will do as well. A bit of a broken evaporating dish will serve for almost every purpose, except when silicon, aluminium, or the alkali metals are to be tested for. 2nd. Never to heat in a platinum vessel the following substances :—

Substances evolving chlorine or sulphur.

Caustic alkalies or barium hydrate.

Cyanides, chlorates, nitrates.

Easily reducible metallic salts, or their corresponding metals, e.g. lead, silver, tin.

When a platinum vessel is dirty, try first to clean it by boiling it in a dish with a little strong hydrogen chloride. If this has no effect, spread over the surface some powdered potassium-and-hydrogen-sulphate, and heat it over a Bunsen's burner until the salt fuses, inclining the vessel so that the liquid salt may flow over every part of it. Finally, boil it with water in a dish.

When a piece of platinum foil becomes creased or wrinkled, place it between folds of glazed writing-paper on a smooth surface, such as a plate of glass, and pass over it with strong pressure a rounded burnisher, such as the handle of a paper-knife.

19. A piece of Platinum Wire, about 25 cm. long and 0.32 mm. in diameter (No. 26 brass-wire-gauge). This is chiefly for use in blowpipe experiments.   •

20. Two pieces of Brass or Copper Wire, about 30 cm.

long, and 1 mm. in diameter. These should be filed to a point
at one end and then bent, the one into the form of Fig. 8,
the other into the form of Fig. 9, and finally a small piece
of wax taper about 3 cm. in length should be stuck upon the

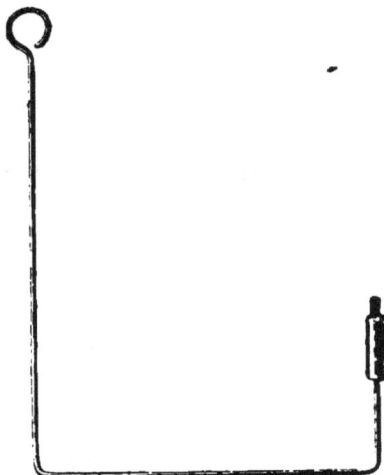

Fig. 8.   Fig. 9.

pointed end of each. One wire may be made to serve the
purpose of the two pieces, if bent into this shape ⌐∟ and a piece
of wax taper stuck on each end.

21. A Deflagrating Cup, see Fig. 10. This consists of
a brass or iron bowl about 1.5 cm. in diameter, screwed to
a piece of stout iron-wire which passes rather stiffly through
a cork stuffing-box attached to a tin flange. It is intended
to hold substances which are to be burnt in gases.

22. A Deflagrating Jar, Fig. 10. This is a wide-mouthed
stoppered jar, open at the bottom, about 28 cm. in height and
15 cm. in diameter. The top should be ground flat, in order
that it may be accurately closed by a glass plate.

23. Two strong cylindrical Glass Jars, for collecting
gases; 10 cm. in height, 3 cm. in diameter, with ground mouths.
They should be made of thick glass, since they are used for
holding mixtures of hydrogen and oxygen gases which are to
be exploded.

24. Three similar Jars, which may be of thinner glass, 20 cm. in height, 5 cm. in diameter.

25. Two circular Glass Discs, 5 cm. in diameter, ground on one side.

26. Two Ditto, 8 cm. in diameter.

27. One shallow Stoneware Tray, for holding gas jars, 8 cm. in diameter.

28. One Ditto, 18 cm. in diameter (see Fig. 10). This, though convenient, is not necessary, as a common dinner-plate may be substituted for it.

29. Four Florence Flasks. These may be procured from any oilman, and should be selected of uniform thickness, free from air-bubbles, and with even mouths not chipped away at one side. In order to cleanse them, put a few lumps of common ' washing soda' into each and heat it gently over a lamp, turning it round so as to bring the salt into contact with every part; finally, rinse it thoroughly, first with common water and then with distilled water, and place it to drain mouth downwards in the ring of a retort-stand.

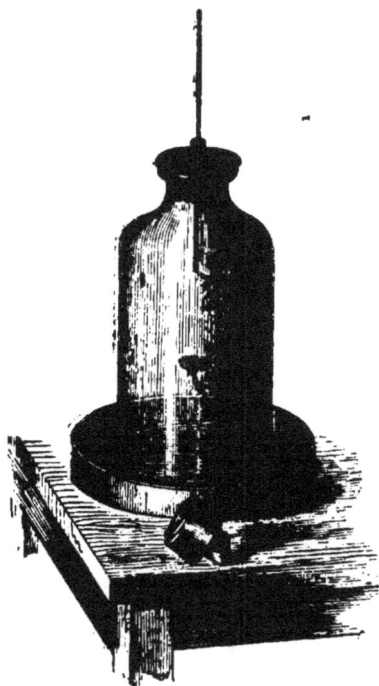

Fig. 10.

30. Three Flasks with flat bottoms, holding respectively 200 C.C., 250 C.C., 400 C.C. [1]

31. One plain Retort, holding about 200 c.c.

---

[1] It will be well to have 2 or 3 more of these, at any rate of the 250 cc. size, in case of breakage.

32. One Stoppered Ditto, of the same size.

33. Two wide-mouthed Stoppered Bottles, of white glass, holding about 700 c.c., for use in experiments on gases.

34. Six Ditto, holding about 200 c.c.[1]

35. Two common corked Bottles, with moderately wide mouths, holding about 200 c.c., for use as washing bottles for gases.

36. Two Ditto, holding about 300 c.c., for containing water which is to be saturated with a gas, such as chlorine, sulphur dioxide, or hydrogen sulphide.

37. One Washing Bottle with tubes (Fig. 11), holding 600 c.c. This is of great use for washing precipitates on a filter, and also for containing a supply of distilled water for general purposes in analysis. Its construction is sufficiently plain from the engraving[2]. When air is blown from the mouth into the upturned tube, a stream of water is forced through the jet at the extremity of

Fig. 11.

the other tube, and may be directed upon a filter, or into a test-tube. If a larger quantity of water is required at once, as in filling an evaporating dish or small flask, the bottle should be inverted so as to bring the blowing-tube lowermost, from which a stream of water will flow while air enters through the other tube.

[1] It is a great advantage to have the stoppers of these gas bottles made much more conical than usual. They are then far less liable to become fixed in their places if the volume of gas in the bottle should contract.

[2] This washing bottle may be easily fitted up by the student himself. Instructions for doing this are given at p. 35.

38. Two Thistle Funnels, *a*, Fig. 12, about 32 cm. in length.

39. Three Glass Funnels, respectively 4 cm., 7 cm., 10 cm. in diameter.

40. Three Glass Beakers, respectively 4 cm., 5 cm., 6 cm. in diameter.

41. Twenty-four Test Tubes, of the following sizes :—

Twelve, 15 cm. in length, 1 cm. in diameter.
Ten   18 cm.   „   1·5 cm.   „
Two   20 cm.   „   2·5 cm.   „

42. Six Watch Glasses, 5 cm. in diameter.

43. One Glass Spirit Lamp.

44. One Glass Mortar, 6 cm. in diameter, with pestle.

Fig. 12.

45. One Porcelain Mortar, 10 cm. in diameter, with pestle[1].

46. Three Porcelain Evaporating Basins, respectively 6 cm., 9 cm., 12 cm. in diameter[2].

47. One Porcelain Crucible, with cover, about 3 cm. in diameter.

48. Two common Cornish Crucibles, respectively 6 cm. and 8 cm. in diameter.

49. One Reduction Tube, *b*, Fig. 12, about 1 cm. in diameter.

50. Two Drying Tubes, *c*, Fig. 12, about 18 cm. in length.

---

[1] This mortar should not be glazed inside.

[2] These should be thin in substance (that there may be less risk of their cracking when heated over a lamp), and highly glazed both inside and outside. The Meissen ware is much the best, both as to shape and quality.

This form of tube is intended to contain calcium chloride, or other hygroscopic substance, in small fragments, for the purpose of removing moisture from gases which are passed through the tube. It is filled in the following way. After removal of the cork, a small tuft of cotton-wool or tow is pushed down into the bulb by means of a glass rod, until it lies across and protects the opening of the narrow tube. The rest of the bulb and the wide tube is then nearly filled with fragments of thoroughly dry calcium chloride, about as large as split peas. Another piece of cotton wool is then lightly pushed in, to keep the calcium chloride in its place, and finally the cork with its short tube is replaced. It is advisable, when the tube is not in use, to keep the ends stopped by little plugs of cork, in order that the moisture of the air may not find entrance.

The calcium chloride for these tubes should not be fused, but only thoroughly dried at a temperature of 200°–300° on a sand-bath. It is in this condition much more porous, and exposes a larger surface to the gas than the fused substance.

51. Twelve 'Ignition Tubes,' Fig. 13, about 6 cm. in length. For directions for making such tubes, see p. 41.

Fig. 13.

52. Two kilogrammes of readily fusible Glass Tubing, free from lead, of different sizes, but chiefly about 6 mm. in external diameter [1].

53. Half a kilogramme of difficultly fusible Glass Tubing, 5 mm. in external diameter, for making 'ignition tubes' (No. 51) and arsenic tubes.

54. Two or three pieces of Combustion Tubing, about 30–35 cm. in length, and 12 or 14 mm. in diameter.

55. Three or four pieces of Glass Rod, free from lead,

[1] The French soda glass is usually very good; but some specimens show a great tendency to devitrify when heated.

about 50 cm. in length, and 4 mm. in diameter: for stirring-rods.

56. A piece of vulcanized India-rubber Tubing[1], about 1 metre in length, and 4 mm. in internal diameter: chiefly for use in connecting glass tubes; for which purpose pieces about 2 or 3 cm. in length may be cut from it, as required.

57. Two or three pieces of similar Tubing, about 60 or 70 cm. in length, and 6 mm. in internal diameter: chiefly for connecting lamps with the gas supply.

58. Three packets of circular Filters, respectively 7 cm., 14 cm., 20 cm. in diameter, suited to each size of funnel (No. 39).

59. One box of Test Papers, containing books of blue litmus, reddened litmus, and turmeric paper.

[Blue litmus is turned red by acids,
Reddened litmus is turned blue by alkalis,
Turmeric is turned brownish red by alkalis.

Their use may be illustrated by laying strips of each side by side on a white plate, and putting on them a drop of (1) dilute hydrogen sulphate (sulphuric acid), (2) solution of potassium hydrate (caustic potash).]

60. Two Brushes for cleaning tubes, one about 3 cm. in diameter, for test-tubes; the other about 5 mm. in diameter for smaller tubes.

61. A light hammer, of the form known as riveting-hammer.

62. A small Anvil, about 6 cm. square, and 2 cm. in thickness.

63. A pair of cutting Pliers.

64. A 'three-square' File, about 12 cm. in length.

65. A round (or 'rat-tail') File, about 20 cm. in length These files should be fitted into handles.

---

[1] Tubing of non-vulcanized india-rubber, which is also manufactured, adheres more closely to glass than the vulcanized tube, and is in many respects preferable to the latter. It has the disadvantage of losing its elasticity in cold weather, but after being warmed and stretched a little, it regains all its good qualities.

66. A Platinum Spatula, about 9 cm. in length, broader at one end than the other. It need not cost more than 6*s.* or 7*s.*, and will be found most useful. Instead of it, an

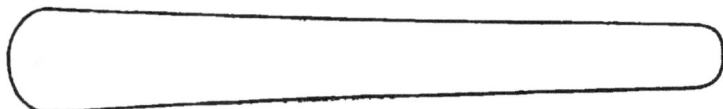

Fig. 14.

aluminium or bone spatula of the same size may be obtained at a much lower price, and will answer for most purposes.

67. A few pieces of Charcoal, for blowpipe experiments. These may generally be selected from the ordinary rough beech-wood charcoal. Sticks about 3 cm. in diameter, free from knots, should be picked out and sawn across the grain into pieces about 3 cm. in length, or rather less. When thus cut, they should show a surface free from cracks and of close, sound texture.

68. A common China Jug, holding about 2 litres.

69. Half a quire of White Blotting-paper, some sheets of glazed writing-paper, and a few cards.

70. Two or three dozen good Corks of various sizes, from 1 to 6 cm. in diameter. Those should be selected which are free from fissures and cavities, and in which the grain runs *across* not *along* the cone.

71. Two or three Dusters.

The following pieces of apparatus are also extremely useful, and access to most of them will be assumed in the exercises :—

72. A Herapath's Gas Blowpipe, provided with a pair of double bellows, or with one of the small French india-rubber blowing machines, Fig. 15. In this form of blowpipe the gas issues from a brass tube about 1 cm. in diameter, in the axis of which a smaller tube is permanently fixed, through which a blast of air is directed into the centre of the gas-

flame.   By attaching a larger or smaller nozzle to the air-tube
and altering the quantity of gas, any kind of flame may be
obtained, from a large brush-like flame 16 or 18 cm. in length,
to a small pointed cone of flame, such as is required for analytical experiments.

The india-rubber blowing machine alluded to, consists of

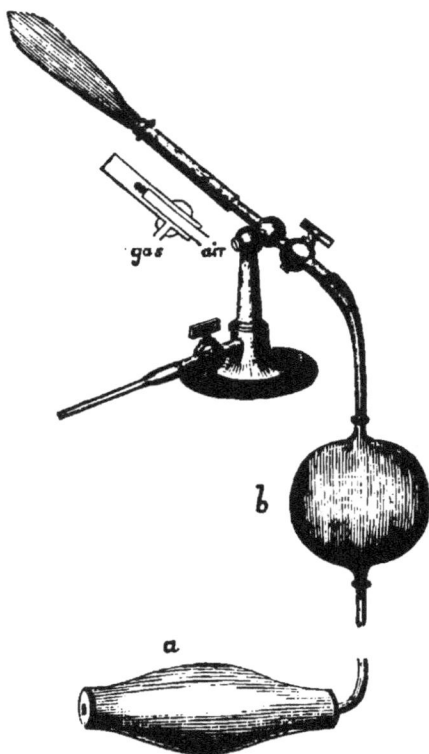

Fig. 15.

two parts: 1st, The blower, *a*—a pear-shaped vessel of strong
vulcanized india-rubber, having a valve fitted at each end
enclosed in a small wooden box; 2nd, The regulator, *b*—
a spherical vessel of thinner india-rubber, with two necks,
one of which is connected by an india-rubber tube with the
valve-box at one end of the blower, while the other neck is
similarly connected with the blowpipe air-jet.   When the

blower is placed on the floor and compressed with the foot, the valve at one end closes, and the air contained in the vessel is forced through the other valve into the regulator, which becomes distended and forces the air through the blowpipe-jet. When the blower is relieved from the pressure of the foot, it recovers its shape, the valve nearest the regulator closes and prevents the return of the air, while a fresh supply of air enters through the other valve. When full, the blower is again compressed with the foot so as to force another supply of air into the regulator.

This blowing machine is much cheaper and more portable than any form of double bellows, and is very effective. The regulator is usually made too small, and thus there is a slight variation in the strength of the blast. The india-rubber must be of the best quality; and it is advisable to enclose the regulator in a net, to prevent its becoming so far distended as to burst.

Fletcher's blower, and a substitute, which can be made without much difficulty, will be described at the end of the book.

Although such a blowpipe as that which has just been described is necessary for some operations in glass blowing, and renders the chemist almost independent of a furnace for fusions on a small scale, yet much may be done by the use of the mouth blowpipe supported in a Bunsen's holder (so that both hands may be free) and directed upon a larger gas-flame than usual.

A very convenient form of Herapath's blowpipe is now made, which can be fitted on a Bunsen's burner. The air-tube is connected with a piece of india-rubber tube ending in a mouthpiece. No bellows are required, as sufficient air can be supplied by the mouth.

73. A Glass-blower's Lamp: when gas is not available. This in its simplest form consists of a flat tin dish, near the centre of which a tin wick-holder is fixed in a slanting direction, sufficiently large to hold a bundle of strands of lamp-cotton about 1 cm. in breadth and 3 cm. in length. The dish is filled with lamp-oil, or melted tallow, and the wick is trimmed, a furrow being formed along the middle of it, in a line with which, and a little above it, the blowpipe-jet is placed.

74. A Clark's Retort and Receiver, Fig. 16; useful for distillations on a small scale.

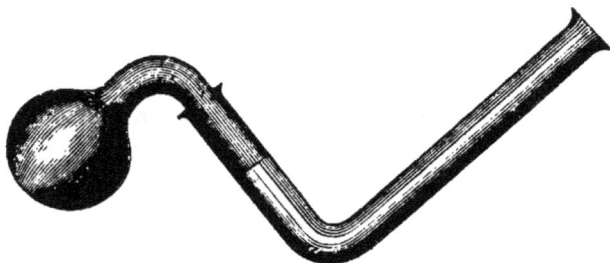

Fig. 16.

75. A Pipette, Fig. 17, for measuring out small quantities of liquid. It consists of a tube drawn out at one end into a jet, and slightly contracted at the other end, so as to be readily closed by the finger. It should hold about 20 c.c. and should be graduated into spaces of 0.2 c.c. To fill it, dip the jet rather deeply into the liquid, and suck the latter up into the tube by applying the mouth to the upper end. When the pipette is nearly full, remove the mouth and immediately press the forefinger (slightly moistened) firmly upon the upper end (see the fig.). Now raise the pipette until the uppermost graduations are on a level with the eye, keeping the jet lightly pressed against the side of the vessel of liquid, but clear of the liquid

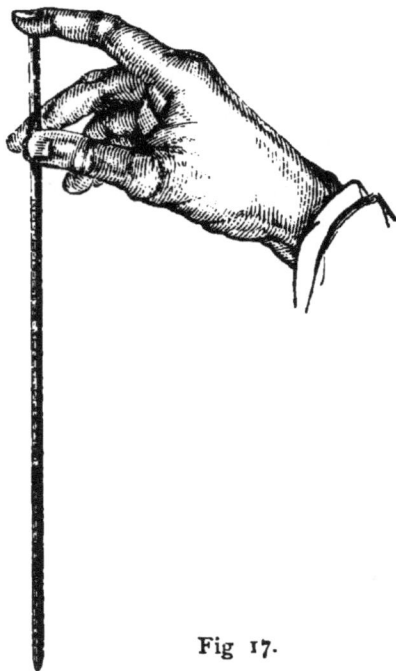

Fig 17.

itself. If the pressure is slightly relaxed, and the finger moved a little sideways, air will enter, and the column of liquid should be allowed slowly to fall until the lowest part of the curved surface of the liquid just touches the top line, when the pressure of the finger must be at once restored, to prevent any more air entering. The pipette may now be steadily removed to the beaker or flask into which the liquid is to be measured, and the desired quantity allowed to escape from the jet, which should be, as before, lightly pressed against the side of the vessel. Be careful always to read from the same point, the lowest part of the curve formed by the surface of the liquid, held at the level of the eye.

76. A Thermometer, with cylindrical bulb, graduated on the stem from $-10°$ to $+150°$ centigrade.

77. A Platinum Capsule : hemispherical, about 3 cm. in diameter. This, which would cost about 5s., will answer almost every purpose of a platinum crucible.

78. A Bunsen's Screw Pinch-cock, Fig. 18. This is a contrivance for regulating the flow of a liquid through an india-rubber tube. The tube is placed, as shown in the engraving, between the two parallel rods, and may be compressed by turning the screws, until the passage through it is entirely obstructed.

Fig. 18.

79. A small round Wicker Basket, with upright sides, about 14 cm. in diameter and 10 cm. in depth; for holding test-tubes.

80. A few sheets or cut filters of Swedish Filtering Paper, for separating precipitates, such as barium sulphate or calcium oxalate, which from their finely-divided condition would pass through the pores of ordinary filtering paper.

A number of bottles will also be required for containing

substances both solid and in solution. Many substances, such as calcium chloride, from their alterability in the air, will of necessity be purchased in bottles. Inclusive of these, the following stock of bottles will be probably sufficient [1]:—

24 wide-mouthed bottles,	with corks,	300 c.c. capacity.
24 ,, ,,	,,	100 c.c. ,,
6 ,, ,,	with glass stoppers,	300 c.c. ,,
12 ,, ,,	,,	100 c.c. ,,
8 narrow-mouthed bottles,	,,	100 c.c. ,,
40 ,, ,,	,,	100 c.c. ,,
3 ,, ,,	,,	2000 c.c. ,,

The common green or blue glass bottles, costing (when stoppered) from 3*s.* to 6*s.* a dozen, will answer quite as well as the more expensive bottles of white glass.

[When a stopper is found to be fixed immovably in the bottle, try to loosen it by tapping it, first on one side then on the other, with a piece of wood such as the handle of a file, pressing the thumb against the opposite side of the stopper and taking care to direct the blows obliquely upwards, rather than directly across the stopper. If this does not succeed, heat the neck of the bottle by passing it to and fro over the flame of a spirit lamp, turning it constantly round. The neck will expand with the heat before the stopper, and if the latter is now tapped again, it will almost certainly be loosened. There is, of course, a risk of cracking the bottle if it is heated too suddenly, but as the success of the method depends upon the difference in temperature between the neck and the stopper, the heat should be applied quickly, and only for a short time. If this method fails put a drop of oil or glycerine round the stopper, and leave the bottle for some time in a warm room. The oil will work its way between the neck and the stopper, and the latter may generally be loosened by tapping. If the bottle contains potassium hydrate, a drop of hydrogen sulphate may be substituted for the oil, and will remove the alkaline cement.]

[1] It is not absolutely necessary for the beginner to start with so large a stock of bottles. Some substances, e. g. marble, sulphur, manganese dioxide, &c., may be kept in boxes. But bottles are far preferable, on the score both of cleanliness and security.

LIST OF SUBSTANCES

Required for the Course of Practical Work contained in Parts I and II of this Book[1].

The numbers in the column on the extreme left of the page are intended to give some idea of the relative quantities of the substances which will be necessary. If 1 be interpreted to mean 10 grammes (about one-third of an ounce), 2 = 20 grammes, &c., the quantities will be sufficient for the purposes of most students, but at least twice the amount should be obtained, if supplies have to come from a distance.

Many of the substances are required for use in solution. In Part I, Sect. 1, Exercise 5, will be found the general method of dissolving a salt in water; and detailed directions as to the strength of solutions &c. are given in Appendix B.

An asterisk is prefixed to the names of those substances which the student may prepare himself.

9	Alcohol, pure, sp. gr. 0.815.
60	„ Methylated, for common purposes.
12	Aluminium and Ammonium Sulphate (Ammonia Alum).
3	Ammonium Carbonate, pure.
3	„ Chloride, pure crystallised.
12	„ „ common.
9	„ Hydrate, solution of, sp. gr. 0.96 (or 0.88[2], Caustic Ammonia).
1	„ Molybdate.
9	„ Nitrate.

[1] Those who are working in a regular laboratory will probably obtain the necessary chemicals from the common stock. This list is more particularly intended to assist those who are working by themselves, in selecting the substances they will require.

[2] This is the most concentrated, and best; but care is required in dealing with it. It must be kept in a cool place, with the stopper tied down, and the bottle must be opened cautiously, especially in warm weather.

3	Ammonium Oxalate.
2	„ Phosphate [1].
2	„ Sodium and Hydrogen Phosphate (Microcosmic salt).
6	„ Sulphide, solution of.
2	Antimony, Metallic.
3	„ Trisulphide.
1	Arsenic Trioxide (White Arsenic).
3	Asbestos, in long loose fibres.
3	Barium Chloride, pure.
2	„ Oxide (Caustic Baryta).
1	Bismuth, Metallic.
1	Cadmium Sulphate.
30	Calcium Carbonate, pieces of white marble.
9	„ Chloride, thoroughly dried, in lumps.
3	„ Fluoride, white fluor spar.
10	* „ Hydrate, solution of (Lime water) [2].
3	„ Hypochlorite (Bleaching Powder).
20	„ Oxide, freshly burnt white Quicklime.
20	„ Sulphate (Plaster of Paris).
6	Charcoal, selected pieces.
1	„ Animal.
3	Carbon Disulphide (Bisulphide of Carbon).
3	*Chlorine, solution of (Chlorine water).
1	Chromium and Potassium Sulphate (Chrome Alum).
1	Cobalt Nitrate.
1	Cochineal, solution of.
20	Copper, Metallic ; strips of sheet-copper, about 0.5 mm thick.
3	„ filings.
	„ pieces of wire, No. 18 and 26 wire-gauge.
6	„ Oxide (Black Oxide of Copper).
12	„ Sulphate.
2	Cotton-wool.

---

[1] This salt is preferable, as a test, to sodium and hydrogen phosphate, but is not necessary.

[2] For the method of making this solution, see p. 58.

Distilled Water, see p. 67.

2   Ether.

Gold, Metallic; a book of gold-leaf.

2   Grape Sugar.

6   Hydrogen Acetate, solution of, sp. gr. 1.04 (Acetic Acid).

6   „   Chloride, pure concentrated (Hydrochloric Acid).

10 * „   Chloride, pure diluted.

15 „   Chloride, common concentrated.

12 Nitrate, pure concentrated (Nitric Acid).

10 * Nitrate, pure diluted.

30 „   Nitrate, common concentrated. '

6   „   Oxalate (Oxalic Acid).

12 „   Sulphate, pure concentrated (Sulphuric Acid).

10 „   · Sulphate, pure diluted.

60 „   Sulphate, common concentrated.

   * „   and Silicon Fluoride, solution of (Hydrofluosilicic Acid).

   * „   Sulphide, solution of (Sulphuretted Hydrogen).

3   Indigo Sulphate, solution of (Sulphindigotic Acid).

2   Iodine.

12 Iron, Metallic; thin wire and one or two strips of thin sheet-iron.

   * „   Perchloride.

2   „   Peroxide (Rouge).

12 „   Protosulphate.

12 „   Protosulphide, in lumps.

1   „   Pyrites.

12 Lead, Metallic; strips of sheet-lead.

3   „   Acetate.

3   „   Nitrate.

3   „   Protoxide (Litharge).

3   Lead, Red Oxide (Red Lead).

3   Litmus, solution of (Archil).

0.5 Magnesium, Metallic; wire or ribbon.

3	Magnesium, Sulphate.
30	Manganese Dioxide (Black Oxide of Manganese).
12	Mercury, Metallic.
3	„ Oxide.
3	„ Perchloride (Corrosive Sublimate).
1	„ Protochloride (Calomel).
1	* „ Protonitrate.
1	Nickel Sulphate.
2	Phosphorus.
1	„ Amorphous (Red Phosphorus).
	Platinum Perchloride.
0.5	Potassium, Metallic.
1	„ Bromide.
20	„ Chlorate.
3	„ Chromate (Yellow Chromate of Potassium).
6	„ Dichromate (Red Chromate of Potassium).
3	„ Cyanide.
1	„ Ferricyanide (Red Prussiate of Potassium).
6	„ Ferrocyanide (Yellow Prussiate of Potassium).
9	„ Hydrate (Caustic Potash).
3	„ Iodide.
12	„ Nitrate (Purified Saltpetre).
2	„ Nitrite.
1	„ Sulphocyanate.
1	*Silicon Dioxide (Silica).
	Silver, Metallic; a book of silver-leaf.
1	* „ Nitrate.
1	Sodium, Metallic.
3	„ Diborate (Borax).
2	„ Carbonate, pure, anhydrous.
20	„ Carbonate, pure, crystallised
12	„ Chloride (Common Salt).
3	„ and Hydrogen Phosphate.
6	Hyposulphite.
3	„ and Hydrogen Tartrate.
12	„ Sulphate.
3	„ Sulphite.

3   Starch.

2   Strontium Nitrate.

6   Sugar (Loaf sugar).

30   Sulphur; roll Sulphur, and Flowers of Sulphur.

12   Tin, Metallic; in strips and foil.

* „ Protochloride (Stannous Chloride), solution of.

2   „ Peroxide (Binoxide of Tin).

2   Turpentine.

12   Zinc, Metallic; pieces of sheet-zinc. The purest form
      of the metal is the Belgian rolled zinc, scraps
      of which may be procured from any tinman.

Advice on the use of the following Exercises.

1. Read over the whole Exercise, or at any rate a complete section of it, before beginning work ; in order that you may understand precisely what you are going to do.

2. Look out all the apparatus, &c. required. It need not be *all* actually on the table before you ; in fact, it is generally better that it should not be so ; but everything should be within reach.

3. Have always at hand a general text-book on Chemistry, and refer to it constantly for explanations of chemical reactions, points of theory, &c. The present treatise is intended as a practical companion to such a book, and not in any way to supersede its use.

PART I.

EXPERIMENTS ON THE PREPARATION AND
PROPERTIES OF SUBSTANCES.

SECTION I.
PRELIMINARY EXERCISES.

EXERCISE 1.

Fusion and Granulation.

Apparatus required—Iron ladle, with bowl about 7 cm. in diameter;
iron spoon; pan or jug filled with clean water; pair of pliers; cloth;
pieces of metallic zinc.

Put a few small pieces of zinc into the ladle, and place
the latter upon a clear fire[1], supporting it on the coals so
that the bowl may rest steadily in a horizontal position. Zinc
requires a rather high temperature ($433°$) for its fusion, but
when the bottom of the ladle becomes heated to faint redness the fragments of metal will sink down into a fluid mass.
When this takes place, add some more pieces of zinc, pressing them down into the fused metal by means of the iron
spoon. If any of the pieces of sheet-zinc are too large to
go conveniently into the ladle, bend them into a more compact
form with the help of the pliers. You will find it much easier
to effect this when the metal is made quite hot by being
held in front of the fire; since a sheet of zinc which is stiff
and unyielding at the ordinary temperature becomes remarkably pliant when moderately heated. Go on adding pieces
of zinc until the ladle is about three-fourths filled, and then

[1] A Bunsen's burner will answer the same purpose, but not so well. If
it is used, the ladle may be supported on an iron tripod, or on the retort
stand, the handle being laid across the largest ring, so that the bowl may
be just outside the ring (not resting in it, lest the brass of the ring should
melt).

leave it on the fire for a minute or two longer, in order that
the metal may become quite fluid. The earthy-looking substance, or dross, which floats upon the melted metal, consists
of a compound of zinc with oxygen, one of the constituents
of the air, and is called zinc oxide. This should, at the
last moment, be skimmed off with the iron spoon, so as
to leave the surface of the melted metal quite bright, like
mercury.   You will notice, however, that no sooner has the
coating of dross been removed than a thin film of it begins
again to be formed, owing to the contact of the air with
the strongly heated metal.   Now take the ladle off the fire at
once, and, holding it about half a metre above the jug of water,
pour the liquid zinc in as thin a stream as possible into the
water.   The steam formed when the hot metal touches the
water blows the particles of zinc asunder, and they fall to the
bottom of the jug in feathery, tumefied fragments, which from
the great surface they expose to the action of a solvent are well
adapted for use in the preparation of hydrogen gas (Sect. 2,
Ex. 2).   The metal in this form is called granulated zinc.

The water in the jug should be poured away, and the
zinc should be collected, dried as far as possible with a cloth,
then completely dried by being placed on a plate in front of
the fire, and kept for use in a wide-mouthed bottle or jar.

EXERCISE 2.

Glass Working[1].

1. To make some elbow tubes, for use in experimenting with gases.

Apparatus required—Pieces of glass tubing, about 6 mm. in external
diameter; pieces of glass rod, rather smaller in diameter; three-square
file; fish-tail gas burner, on iron foot; Bunsen's burner; Herapath's
blowpipe.

Select a piece of readily fusible glass tubing about 6 mm.

---

[1] In this and the succeeding Exercises only the more elementary operations in glass blowing are treated of, such as must be learnt in order to
fit up the apparatus required in Sect. 2.

in external diameter. Lay it down on the table before you, holding it down with the thumb and forefinger of the left hand, placed 15 cm. from one extremity. Make a small notch in the tube close to this point with a three-square file, slightly pressing the side of the file against the left thumb, which will thus serve as a guide to prevent the edge of the file slipping along, instead of cutting across, the glass. The notch should be made, not so much by a repeated to-and-fro motion of the file, as by one, or at most two, short forward strokes combined with as much downward pressure as the tube will bear, the hand being raised a little as the file goes forward, so that it may follow the curve of the surface of the glass. By this means the file will cut deeper into the glass with less injury to itself, than if the edge were drawn to and fro from point to handle as is usually done ; the effect being then rather a rub than a cut. Now take up the tube, holding it with both

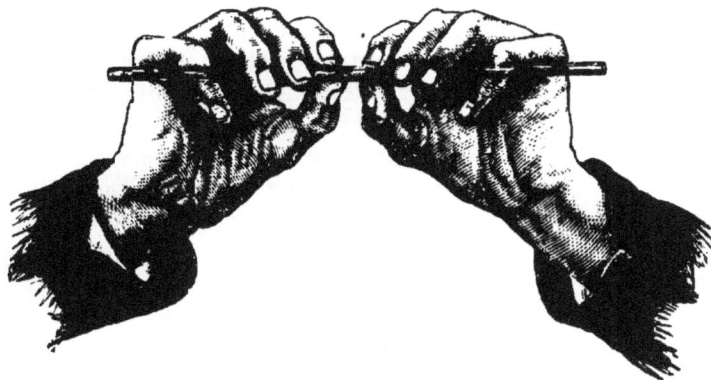

Fig. 19.

hands thus, Fig. 19 (one hand being on each side the notch, and the thumb-nails pressing against the glass on the side *opposite* the notch), and break it asunder precisely as a stick is broken. The edges of the freshly-cut glass are extremely sharp and must next be rounded off by holding the end of the tube just within the edge of the flame of a Bunsen's burner a little below the top of the flame, turning it constantly round, and keeping the outer end lowermost, lest vapour should condense in the tube, and run down to the hot part, so as to crack the glass. Do not heat the glass so long as to cause the end to sink in, and thus contract the bore of the tube; as soon as the edge is observed to be fairly rounded, the tube should be removed from the flame and allowed to cool, when the other end may be rounded in the same manner.

[It should be noticed, once for all, that in all kinds of glass working, the process of annealing is of the utmost importance. Glass must never be either heated or cooled suddenly, unless the special object is to produce a crack. The material is such a bad conductor of heat, that the end of a piece of glass may be raised to a red-heat, while at a distance of 3 cm. from this portion it remains for some time sufficiently cool to be held in the fingers. In consequence of this low conductivity, when heat is applied suddenly to a piece of glass, the parts immediately in contact with the source of heat expand before the heat is communicated to the neighbouring parts, and thus tend to tear the latter asunder. Again, when glass is suddenly cooled, the surface contracts at once, and is torn asunder by the still expanded adjacent portions, which have not had time to lose their heat. The useful applications of this property will be alluded to presently; but it should be a rule—

1st, *Never to bring a piece of glass into a flame suddenly,* but to hold it for half a minute, more or less according to the size and thickness of the glass, in the current of hot air above the flame, constantly turning it round, and heating more of it than is intended ultimately to be brought to a red-heat.

2ndly, *After the work is done, to withdraw the glass very gradually from the flame,* occupying a minute or so in removing it to a distance of 12 or 13 cm. above the flame; then to leave it to cool very slowly in a position protected from currents of air. It will sometimes be found useful to have at hand a dish of strongly heated sand, into which the hot glass may be plunged and left to cool slowly; and in all cases it is better to err on the safe side, than to risk the breaking of a tube owing to its particles being in a state of tension from deficient annealing.

3rdly, *Always to keep the glass turning while it is in the flame.* The heat of a lamp or blowpipe is mainly applied to one side of an object, viz. that which is turned towards the wick; while almost all operations in glass blowing require that the piece of glass should be uniformly heated on all sides. If this simple rule be not attended to, it will be found impossible to blow a good bulb and even to make a good bend in a tube. Practice alone, however, can give that steadiness of hand, and *adhesiveness*, as it were, of the fingers to the glass, which will enable the student to rotate a piece of tube which is heated to fusion in the middle, without sensibly distorting the softened part. The portions of the tube on either side of the centre may be regarded as two distinct tubes united by a flexible material; and the object should be, to keep these two tubes in the same straight line, and to rotate them continuously at the same rate, without laying any stress on the connecting portion. It is generally best to hold the hands under the glass, the up-turned forefinger and thumb being chiefly employed in rotating the tube, while the other fingers sustain it at such points that the portions of glass on either side of the heated part are pretty evenly balanced, and have no great tendency to tilt in either direction.]

For bending the ordinary fusible glass tubing, if its external diameter does not exceed 1 cm., a blowpipe-flame is neither required nor so suitable as the flame of a common fish-tail or bat's-wing gas-burner; but if gas is not at hand, a spirit lamp with a large flame may be used. Light the gas, or spirit lamp; then holding the piece of tube by its extremities, bring it about 7 or 8 cm. above the flame, turning it constantly round and moving it laterally so as to heat about 5 cm. of it equally on all sides. The flame of a fish-tail burner is flat, and the glass must be held *along*, not across it: the object being to heat a considerable length of the tube, so as to make a gradual bend. After a few seconds, lower it gradually into the flame, still constantly turning it round. If the gas-burner be used, the glass will become covered with soot when .immersed in the flame; but this is of no consequence, as the heat of such a burner is never high enough to incorporate the carbon with the glass. When the heated portion becomes soft and yielding, which will take place even before it has acquired a visible red-heat, withdraw it from the flame, and gently bend it to a right angle, avoiding the use of much force. You will probably find some difficulty at first in making the bend in one plane, i. e. so that the bent tube when laid on a flat surface may touch it in every part of its length. The best method of accomplishing this is, to support the tube lightly by its extremities, so that the direction of the bend may be determined mainly by the weight of the tube itself; then, holding it before you so that a line drawn from the eye may pass through both its extremities, gradually approach the hands to each other, as if you were endeavouring to snap the tube in two. Do not attempt to use much force, or to make the bend suddenly, or you will inevitably either flatten the glass on the outer side or wrinkle it up on the inner side ; either fault being fatal to the strength of the bend. It will usually be found necessary to heat the tube again in order to complete the bend; and it is better, if there are any signs of wrinkling or flattening, not to attempt to bend it further in that part, but to heat another portion of it a little on either side of the partially-made bend, and to complete the curve in that portion.

Another method of bending small tubes (not more than 6 or 8 mm. in diameter), which practically succeeds very well, is the following :—Heat the tube as above directed until it becomes soft ; then hold it steady, just at the top of the flame, withdraw the left hand, and allow the tube to bend by its own weight as far as necessary. It is essential that the tube should be held perfectly still during the bending, and not rotated in any way by the right hand.

The correctness of the angle may be judged of by holding the tube close to, but not actually touching, an ordinary square, or the corner of the table or of a book. When the proper bend is completed, lay the tube on a bit of glass in such a position that the heated portion does not come into contact with any cold surface, and leave it to cool slowly.

While it is cooling you may cut off another portion of the same tubing about 24 cm. long, and after rounding the ends, bend it in a similar way, making the bend, however, not in the middle of the tube, but about 7 cm. from one end. If that end becomes too hot to hold in the fingers it may conveniently be inserted in a hole made in a small cork, which will then answer the purpose of a handle. The tubes thus bent will serve to fit up a washing bottle or generating flask for gases; and several similar ones with the branches varying in length may be made at leisure moments from any waste bits of tubing, and will be found generally useful, saving much time in fitting up apparatus for any particular experiment.

2. **To make a glass jet and a dropping-tube, or pipette.**

· Cut off a piece of glass tubing (the less fusible the better) about 25 cm. in length and 5 mm. internal diameter. Round off the ends of the tube as directed in p. 29, and heat a portion of it, about 7 cm. from one end, in the flame of a Bunsen's burner[1] (remembering to turn it constantly round), until it becomes quite soft and begins to thicken and contract in diameter : then withdraw it from the flame, and pull the two ends apart by slightly separating the hands, until the drawn-out portion is contracted to an external diameter of 2 mm. Be careful not to use much force in drawing out the glass, or it will be contracted to a fine thread so thin as to be useless. When it is cool, make a fine scratch with a file at the middle, *a*, of the drawn-out portion, and break the tube at this point. You will then have two tubes, each ending in a jet, the edges of which should be slightly rounded in the lamp flame, and the aperture, if necessary, reduced by holding it rather longer in the flame, until a large needle would just pass through it. The shorter of these tubes may be kept for use as a jet, in

Fig. 20.

Sect. 2, Ex. 2. The longer one will form a very useful 'pipette' for delivering small quantities, such as single drops of a liquid, in testing. For this purpose the narrow end is dipped into the liquid, and the forefinger (moistened slightly) is pressed firmly on the top of the tube (see Fig. 17): it may then be withdrawn, and the pressure of the air will prevent any liquid escaping, but by relaxing the pressure of the finger one or more drops may be allowed to fall, as required.

[1] If the glass is very infusible the blowpipe flame must be used, but a Bunsen's burner will generally be found to give heat enough.

D

3. To make a stirring rod.

For this purpose the heat of a gas or spirit lamp is scarcely sufficient; it must be intensified by the use of a blowpipe. The best form of this instrument is fully described at p. 16, and it will only be necessary here to explain the method of using it so as to produce the greatest effect.

If a Herapath's blowpipe is at hand, the process is simple. Turn on the gas full, and light it at the mouth of the blowpipe. Then work the bellows gently and uniformly, and gradually reduce the supply of gas, until you have a small brush of blue flame about 10 or 12 cm. long.

If you have one of the mouth Herapath's blowpipes, described in the list of apparatus p. 17, it should be fitted into the tube of the Bunsen's burner and the whole raised on blocks to a convenient height above the table. You should then refer to the Exercise on the Use of the Blowpipe, p. 85, and practise the method of keeping up a continuous stream of air, as there described; remembering not to turn on more gas than is necessary to give a small brush of blue flame.

[If a spirit or oil lamp must be used, the wick should be trimmed flat and pulled up sufficiently to give a large flame; then with the trimming scissors separate the wick in two parts, and bend each portion sideways so as to leave a clear passage for the air between them. Arrange the lamp and jet so that the latter may lie, pointing slightly upwards, in the line of the passage just formed in the wick, and on a level with the top of it. Light the lamp, and, introducing the jet just within the flame, commence the blast of air. The flame of the lamp will now be almost entirely deflected in the direction of the stream of air; and by pushing the jet a little further into the flame, or drawing it back beyond the margin, any kind of flame may be produced, from a quietly-burning pointed blue cone to a large roaring brush of flame; the former being most suitable for the present purpose. If the flame is ragged or irregular, see whether any filaments of the wick stand in the way of the blast, and if so remove them with the scissors. If this does not cure the defect, the blowpipe-jet is dirty or not truly circular, and must be cleaned out by a large needle, or, better, by a small broach.]

Having obtained a satisfactory flame, select a piece of glass rod about 5 mm. in diameter, and cut off a piece about 18 cm. long, as directed in the case of the glass tubing. Hold one end of this piece about 5 cm. in front of the visible flame, turning it constantly round by a twisting motion of the fingers, and gradually bring it just within the apex of the flame, which is its hottest portion. The edges of the glass will soon begin to melt, and the sharp angle will disappear as the glass approaches the liquid condition. The heat should be continued until the end is perfectly round, and then the rod should be gradually withdrawn from the flame and allowed to cool slowly, resting on another fragment of glass or on the table so that the hot end may project over the edge. As soon as it is cool, the other end may be drawn out so as to form a blunt point (in shape resembling the pipette) by heating it in the flame until it becomes soft, then pressing upon it the end (previously heated) of a short bit of glass tubing or rod, so as to weld them together, and lastly directing the flame on the rod close to the junction, and drawing it out, when soft, precisely as was done in making the pipette. The rod should then be cut at the narrowest part and the sharp edges slightly rounded by fusion.

EXERCISE 3.

Glass Working (*continued*).

Apparatus required.—Piece of glass tubing, about 5 or 6 mm. external diameter; pieces of less fusible glass tubing, of the same diameter (the ordinary German glass will, however, do); three-square file; Bunsen's burner; fish-tail burner; corks; cork-borers; cork-squeezer, or pliers; narrow-mouthed bottle, about 600 c.c. capacity; rat-tail file; piece of charcoal, about 3 cm. in diameter, and 6 or 7 cm. long.

1. **To fit up a washing bottle, as shown in Fig. 21.**

In doing this you will have to apply the experience in glass working which you have already gained, and in addition to learn the method of boring holes in corks.

Cut off a piece of glass tubing about 5 mm. in external diameter and 50 cm. in length; hold it horizontally in the hands and heat it (observing the precautions given in p. 30) in the flame of a Bunsen's burner at a point about 15 cm. from the right-hand extremity. As the tube is somewhat long, you will find an advantage in supporting it near the other extremity on a ring of the retort-stand, or other support, fixed at the same height as the burner. This will render it easier to keep the tube turning between the fingers (p. 30), and to maintain it in the proper position when it becomes soft.

Fig 21.

Allow the heated portion of the glass to become slightly thickened, then raise the tube out of the flame and draw it out slowly and carefully (waiting until the glass has so far cooled that a little force is required for the purpose), as was done in making the pipette, p. 33. Allow it to cool slowly, and then cut it in two at the middle of the contracted portion. You have now two tubes, the one 35 cm. an l the other 15 cm. in length, each terminating in a jet. Lay aside the shorter tube for the present, and heat the extremities of the longer tube just sufficiently to round off the sharp edges; then (using the fish-tail burner) bend it to an acute angle equal to that of the left-hand tube in Fig. 21, making the bend about 8 cm. from the jet. While it is cooling take the other tube, cut off the contracted portion at the end, round off the sharp edges of each end in the Bunsen flame, and bend it (using the fish-tail burner) near the middle to an obtuse angle equal to that of the right-hand tube in the figure.

It now only remains to fit these tubes to the bottle by means of a cork perforated with two holes. It will be best however to begin by practising the method of making a single

hole in the centre of a cork. Take a good sound cork about
2 cm. in diameter, squeeze it until it becomes soft and elastic
(a pair of pliers or nut-crackers may be used instead of a
regular cork-squeezer), then take it up thus, Fig. 22, between
the second finger and the thumb of the left hand, and place
the sharpened end of the smallest cork-borer against it, as near
the centre of one end as you can judge. Urge the cork-borer
into the cork with a twisting motion, as if you were using
a cork-screw. Some care will be required to make the hole

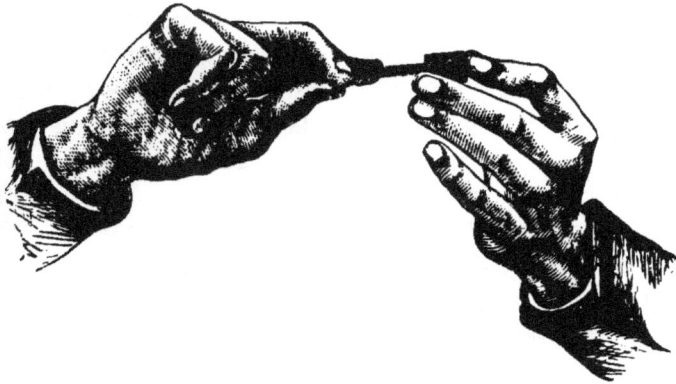

Fig. 22.

straight through the cork, so that it may be truly central. Of
the proper direction the eye will be the best judge: and when
the cork-borer has penetrated some little way, it will be advisable to turn the cork a quarter round, in order that it may
be seen whether the axis of the cork-borer and of the cork
are still in the same straight line. If not, a slight pressure
on the cork-borer in one direction or the other, while the
boring is continued, will set it straight. When the borer has
penetrated quite through the cork it may be withdrawn with a
twisting motion, and will bring with it a cylindrical plug of
cork, leaving a hole, the sides of which should be smoothed
with the round or rat-tail file. The plug of cork remaining in
the borer may be pushed out by means of a wire which is
usually sold with the set of borers for that purpose. It should

not be thrown away, as such small corks are often very useful for stopping the ends of drying-tubes, and other purposes.

When you have practised in a similar way on one or two other corks, and have learnt to control the direction of the borer, you may complete the fitting up of the washing bottle. Take an ordinary narrow-mouthed bottle holding about 500 or 600 c.c.; choose a cork slightly too large to fit it (since the cork is reduced in size when it is squeezed), and render it soft and elastic by squeezing it. You have now to make two holes in it in the position of Fig. 23, on opposite sides of the centre, and about midway between the centre and circumference. Take for the purpose a cork-borer rather smaller than the tubing which you have been using, and bore the two holes, with particular care that each hole does not run into the other or pierce the side of the cork. The cork-borer may be slightly oiled, if thought necessary, but this will be seldom requisite if the end is kept properly sharp. The holes should next be smoothed and slightly enlarged by the rat-tail file, until the end of one of the tubes will just enter them when some little pressure is used. Now pass the longer branch of the longest of the two tubes through the cork, with moderate pressure[1] and a twisting motion, until it projects so far as to reach, when the cork is fitted into its place, nearly to the bottom of the bottle. When this is done, pass one of the branches of the other tube through the other hole in the cork, until it projects 3 or 4 mm. on the other side. Fill the bottle with distilled water, fit the cork carrying the tubes tightly into the neck of it, and your washing bottle is ready for use. Blow gently through the up-turned end of the shorter tube, and see whether a fine stream of water issues from the jet of the other

Fig. 23.

---

[1] If much pressure is used the tube is not unlikely to break, and the splinters of glass may cause a serious cut. The hole should never be so much smaller than the tube as to make it necessary to use much force in passing the latter through it. It is a good plan, also, to wrap the tube in a cloth or handkerchief while it is being passed through the hole in the cork.

tube. If the jet is found to be too large, it may be easily reduced in size by holding the tip of it for a second or two in the flame of a Bunsen's burner. Care should be taken not to allow any water to flow down into the jet, while it is being heated, which would infallibly crack it.

If no water issues from the jet when you blow air into the bottle, either the aperture is closed up, in which case a small portion of the tip of the jet may be cut off with the file, or there is a leakage of air at the cork. Place a drop or two of water on the cork, and observe whether, on blowing as before, bubbles rise through it. If they do, you may remedy the fault either by pushing the cork more tightly into its place or by melting a little sealing-wax over the top of it, or, if these fail, by taking a new cork altogether, and boring the holes with more care. Never be satisfied with an imperfect apparatus.

2. To seal a glass tube.

This is the simplest operation in glass blowing strictly so called, i. e. in which the assistance of the breath is called in, to mould the glass into shape; and it is one which the student is continually called upon to practise for mending broken test-tubes and making arsenic reduction-tubes.

Test-tubes, however, are made of such thin glass that it is by no means so easy as it appears to seal them neatly ; and you will find it best to commence work on a piece of soft glass tube about 5 mm. in external diameter. Cut off a piece of this tubing about 15 cm. in length, round the ends in the Bunsen flame, and having arranged the blowpipe so as to give a large steady flame, hold the tube horizontally and bring it gradually into the hottest part of the flame, so that about 1 cm. near the middle may be thoroughly heated. When the glass becomes quite soft, remove it from the flame and draw it out a little by separating the hands, until it assumes the form of Fig. 24. Now direct the flame against the part which lies a little to the left of the most contracted portion, and draw it out further until nothing but a thread

of glass remains to connect the two portions of the tube,
Fig. 25.   This thread of glass should next be heated just at
the point where it joins the tube on the left hand, when it
will fuse and divide, running up into a small knob against
the thicker portion of the tube.   Lay down the right-hand
piece of tube (resting it on another bit of glass that it may
not burn the table), and proceed to make the end of the
other piece smooth and round.   This is done by directing
the flame upon the small projecting knob, which will soon
fuse and partially incorporate itself with the surrounding glass.
The whole end of the tube will, however, have become contracted and thickened, and must be expanded a little by
removing it from the flame and immediately forcing air very

Fig. 24.

Fig. 25.

Fig. 26.

gently into it from the mouth, until it takes the shape of
Fig. 26.   Do not blow too hard at first, or heat more of
the tube than is necessary, or you will probably expand the
sides of the tube into a bulb, which is not your present
object: your intention being simply to distribute evenly the
thickened glass at the extremity by driving it forwards rather
than outwards, and to mould the end of the tube into the
regular round form of Fig. 26.   If this is not accomplished
at the first trial, bring the end of the tube again into the blowpipe-flame until it contracts, and blow it out as before, keeping
the attention, while blowing, fixed on the heated glass so as
to be ready to moderate or stop altogether the supply of air
from the mouth, if the glass shows signs of yielding too
much.   Now anneal the tube as already directed, and finish
the other, or right-hand piece of tube, in a similar way, drawing
it out and rounding it exactly as before.   If the bit of capillary
tube remaining attached to it is too short or too slight to be

used as a handle in drawing out the glass, it should be cut off, and while the end of the tube is heated in the flame, a bit of waste glass held in the right hand should be heated and pressed against it, and the flame directed upon the point of junction. The two pieces will then adhere, and by applying the flame a little more to the left the superfluous glass may be melted and drawn off, attached to the bit of waste glass. Moreover, if the remaining knob be too large to be neatly melted into the bottom of the tube, it may, when soft, be touched with a bit of heated glass and drawn off in a similar way.

After a few small tubes have been thus sealed, larger tubes may be operated on in the same way, a larger blowpipe-flame being employed. Test-tubes broken at the bottom will be found good materials for practising on, and may be mended by drawing off the broken portion and sealing them as above. The glass of which these tubes are made is so thin that it is not easy at first to avoid over-heating them in some one spot, thus producing a mis-shapen end, which is quite inadmissible in vessels which, like test-tubes, are exposed to comparatively sudden changes of temperature. The softened portion should not be drawn out much at first, but allowed to sink in and contract of its own accord, in order to thicken it a little: and especial care must be taken that the bottom is of uniform thickness and well annealed. If there is a crack in the tube, the piece must be broken off by a slight blow; or, better, the crack may be led round the tube as described in the next section, otherwise it will extend itself when the tube is heated.

Another very useful form of sealed tube, which may be made by the student himself, is that known as the 'ignition tube,' which serves for heating substances either *per se*, or with a flux, in the preliminary examination in qualitative analysis. These tubes are of hard infusible glass, about 5 or 6 cm. long and 5 mm. in external diameter; and the manufacture of them will be a moderate test of proficiency, as it will require some dexterity to get rid entirely of the small knob of glass already mentioned. The end must be very strongly heated, and the breath thrown in with some force, the moment that the glass is removed from the flame. These tubes are shown of the actual size in Fig. 27. It will be seen that the sealed end is slightly expanded so as to form a small bulb. This is done after the knob has been got rid of, by strongly heating a portion of the sides as well

Fig. 27.

as the end of the tube, and then immediately blowing into it with considerable force. The bulb should not, however, be larger than is shown in the figure, or it will be too thin to be of much use.

3. To divide glass by leading a crack along it.

The low conducting power of glass for heat renders it, as has been already noticed, very liable to crack from sudden changes of temperature. From the same cause, however, it is easy to extend a crack, when once begun, in any desired direction by heating the parts of the glass which lie just in front of the crack. The method is extremely simple, and will be found especially useful in cutting off the necks of flasks when they are chipped or uneven, and in making evaporating basins or capsules out of broken flasks or retorts.

Suppose, for instance, a Florence flask has, as is often the case, a neck too uneven to allow of a cork being fitted into it. Arrange the blowpipe to give a rather large flame; take any waste bit of glass rod or tube about 6 mm. in diameter and 7 or 8 cm. in length; draw it out in the flame as directed in the last section, leaving the end somewhat pointed. Now take the body of the flask in the left hand, the neck pointing upwards, and, having heated the pointed end of the rod to full redness, apply it to the outside edge of the neck, and hold it there for a second or two. It is very probable that a small

crack will be thus started, but if not, it may with certainty
be produced by removing the heated rod, and immediately
touching the spot lightly with a moistened finger or splinter
of wood[1]. The crack once begun, press the red-hot end of
the rod on the glass a little in front of it, Fig. 28, when it
will at once extend itself to
the heated spot; and by slowly
drawing the rod in the direction required, re-heating it
from time to time in the
blowpipe-flame, the crack may
be led at first a short distance
downwards, and then, turning
at right angles, horizontally
round the neck so as to cut
off a ring of glass including
the uneven portion. It is
generally not possible to carry
the crack entirely round the
glass until it returns into it-

Fig. 28.

self; a small portion of the glass will remain undivided, but
after laying down the rod the ring may be readily pulled off,
a very slight inequality marking the point where the crack was
incomplete.

A rather better form of termination for the rod, especially
when the glass to be cracked is somewhat thick, is the

Fig. 29.

following, Fig. 29. It is made by heating strongly about
5 or 6 mm. of the end of the rod, and then, the rod being
held in a slanting direction, pressing it down on any flat
metallic surface. The glass will spread out laterally, and by

---

[1] By making a notch with the file in the place and in the direction in
which the crack should be, and applying the heated rod to the end of the
notch, it is easy to start a crack at once and in the desired direction.

turning the rod half round and pressing it down again in the same way the desired chisel-shape will be obtained. In using it the edge should be kept in the line of the crack required, and moved along the surface with slight pressure. The advantages of this form are: 1st, That more of the glass in the desired line is heated at once, so that the crack extends more rapidly, yet under perfect control; 2ndly, That it is easier to lead the crack in an unswerving straight line, the eye being guided by the line of the edge of the tool, than when a blunt point only is used, precisely as a carpenter finds it easier to cut the side of a mortice by a broad chisel than by a narrow one.

Having thus rendered the neck of the flask even, it will be desirable not only to fuse the edge in the flame, but to turn it slightly outward so as to form a border like that of a bottle, which will give it much greater strength when a cork is to be fitted to it. To do this a very simple tool must be first prepared. Select a piece of sound charcoal, free from fissures, about 3 cm. in diameter, and 6 or 7 cm. long. Cut and rasp the end of this to a point like that of a pencil, but more obtuse, the angle at the apex being nearly a right angle. Take the flask in the left hand, and cautiously heat the extreme edge of the neck in the blowpipe-flame until it softens, turning it constantly round and holding it at right angles to the flame so that the latter may play across the mouth and heat two opposite sides of the glass at once. When the glass begins to soften and sink inwards, take the pointed charcoal in the right hand and press it gently with a twisting motion into the neck. The edge will spread out, and by repeating the operation a border shaped thus, Fig. 30, will be obtained, which should be annealed with especial care.

Fig. 30.

Test-tubes, &c. may be bordered in the same way, but it is best to use a spirit lamp or Bunsen's burner, and not the blowpipe, for heating the edges, on account of the thinness of the glass.

EXERCISE 4.

Weighing and Measuring.

Apparatus required.—Balance, and set of weights; Bunsen's holder; test-tube, selected of stout glass, about 15 cm. long, and 1.5 cm. in diameter; beakers; washing bottle; pipette; three-square file; writing diamond. Alcohol (methylated spirit); strong hydrogen sulphate (sulphuric acid); piece of glass rod, and strip of lead or copper.

General rules to be observed in weighing.

1. Handle the weights very carefully, using a pair of light brass forceps for the smaller ones.

2. Never allow the scales to swing violently about, but keep them steady with the hand while the weights are being put in.

3. Try the weights in a definite order, beginning with the largest, then taking the next smaller, then the next, and so on.

4. Count the weights over twice,—once in the scale-pan, and again as they are being put away in their places,—and put down results at once in your note-book.

Support the balance in the Bunsen's holder [1], as shown in

Fig. 31.

Fig. 31, at such a height that the pans may be about 3 cm. above the table. Place under the pans a flat ruler (a half-

[1] It will be advisable to put a weight (such as the Argand lamp) on the base of the Bunsen's holder, to prevent the possibility of its falling forward when heavy weights are put in the scales.

metre rule will do very well) so that by tilting it on its edge they may be steadied when required.

In the first place see that the beam, when free to swing, comes to rest in a horizontal position. If it does not do so, one of the pans may be dirty, or not hanging fairly from the beam; but if nothing of this kind is noticeable, the construction of the balance may be at fault, and the equilibrium must be adjusted by putting a bit of paper, or tin foil, in the lighter scale; this must, of course, remain there throughout the exercise.

Next, examine the weights to see if they correspond with, and are correct multiples of one another. If you have the complete set of 100, 50, 20, 10, 10, 5, 2, 2, 1, ·5, ·2, ·2, ·1 grm.[1], the following trials may be made.

1. Place a 10 grm. weight in each scale and see whether they balance each other, as they should do. Absolute accuracy can hardly be expected in common balances and weights, but there should be no error so great as 1 decigramme.

2. Transpose the weights from one scale to the other. They ought still to balance each other; otherwise the arms of the balance are not of equal length. If the discrepancy is serious, the balance should be rejected altogether.

3. Weigh out carefully 1 grm. of writing paper; then put the 1 grm. weight in the same scale as the paper, and see if the 2 grm. weight balances the whole, as it ought to do.

4. Try whether the 2 grm. weight which you have just used balances the other 2 grm. weight.

5. Try whether the 2 + 2 + 1 grm. weights balance the 5 grm. weight.

6. Try whether the 5 + 2 + 2 + 1 grm. weights balance the 10 grm. weight.

7. Try whether the two 10 grm. weights balance the 20 grm. weight.

8. Try the 50 grm. weight against the 20 + 10 + 10 + 5 + 2 + 2 + 1 grm. weights.

9. Try the 100 grm. weight against all those just mentioned.

---

[1] Suggestions for making weights to fill up any deficiencies in a set are given in Appendix A.

If you have not a complete set of weights you may still be able to make some of the above trials, and others will suggest themselves. For instance, if you have 1, 2, and 5 grm. weights, 4 grms. of a substance may be weighed out by putting the 5 grm. weight into one scale, and the 1 grm. weight into the other, and then placing in the latter scale enough of the substance to restore equilibrium. This method is called 'weighing by subtraction,' and is constantly useful.

1. Graduation of a test-tube as a measure of cubic centimetres.

The original gramme weight was obtained (see Appendix E) by measuring 1 cubic centimetre of pure water (at 4° C.) and making a piece of metal of such a weight as to balance it. Hence it is easy to obtain a correct measure of 1 c.c. by putting into a tube enough water to balance the 1 grm. weight (the small correction for temperature may be neglected) and making a mark at the level of its surface.

Select a test-tube of stout glass, about 1·5 cm. in diameter and 15 cm. in length. Place a beaker in one scale of the balance, and support in it the empty test-tube. Counterpoise them by a beaker partially filled with water (or by small shot) placed in the other scale. When you have seen that the scales are in equilibrium, put into the scale containing the counterpoise a 1 grm. weight, and pour distilled water into the test-tube from the jet of the washing bottle until equilibrium is restored. If too much water is added, the excess may be removed by a glass rod, or by the pipette which you have already made (p. 33).

Bring the test-tube up to the level of the eye, holding it vertically against the wall, or some other upright support; place the thumb-nail just at the level of the *lowest* part of the curve formed by the surface of the water[1], and make

---

[1] This is the best and most sharply-marked level at which to read off a volume of such a liquid as water, and it may be made even clearer by holding behind the tube a white card over the lower half of which has been pasted a piece of black paper; the boundary line between the white

a slight mark with a file at this level. The mark will then, as already explained, denote a volume of 1 c.c.

Weigh into the tube in a similar manner, 2 grms., 5 grms., 10 grms., and 20 grms. of water, making a file-mark at each point. Finally, empty out the water and make a shallow scratch with the file at each of the marks, guiding the edge of the file by a strip of card held firmly round the tube. The lines may extend round one-fourth of the circumference of the tube, and the figures 1, 2, 5, 10, 20 c.c. may be scratched on the glass close to the proper mark with a sharp point of the file, or, better, with a writing diamond. The tube is now ready to be used whenever a measure for small quantities is required, and also for determining densities, in the manner next to be explained.

2. Determination of Densities.

'Density' means, strictly speaking, 'quantity of matter'; and equal bulks of different substances contain very different quantities of matter. This is generally found out by weighing them. Thus the weight of 1 c.c. of water is, of course, 1 grm.; the weight of 1 c.c. of silver is found to be 10·5 grms. Hence we say that silver has $10\frac{1}{2}$ times the density of water[1].

In practice, water is usually taken as the standard of comparison for liquids and solids, and the 'density' or 'specific gravity' of a substance is the number which expresses how many times a certain bulk of it is heavier or lighter than the same bulk of water.

A. Density of Liquids.

In the case of liquids the process is very simple. A vessel is (1) counterpoised so that its own weight may be neglected; (2) filled up to a particular level with water, and weighed;

and black being made to coincide nearly with the lowest part of the curve. The latter then appears black and perfectly sharp against the white background.

[1] In illustration of this, a piece of good cork may be cut with a sharp knife into a cube each side of which measures 1 cm., and then weighed: it will be found to weigh nearly one-fourth of a gramme. A cube of lead of the same dimensions (filed and scraped as true as possible) will weigh between 11 and 12 grms.

(3) filled up to the same level with the liquid and again weighed.

Then we have only to find the proportion between the weight of the water and the weight of the liquid by the rule-of-three sum ;

Weight of water : weight of equal volume of liquid : : 1 : density of liquid.

The measure you have made is particularly convenient for this purpose, since it is easy to obtain with it a definite volume of any liquid, and it is clear from the way in which you have graduated it, that the number of cubic centimetres of the liquid taken will express the weight in grammes of the same volume of water, thus saving the trouble of weighing the latter.

Thus if the volume of the liquid is 5 c.c. we know at once without weighing that this volume of water weighs 5 grms.

Density of Alcohol, &c.

In the first place see that the tube is properly counterpoised : then measure in it 10 c.c. of common alcohol, and weigh it carefully. Put down the weight at once in your note-book.

Then, since it is clear that 10 c.c. of water weigh 10 grms. (neglecting the slight correction for temperature) we shall have the proportion,

Weight of 10 c.c. of water : wt. of 10 c.c. of alcohol : : 1 : density of alcohol.

After rinsing out the tube with water, and drying it, you may find the density of strong hydrogen sulphate (common sulphuric acid) in a similar way; but it must be borne in mind that the substance is very corrosive, and should be handled carefully. A strong solution of common salt may be substituted.

B. Density of Solids.

In the case of solids there would seem, *primâ facie*, to be a great difficulty in getting a volume of the substance exactly equal in size to a given volume of water. It may be done readily, however, with the measuring tube in the following way.

1. *Density of glass.*

Dry the tube, and, after seeing that it is properly counterpoised, put into it a bit of glass rod about 8 or 9 cm. in length and about as thick as a pencil.

Find the weight of the glass, and write down this weight at once on a piece of paper. Thus:—

(i)   weight of glass in air =

Then fill up the tube (the glass still remaining in it) with water, to the level of the 20 c.c. mark, see that no air-bubbles cling to the glass, and weigh the whole.

By this last weighing you have found the weight of a mixture of the solid + water measuring 20 c.c. And if you subtract from this the weight of the solid, the remainder will be the weight of the water you have added to make up the total volume to 20 c.c. Thus:—

> weight of solid + water =
>
> — weight of solid alone   =
> _____
>
> weight of water added =

Now, if the solid was not there, but its place supplied by water, we should have 20 c.c. of water, which would, of course, weigh 20 grms. And it is easy to see that the difference between 20 grms. and the actual weight of the water you added will be the weight of the water which would be there if the solid was not, i.e. the weight of a volume of water equal in size to the solid.

Hence, the next step in calculation will be

> weight of 20 c.c. water =
>
> — weight of water added   =
> _____

(ii)   weight of water equal in ⎫
      volume to the glass ⎭ =

Then, since you know already the weight of the glass in air, you have lastly to work out the proportion,—

weight of water (ii) : weight of equal vol. of glass (i) : : 1 : density of glass.

In the same way you may ascertain the density of a piece of copper wire, or of a number of shot, remembering that it is advisable to take a good large piece of the substance; otherwise the errors of the balance and weights will be so relatively large as to make the results uncertain.

The density of many solids is, perhaps, more usually taken by another method, in which recourse is had to Archimedes' principle, that 'a solid when immersed in a liquid loses a weight which = that of the liquid which would occupy its place, i. e. the weight of a volume of the liquid equal to it in size.'

All that is required, therefore, is (1) to weigh the solid in air, (2) to immerse it in water and weigh it again in that position.

The difference between the two weights will be the weight of a volume of water equal in size to the solid, and the density can then be found by the usual proportion sum.

In illustration of this method, the density of a bit of stone or glass (a marble or a solitaire ball), or of a copper coin may be taken as follows :—

Arrange the balance as usual; except that, since the chains or cords by which the scale is hung are likely to get in the way, it will be well to keep them farther out by attaching to them, about 2 or 3 cm. below the ring from which they are hung, a wire frame of the shape shown in Fig. 32, made of three pieces of brass or copper wire about 9 cm. long twisted together; each arm being about 4 cm. long. Such a

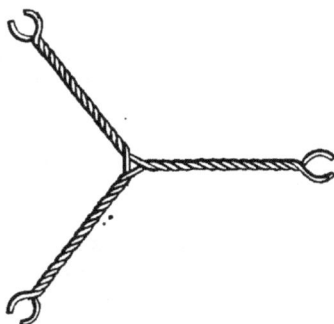

Fig. 32.

frame may be made for each scale, and may remain permanently attached to the balance, as it will be found of great use in preventing the chains getting entangled.

Support over one scale a beaker upon a glass plate resting

on a ring of the retort stand, as shown in Fig. 33, care being taken to leave just room enough for the scale to rise and fall freely below the retort ring.

Fig. 33.

Hang the solid, of which the density is to be taken, in a secure noose of the finest cotton or silk, and tie it to the centre of the three-armed frame above mentioned, so that it may hang freely in the middle of the beaker. Weigh it in this position, and note down the weight at once. Then fill the beaker with water, brush away with a piece of wire any air bubbles which may cling to the solid, and weigh it again. Put down the weights thus :—

weight of solid in air   =
weight of solid in water   =   _____
weight of equal volume of water =

Then calculate the density by the usual proportion sum.
The calculation may be expressed in the following rule

$$\frac{\text{Weight of solid in air}}{\text{Loss of weight in water}} = \text{density of solid.}$$

The principle of Archimedes suggests a neat method of taking the densities of liquids, which may sometimes be found useful. It is simply, to weigh a piece of glass or other suitable solid; (i) in air, (ii) in water, (iii) in the liquid of which the density is required.

Then, as already explained,—

Loss of weight in water = weight of volume of water equal to the solid.

Loss of wt. in the liquid = wt. of vol. of the liquid equal to the same solid.

Thus we have ascertained the weights of equal volumes of water and the liquid, from which the density of the latter is found in the usual way. You may, in illustration, take the density of alcohol by this method, using the solid of which the density has just been found; and of which, therefore, the weight in air and water is already known. Compare the result with that which you previously obtained.

EXERCISE 5.

Solution, Evaporation, and Crystallisation.

Apparatus required—Porcelain mortar and pestle; scales and weights; Bunsen's holder; flask, with flat bottom about 300 c.c. capacity; glass measure; retort-stand; sand-bath, with sand; argand, or spirit lamp; funnel, 7 cm. in diameter; filters, 14 cm. in diameter; beaker; porcelain dish, 12 cm. in diameter; watch-glass; glass rod; washing bottle, filled with distilled water; spatula; cloth; writing-paper; blotting-paper; alum; sodium sulphate.

Place the mortar on a clean sheet of paper, and put into it two or three lumps of common alum. Reduce these to a coarse powder by first striking them with the pestle until they are broken up into small pieces and then completing the pulverisation by rubbing these fragments with a circular movement of the pestle (not unlike the motion used in stirring a liquid, but combined with downward pressure), occasionally shaking down to the centre portions of the salt which adhere to the sides of the mortar. When this is done, take two half sheets of writing-paper, turn up two opposite sides of each, so as to form a trough, and place one in each scale of the balance; they should then be found to counterbalance each other. Into one scale-pan put weights amounting to 30 grms., and into the other bring some of the powdered alum, using a card or spatula for transferring it. Continue to add the salt until the beam turns, then remove any excess of the powder little by little, until the beam remains level, showing that there is the same weight in each scale. Now take out of the scale the paper with the powder in it, and transfer the latter to a flask, as in Fig. 34.

Place the measure on a level table, and pour distilled water into it until the lowest part of the curved surface of the water is seen to touch the division which marks 150 c.c. The eye should be placed on a level with this division and neither above nor below it, or it will be impossible to obtain a correct measurement. Pour the measured water into the flask and

set the latter on the sand-bath, in which should be placed
enough sand to form a stratum about 1 cm. deep, or rather
less. Support the sand-bath on the largest ring of the retort-

Fig. 34.

stand, and place under it the argand burner, the flame of
which should nearly, but not quite touch the bottom of the
sand-bath.

While the solution of the alum is going on, you may get
ready a filter, or strainer, to separate any particles of dirt
which may be in the liquid.
Take a circular piece of filtering paper about 14 cm. in
diameter, fold it in half, and
then again at right angles to
the first fold, so that the circle
is reduced to a quarter-circle
consisting of four thicknesses
of paper. Open this out, so
as to form a conical cavity,
having three folds of the paper
on one side and one on the

Fig. 35.

other; and place it in a funnel slightly larger than the filter
thus folded. Fig. 35 will serve to explain the mode of folding
the filter. The filter should fit the funnel pretty accurately,
and may require to be opened out a little more, or, on the

other hand, contracted so as to form a more acute cone; but in all cases care should be taken not to injure the point, which, although it requires most strength, is generally the weakest part, since all the creases meet there.

Set the funnel in a ring of the retort-stand, and place under it a beaker or other vessel. Pour a little distilled water upon the filter, directing the stream not into the point, but down the thicker part of the side, and allow the water to drain off into the beaker. Meanwhile see if the whole of the alum has by the aid of the heat dissolved in the water, and if not, shake the flask to bring fresh portions of the liquid into contact with the undissolved portion. When the solution is complete, take away the lamp; take the flask out of the hot sand and put in its place a porcelain dish; turn round the ring holding the funnel, raising or lowering it if necessary, until the tube of the funnel just touches the side of the dish near its rim. Now, grasping the neck of the flask with a cloth, pour its contents along a glass rod, Fig. 36, so as to fall on one side of the filter; pouring slowly at first, until the filter becomes fully saturated with the solution, but afterwards keeping the filter nearly full. When all the liquid has run through, remove the funnel and filter, and replace the lamp in order to evaporate the solution; i.e. to drive off the water until only enough is left to retain the salt in solution at a temperature near the boiling-point. The liquid is then said to be 'saturated' at that temperature. To see when this is the case, dip from time to time a clean glass rod into the solution, place a drop of the liquid on a glass plate or watch-glass and stir it with the rod as it cools. If it deposits minute crystals on cooling, the proper point has been reached. The lamp should then be removed, and the porcelain dish taken carefully from the sand with the fingers (protected by a glove or cloth), and deposited in a cupboard or on a folded cloth laid on the table. Cover it loosely with a piece of paper supported under-neath by a short glass rod or tube laid across the dish. The whole should now be left quite undisturbed for two or three hours at least, in which time a good crop of crystals ought to be formed by the slow cooling of the solution. Meanwhile, if time permits, you may crystallise some sodium sulphate in exactly the same way, weighing out 60 grms. of the salt, and

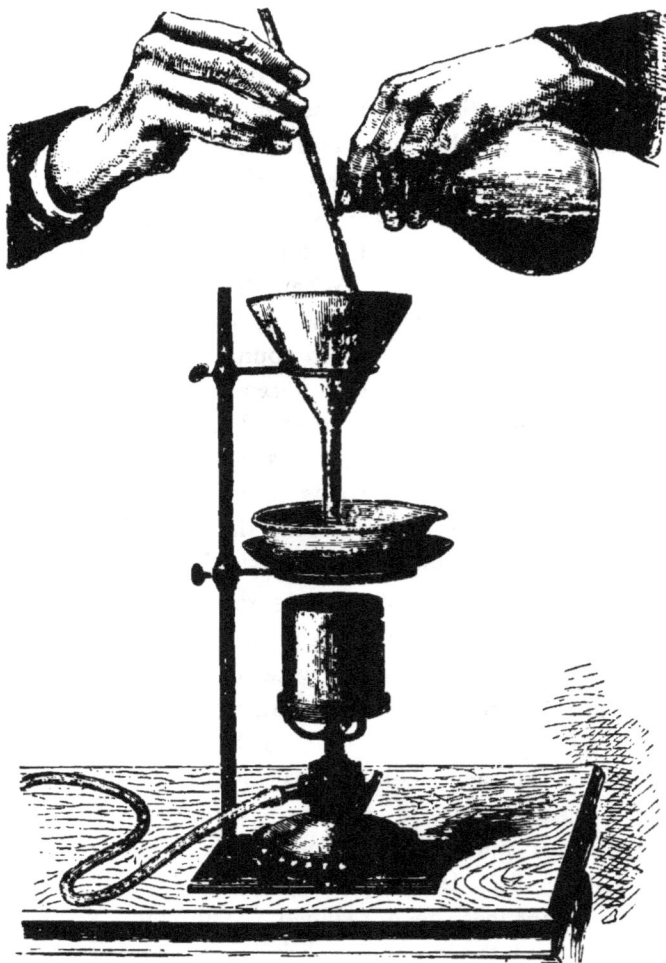

Fig. 36.

dissolving it in 100 c.c. of water. (The solution may be left to crystallise in a beaker, if no other large porcelain dish is at hand.)

When the dish containing the crystals is quite cold, pour

off the remaining solution, or 'mother liquor' as it is called, into a beaker, holding a glass rod in contact with the lip of the dish, so that the liquid may run down it and not down the outside of the dish. After the last drops have drained away, shake out the crystals on a folded sheet of white blotting-paper, dry them by pressing them gently and repeatedly with fresh pieces of blotting-paper, and examine their shape.

[If no good, well-defined crystals are obtained, put the whole of the crystals into the beaker containing the mother liquor, and heat until they are re-dissolved; then hang in the solution a loop of lamp-cotton or worsted, or a small splinter of coke, to serve as a nucleus upon which crystals may form and be free to grow on all sides.]

Notice the totally different form of the crystals of the two salts; the alum crystallising in shapes derived from the octohedron, a solid figure obtained by joining two four-sided pyramids base to base; the sodium sulphate crystallising in long four-sided prisms, like flattened rods.

Notice also the different behaviour of the salts when a crystal of each is exposed to dry air for some time [1]: the alum remains nearly unaltered, while the sodium sulphate 'effloresces,' as it is termed, or becomes converted into a white opaque substance, by giving up a certain amount of water which it contains, and which seems essential to its crystalline form. The water thus combined with the salt is called 'water of crystallisation.'

Mix in a clean porcelain dish the solutions of alum and sodium sulphate which were poured off the crystals, evaporate the mixture, as before, on the sand-bath, and leave it again to crystallise. You will now obtain a crop of crystals of both salts, and each will be found to have crystallised in its own characteristic form. If, however, the evaporation has not been carried very far, the alum (which is the least soluble of the two salts in cold water) will crystallise alone. This method, of

---

[1] The effect may be quickly shown by putting one or two of the crystals of each substance into a small wide-mouthed gas bottle, together with a small lump of quicklime loosely wrapped in paper, which will by its affinity for water maintain the dryness of the air.

partial crystallisation, is continually employed for the purpose of, separating a crystallisable salt from impurities which are more soluble than itself.

EXERCISE 6.

Solution of Calcium Hydrate (Lime water).

Apparatus required—Retort-stand; sand-bath and sand; argand, or spirit lamp; Bunsen's burner; scales and weights; porcelain dish, 8 cm. in diameter; stoppered bottle, holding about 200 c.c. (a small wide-mouthed gas bottle will do); glass rod; two watch-glasses; test-tube, about 1.5 cm. in diameter; litmus and turmeric paper; writing-paper; cloth; washing bottle, filled with distilled water; lumps of quicklime.

The salts which you have selected for the last experiment are not only very soluble in water, but also more soluble in hot than in cold water [1]. You may take calcium hydrate as an example of a substance which dissolves only in small proportion in cold water, and is still less soluble in boiling water.

Place about 5 grms. of good quicklime, in lumps, in a porcelain dish, and pour over it 4 or 5 drops of distilled water from the washing bottle. The lime, if it is freshly burnt, will become very hot and fall to pieces [2], forming a white impalpable powder. The water disappears entirely, its elements having united with the elements of the quicklime to form a single substance, calcium hydrate [3].

at 0°, at 33°, at 100°,

[1] 100 c.c. of water dissolve 5.22; 22; 421 grms. of alum.
    ,,   ,,   ,,   12.0; 322; 244 ,, of sodium sulphate.
Hence we see that, while sodium sulphate is more soluble in hot than in cold water, its solubility does not increase regularly with the temperature, reaching a maximum at 33°, and then slightly decreasing.
[2] If it does not become hot in the course of a minute, the reason is that it has absorbed moisture and carbon dioxide by exposure to the air. A lump or two may be put into the hottest part of a common fire, until it has become thoroughly redhot, then taken out, covered with dry sand and allowed to cool. It must be kept in a well-stoppered bottle.
[3] The evolution of heat is due to two causes :—(1) The chemical combination which is taking place. It is a good example of the universal law that heat is evolved during chemical combination. (2) The fact that a liquid (water) is entering into combination, while a solid (calcium hydrate) is the sole product. For in all cases in which a liquid becomes a solid, heat is evolved.

Add more water in successive small portions, until all the lumps of lime have been broken up, and the mass is thoroughly moist: then transfer it with the help of a glass rod or spatula to a stoppered bottle holding about 200 c.c. The last portions may be rinsed in by the aid of a stream of water from the jet of the washing bottle. Fill the bottle nearly to the neck with distilled water, and, after inserting the stopper, shake the liquid for a minute or two, then leave it undisturbed for a quarter of an hour. You will notice that the salt does not dissolve, like the alum, to a clear fluid (although you have added about forty times its weight of water), but that the greater part of it subsides to the bottom of the bottle. Leave the bottle until the next day, occasionally shaking it thoroughly. You will find that even then there is a large quantity of calcium hydrate remaining undissolved; indeed, you will require some positive proof that any of it has been dissolved.

Take out the stopper, wipe the inside of the neck of the bottle with a clean cloth, and pour some of the liquid into a test-tube, inclining the bottle very gently so as to avoid disturbing the sediment. Hold the test-tube up to the light, in order to see if there are any solid particles floating in the liquid. If such is the case, the liquid may be filtered into another test-tube. The following experiments should be tried with the clear solution :—

1. Dip a clean glass rod into the liquid, and taste what adheres to the rod. You will find that it is not tasteless, like pure water, but has acquired a sharp caustic taste.

2. Try its action on test-paper.

Lay strips of blue litmus, reddened litmus and turmeric paper on a white plate, and place on each, with a glass rod, a drop of the solution. It will not alter the blue litmus, but will turn the reddened litmus blue and the turmeric a brownish red. A substance which does this is said to have an 'alkaline reaction.'

3. Pour a few drops into a watch-glass; place the latter upon the sand-bath, supported as before on a ring of the retort-stand, and evaporate the liquid to dryness. For the sake of

comparison you may evaporate in another watch-glass (placed on the same sand-bath) a little of the distilled water which you have been using. When all the water has been driven off, you will find that in the former case a solid white residue is left on the watch-glass, in the latter case there is no residue whatever, or at all events a mere trace. The occurrence of such a residue is a conclusive proof that something has been dissolved by the water.

4. While the evaporation is proceeding, you may heat over a Bunsen's burner the test-tube containing the remainder of the solution of calcium hydrate.

[The operation of heating a test-tube in the naked flame requires some care, otherwise the tube may crack, or its contents may be thrown out, owing to the sudden formation of large bubbles of vapour at the bottom of the tube. Hold the tube in a slanting position, turning it round between the fingers and moving it to and fro in the flame [1], so as to distribute the heat over as large a surface as possible. Remember, moreover, that the test-tube should not, as a rule, be more than *one-third* full, and that no heat must be applied to the part of the tube above the level of the liquid; otherwise, if the colder liquid should reach this part, the tube will be almost certain to crack. When the liquid has nearly reached the boiling-point, remove the tube from the direct flame and hold it at the side or above the flame, occasionally shaking it to promote the formation of bubbles of vapour.]

Observe that the liquid, as the temperature rises, becomes milky, a solid substance being formed in it. This is owing to the fact that calcium hydrate is much less soluble in hot than in cold water [2]. A solution, therefore, of calcium hydrate, which is saturated at the ordinary temperature, will deposit a portion of the substance when it is heated. Cork the tube

---

[1] If the tube becomes too hot to be held between the fingers, a simple and convenient holder is obtained by folding half a sheet of writing-paper into a band about 2 cm. in breadth. This should be passed round the tube near its mouth, and held (close to the tube) between the forefinger and thumb. The jaws of a Bunsen's holder, Fig. 3, taken out of the socket, also form an excellent holder.

[2] 100 c.c. of water dissolve at 15°, 0.173 grms.; at 100°, 0.083 grms. of calcium hydrate.

and cool it by immersing the lower portion of it in cold water, shaking it occasionally. The greater part of the turbidity will, after some little time, disappear : the calcium hydrate being again dissolved.

5. This solution of calcium hydrate (or ' lime water,' as it is often called) is chiefly used to detect the presence of carbon dioxide (carbonic acid) since when it is brought in contact with this gas, it becomes cloudy, owing to a white insoluble substance, calcium carbonate or chalk, being formed. Hence it is called a ' test '[1] for carbon dioxide.

You may illustrate this use of it, and at the same time prove that carbon dioxide is present in the breath, by putting a little of the solution into a small beaker, and blowing into it for a few seconds through the long branch of an elbow tube, when the clear liquid will gradually become milky.

Pour off the rest of the solution of calcium hydrate into a clean bottle, inclining the bottle which contains the solution very gently, in order to avoid disturbing the solid substance at the bottom. This operation is called ' Decantation,' and will be again alluded to in Exercise **9**. Be careful to cease pouring before any turbid solution finds its way into the second bottle (if any floating particles are noticed, the whole must be filtered) ; close the latter with a good cork or stopper, and, after placing a label on it, put it away for use in future experiments.

---

[1] A test means an experiment, or a material for an experiment by which we examine an unknown substance to find out what it is.

EXERCISE 7.

Distillation.

Apparatus required—Stoppered (or plain) retort, about 200 c.c. in capacity; retort-stand; sand-bath with sand; argand, or spirit lamp; thistle funnel (Fig. 12, *a*); small flask; porcelain mortar; Bunsen's holder; funnel, 10 cm. in diameter; beaker; test-tube-stand; test-tubes in basket; watch-glass; washing bottle with distilled water; wooden blocks; blotting paper; lamp-cotton, or tow; cloth; solutions of barium chloride, silver nitrate, ammonium oxalate, hydrogen nitrate (dilute), ammonium hydrate, calcium hydrate.

Distillation is the process of converting a volatile liquid into vapour, condensing the vapour again into a liquid, and collecting the product or ' distillate.'

It is very frequently used in chemical work for separating a more volatile from a less volatile substance. For example, alcohol may be thus separated from water, and water itself may be, as will be seen in the following exercise, separated from earthy impurities which unfit it for use in the laboratory.

An apparatus for distillation consists of three parts;—

1. A boiler; for which, in laboratory work, a glass retort is generally used.

2. A condenser; for which the long neck of the retort, kept cool by being wrapped in paper soaked in cold water, will serve sufficiently well[1].

3. A receiver; such as a glass flask.

The arrangement of the apparatus is shown in Fig. 37.

Select a stoppered retort with a long neck; place it upon the sand-bath (containing a little sand), supported upon the largest ring of the retort stand at such a height as to allow the lamp to be placed under it. Put over the tubulure the smallest retort-ring, to hold the retort steady in its place. This will be most conveniently done if the rod of the retort-stand is placed behind the retort, as shown in the figure. Pass the

---

[1] A preferable form of condenser, known as 'Liebig's condenser,' is described in Appendix A.

ʹbeak of the retort into a clean flask, placed in a slanting position in the porcelain mortar, and support the latter on wooden
blocks at such a height that there may be a gradual fall from
the point at which the neck joins the body of the retort to
its outer extremity.

Now fill the retort about half full (not more) of common

Fig. 37.

water, using a funnel passed through the tubulure and taking
care that no water passes down the neck of the retort.

[If you have to use a plain retort, the water must be put in
before it is arranged in its place. For this purpose, lay a folded
cloth upon the table and place the retort upon it, supporting its
neck in a vertical position by a Bunsen's holder; then pass a thistle
funnel down the neck and pour through this the proper quantity
of water.]

Now light the lamp and put it under the sand-bath. While
the water is being gradually heated, the arrangement for condensing the steam should be set up in the following way.

Take a piece of blotting-paper about 14 or 15 cm. broad and
about three-fourths as long as the neck of the retort ; fold two
of its adjacent corners inwards so as to give the paper the following outline ; wet it thoroughly, and place it
over the neck of the retort with the folded corners underneath, so that while the narrower part
of the paper lies upon the broader part of the neck, the narrower
part of the neck may be completely enveloped by the paper,
the two unfolded corners (pressed close together) hanging down
some way below it.   Next, take a strip of blotting-paper about
10 cm. in breadth and somewhat shorter than the first strip,
fold it, in the direction of its length, into the shape of the
letter Λ, and place it like a saddle upon the first wrapper.
Just beyond the lowest end of the folds of paper, twist round
the neck of the retort a few strands of lamp-cotton or of tow,
thoroughly wetted, the end of which should hang down 10
or 12 cm.   The object of this is to prevent the water which
trickles down the paper from passing any farther down the
neck, of the retort and entering the receiver.   Underneath the
end of this piece of cotton is to be placed a beaker or basin
to catch the waste water.   Directly over the upper extremity
of the paper-wrapping support in the Bunsen's holder a large
funnel, the neck of which has been partially plugged up by
a bit of paper or tow, so that water will only issue from it
in drops.   This should be tried before the funnel is arranged
in its place, by holding it over a basin or sink, pouring in
some water, and pushing in or pulling out the plug of tow
until the drops follow each other not too quickly to be
counted.

After fixing the funnel in the Bunsen's holder (which should
have a weight put on the wooden base behind to prevent any
chance of its toppling over) fill it with common water, and
see that the drops fall slowly and regularly from its beak upon
the upper end of the blotting-paper, and that the water, after
saturating the paper, finds its way into the beaker, and not
into the receiver.   You will now see the use of the upper
strip of paper ; it serves as a channel to guide the drops of

water and to distribute them regularly over the paper-wrapping underneath.

The water in the retort will soon begin to boil, and the steam, passing over into the colder parts of the neck, will be reconverted into water which will collect in drops and run down into the receiver. When twenty or thirty drops have thus distilled over, withdraw the receiver (holding the neck of the retort steady with one hand), and pour away the liquid which it contains ; since the first portions which come over are liable to contain impurities derived from the neck of the retort. When the last drop has drained away, replace the receiver in its former position and continue the distillation until about three-fourths of the water in the retort has passed over. Pour, from time to time, some cold water into the large funnel so as to keep it nearly full. Be particularly careful to regulate the heat so that the water may not boil so violently as to splash over into the neck of the retort, and thus be carried down into the receiver. If, in spite of care, it boils unsteadily, with 'bumping,' owing to the sudden formation of large bubbles of steam in contact with the more strongly heated portions of the retort, it will be advisable to remove the lamp, to wait until the temperature of the water has sunk a few degrees, and then to introduce into the body of the retort a fragment or two of charcoal, or a bit of crumpled platinum foil. This will, in consequence of its numerous points and edges, materially assist the formation of smaller bubbles of steam, and when the lamp is replaced the boiling will go on more quietly.

While the distillation is going on, you may examine the qualities of the water which you are using, in the following way :—

Place four clean test-tubes in the stand, and fill each about one-third full of the same water which you have taken for distillation.

[In pouring a solution from a bottle into a test-tube, it should be a rule never to allow drops of the liquid to run down the side of the bottle. Apart from the unsightly appearance of a bottle encrusted with crystallised salts, the labels are likely to be obliterated or washed off, and shelves and tables will be damaged. These inconveniences may be entirely avoided by the simple method of pouring

F

which is illustrated in Fig. 38. The test-tube, into which the liquid is to be poured, is held near the top between the third finger and little finger of the left hand. The stopper, grasped by the forefinger and thumb of the same hand, is loosened and wetted with the solution by slightly inclining the bottle; it is then taken out, drawn across the neck, so as to leave a wet track, for the liquid to follow, and held against the neck of the bottle, while the test-tube is

Fig. 38.

brought directly under it. The stopper then (like the glass rod in Fig. 36, p. 56) forms a prolongation or lip down which the liquid runs, not a particle finding its way down the side of the bottle. Besides the greater facility with which drops may be measured out, this method has the further advantage, that the stopper is never out of the hands, and hence there is no danger of the solution being contaminated by impurities taken up by the stopper from a dirty table.]

(*a*) Add to the contents of one test-tube five or six drops of dilute **hydrogen nitrate** (nitric acid), and then one drop of solution of **barium chloride**; shake the test-tube, and hold it up to the light. If a white cloudiness or precipitate is produced, it is a proof that the water contains a SULPHATE.

(*b*) Add to the contents of another test-tube five or six drops of dilute **hydrogen nitrate**, and then one drop of solution of **silver nitrate**. If a white cloudiness or precipitate is produced, the water contains a CHLORIDE.

(*c*) Add to the contents of the third test-tube three or four drops of solution of **ammonium hydrate** (ammonia), and then one drop of solution of **ammonium oxalate**. If a white cloudiness or precipitate is produced, the water contains a CAL-CIUM salt [1].

(*d*) Add to the contents of the remaining test-tube about 1 cc. (a small tea-spoonful) of solution of **calcium hydrate** (lime-water). If a white cloudiness or precipitate is produced, it is, as you have already seen, p. 61, a proof that the water contains CARBON DIOXIDE.

By this time sufficient distilled water will have collected in the receiver to be examined by the same tests, in order to see whether the process of distillation has freed it from impurities (if any have been found).

Place four clean test-tubes in the stand, pour into each some of the distilled water, and repeat (*a*), (*b*), (*c*), (*d*) experiments. The water should give no cloudiness or precipitate in any of the above experiments, and if a few drops are poured into a clean watch-glass and evaporated to dryness on the sand-bath no solid residue should be left; otherwise it is most likely that the water in the retort has boiled too rapidly, and a portion of it has splashed over, or has been carried over as spray into the receiver.

Before the water in the retort has wholly evaporated, you should stop the distillation by removing the lamp, and withdraw the receiver, resting the beak of the retort temporarily upon the edge of the mortar. The retort and the rest of the apparatus should now be cleaned and put away for future use. If there is any deposit in the retort which cannot be removed by water alone, a few drops of dilute hydrogen chloride (hydrochloric acid) will at once dissolve it [2].

---

[1] If the water which you have used is fresh spring water, and if, from experiment (*c*), you have discovered that a calcium salt is present, you will probably find that, as the distillation proceeds, the water in the retort becomes turbid, and a white deposit is formed. The reason of this will be explained in the exercise on Carbon Dioxide.

[2] It will be scarcely worth your while to attempt to distil for yourself all the water you will require. You will in all probability be able to obtain it at the nearest chemist's shop at a cost of 2*d*. or 3*d*. per quart. But the water which you buy should in all cases be examined by the above tests, and rejected at once if any impurity is found in it.

EXERCISE 8.

Chemical Action.

Apparatus required.—Bunsen's burner; test-tubes in basket; test-tube
stand; cedar matches; glass rod; platinum foil; porcelain dish 5 cm.
in diameter; gas tray; crucible tongs; glass mortar; porcelain mortar;
china plate.

Mercury oxide; solution of copper sulphate (blue vitriol); granulated zinc;
dilute hydrogen chloride; phosphorus; iodine; copper filings; flowers
of sulphur; mercury perchloride ('corrosive sublimate'); potassium
iodide: solution of ammonium chloride ('sal ammoniac'); solution of
silver nitrate ('lunar caustic').

ANALYSIS.

The first and chief work of the chemist is, to examine
the composition of all kinds of matter. When anything is
brought under his notice, he tries by various forces to break
it up into two or more substances unlike itself and each other,
and to break up these again into others, until at last he has
dissected things, as it were, into a number of substances which
he is unable (with his present means) to separate further.
These simple substances are called 'Elements,' and the process
of separating bodies into their elements is called 'Analysis.'
In the work of analysis several forces are used, and the following experiments will illustrate their employment.

I. Heat.

Place about 0.5 grm. (as much as will lie on the end of a
penknife) of the red substance called 'red oxide of mercury' in
a moderate sized test-tube, and heat it strongly over a Bunsen's
burner, closing the mouth of the tube *loosely* with a cork along
the side of which a small notch has been cut. The substance
will become darker in colour, and, as the heat approaches
redness, a mirror-like ring will be formed in the cool part of

the tube. Continue the heat until the whole of the red substance has disappeared, then light a cedar match, blow it out,
and while it is still glowing plunge it into the tube. It will
burst into flame. This property of re-kindling a glowing
match is characteristic of
oxygen, and on trying
other experiments we
should find that the pro-
perties of the gas in the
tube agree altogether with
those of the substance
which has been called
oxygen. Next, examine
the mirror-like deposit in
the tube; you will see that
it is a cluster of small
globules which, on being
touched with a glass rod,
run together into a larger
globule possessing a bright
metallic lustre and con-
siderable weight. These
properties, together with

Fig. 39.

others which you would discover on further examination of the
deposit, coincide with the properties of the substance called
mercury. Moreover, if we weigh (a) the red substance before
heating it in the tube, (b) the gas evolved, and (c) the metal
deposited, we should find that the weight of the original substance was equal to the sum of the weights of the gas and of
the metal obtained from it.

Further, on making other experiments with the gas and the
metal, we should find it impossible, with our present means, to
break them up into simpler forms of matter.

Our inference, then, is that the red substance is composed of
the element oxygen and the element mercury. In what state
these substances exist in the compound we cannot say; all that
we know is, that they, and they only, can be obtained from it.

Especial notice should be taken of the difference between the *physical* action of such a force as heat, and the chemical action of the same force.

When heat was communicated to the red mercury oxide, the substance changed its size, becoming larger than before: and on further heating it became altered in its power of reflecting rays of light: in other words, it changed its colour. But these were only *temporary* changes: when the heat was withdrawn the substance recovered its former size and its former colour. Such an action is a physical one only.

But when the communication of heat was carried further, a change of a totally different character took place; the original red powder disappeared, and two totally different substances appeared,—a colourless gas, and the vapour of a brilliant metal. Further, on withdrawing heat, the original red substance was not again obtained.

We have, in fact, here a *permanent* change, and this is characteristic of a chemical action.

[Keep the tube with the deposit of mercury for another experiment, p. 73.]

II. ELECTRICITY.

Lay a piece of platinum foil in a porcelain dish, and pour upon it about 20 c.c. of solution of copper sulphate (blue vitriol). The surface of the platinum foil will be quite unaltered; but if a bit of zinc is dropped into the solution and pressed closely into contact with the platinum, a coating of a red metal, which can be no other than copper, will be in the course of a few seconds deposited upon the platinum[1].

The reason is, that an electric current has been developed by the contact of the zinc and the platinum; and this, in passing through the solution, has separated the copper from its combination with sulphur and oxygen. The metal is thrown down

---

[1] This will scarcely be seen until the platinum is taken out of the solution; since the blue liquid absorbs nearly all the red rays, which are the only ones the copper can reflect.

on the platinum : the sulphur and oxygen are liberated at the surface of the zinc, with which they combine [1].

III. CHEMICAL AFFINITY.

This is, perhaps, more generally used than any other : the principle being to overcome the attraction between two elements, A and B, which form a compound, by acting upon the compound with some substance, C, for which A has more affinity than it has for B. Then A combines with C, and B is left uncombined (always supposing that it has itself no tendency to combine with C).

Put about 10 c.c. of dilute hydrogen chloride (hydrochloric acid) into a test-tube supported vertically in a Bunsen's holder, drop in a piece of zinc, and cover the mouth of the tube with a bit of paper or card. An effervescence will begin at the surface of the zinc, and a gas will be given off which will catch fire when a light is held to the mouth of the tube, and may be proved to have all the other properties of hydrogen.

The reason is this : hydrogen chloride is composed of two elements, hydrogen and chlorine ; chlorine has a greater affinity for zinc than it has for hydrogen ; therefore it leaves the hydrogen and combines with the zinc, while the hydrogen (having itself no affinity for zinc) is liberated as gas.

If you leave the liquid until all action is over, and then, after filtering it from any undissolved particles, put it in a porcelain dish and evaporate it to dryness, you will obtain a white residue which is the combination of zinc and chlorine (zinc chloride).

SYNTHESIS.

Having obtained these simple substances or elements, the chemist proceeds to examine the way in which they combine with each other to form compounds, and the action which these compounds have on one another to produce other compounds.

---

[1] The platinum may be cleaned from the film of copper by pouring on it a few drops of strong hydrogen nitrate, which will dissolve the copper. It should then be washed thoroughly with water.

And here, in the first place, we have to distinguish between a mere mixture and a true chemical compound.

If, on putting two things together, we find,

1. That heat is given off,
2. That there is a sudden change in appearance, more or less complete,
3. That the two substances originally taken disappear permanently, and can no longer be distinguished as lying side by side in the mixture, while the resulting product is to all appearance *one* substance, alike in every part,

then we may be sure that chemical action has taken place, and that a true chemical compound has been formed by the union of the elements.

Examples.

A. Action of elements on one another.

1. Take a piece of phosphorus out of the bottle with a pair of crucible tongs, and place it in a gas tray filled with water [1]. Cut off a small piece, not larger than half a pea, dry it carefully by pressing it gently between folds of blotting-paper, and put it on a plate. Then place in contact with it a few crystals of iodine. The two substances will almost immediately combine ; so much heat will be given off as to cause inflammation of the mass, and a scarlet compound (called phosphorus iodide) will be formed, totally unlike either of the elements, and in which it would be impossible to distinguish the separate particles of iodine and phosphorus, even by a powerful microscope.

This is an example of a combination which takes place without the aid of any other force than chemical affinity. But in many cases the substances must be heated before they will combine.

---

[1] In all experiments where phosphorus is made use of, great care must be taken to prevent it catching fire. It must never be allowed to remain for more than a few seconds out of water, and should always, if possible, be held with the forceps, not with the fingers. A jug of water should always be within reach to extinguish it in case it should accidentally inflame, as may happen in a hot room.

2. Weigh out 1 grm. of copper filings and 0.5 grm. of sulphur; mix them together in a mortar and put the mixture into a moderate-sized test-tube. No combination takes place at once, as in the first experiment, and it is easy to distinguish the red particles of copper and the yellow particles of sulphur lying side by side in the mixture.

Now heat the tube over a lamp; the sulphur will melt and get darker, but nothing further will happen until it begins to boil, when the mixture will suddenly glow brightly. Take the tube away from the lamp and observe that the glow continues, showing that heat is being evolved. Combination has, in fact, taken place; and after allowing the contents of the tube to cool, you will find that the sulphur and the copper have both disappeared, and a black substance (copper sulphide) is formed, which is composed of the two elements, but is wholly unlike them in appearance and properties. Detach a little of it from the tube, and prove its brittleness (unlike copper) by powdering it in the mortar.

It is to be noticed that the weights you have taken represent almost exactly one of the two proportions in which alone copper and sulphur will combine. The exact proportion is 63.5 parts by weight of copper and 32 parts by weight of sulphur. Another compound is known containing 127 (or twice 6.35) parts of copper to 32 parts of sulphur; but if you take weights of the substances in any other proportions, the excess of one or the other will remain uncombined.

This is an important characteristic of chemical combination; you may *mix* substances in any proportion whatever, but they will only *combine* in a few definite proportions.

3. Take the tube containing the deposit of mercury which was obtained in a former experiment, p. 70, put into it one or two crystals of iodine, and heat gently first the deposit of mercury on the sides, and then the iodine at the bottom of the tube. The iodine will volatilise as a splendid purple vapour and will combine with the mercury forming a bright scarlet substance called mercury iodide, which is often used as a paint.

B. Action of compounds on one another. Double Decomposition.

In the above two cases we have taken elements and made them combine. But many chemical compounds will act upon each other, either

(*a*) combining just as if they were elements; or,

(*b*) decomposing each other and exchanging the elements they contain.

You have had a good example of the first case already in the action of water upon quick-lime (p. 58). The calcium oxide (quick-lime) and the hydrogen oxide (water) combined to form calcium hydrate, while considerable heat was evolved.

The second case is a very common form of chemical action, and is called 'double decomposition.'

Examples.

1. Weigh out 2 grms. of potassium iodide, and 3 grms. of mercury perchloride; mix them together in a porcelain mortar, and grind them to powder.

As the grinding goes on, and the particles of the salts are brought closely into contact with each other, chemical action begins and the white mixture turns scarlet.

What has happened is this: the potassium and mercury have changed places, the potassium combining with the chlorine and the mercury with the iodine; the latter compound forms a bright scarlet substance (mercury iodide) the same, in fact, which you formed in the last experiment by direct synthesis.

The change may be conveniently represented thus :—

Before Decomposition.		After Decomposition.
Substances taken.	Composition.	Substances obtained.
Mercury perchloride, 271 parts.	{ Chlorine, 71 parts { Mercury, 200 parts	} Potassium chloride, 149 parts.
Potassium iodide, 332 parts.	{ Potassium, 78 parts { Iodine, 254 parts	} Mercury iodide, 454 parts.

[Set aside the mixture in the mortar for use in the next exercise.]

Radicles—Simple and Compound. Salts.

The above examples have been selected to show the action which occurs between elements whether isolated or in combination. But it is very often the case that *groups of elements* take part in a chemical action and are transferred from one compound to another just as if they were single elements. Such closely associated groups, as well as the elements themselves, are called ' radicles,' as being the common root (radicula) or basis of a series of compounds.

Thus we find that the class of substances called ' nitrates ' all contain nitrogen and oxygen associated in the proportion of 14 parts by weight of nitrogen to 48 parts by weight of oxygen. This group is called the ' nitrate' radicle.

Similarly the substances called 'ammonium salts' all contain nitrogen and hydrogen associated in the proportion of 14 parts of nitrogen to 4 parts of hydrogen. This group of elements is called the ' ammonium' radicle; and the elements thus connected may be transferred from one compound to another without apparently losing their hold on one another.

The compounds formed by the union of two or more radicles, one being more electropositive than the other (i. e. separated by electricity at the negative pole of a battery), are called ' Salts ' in the most general sense of the term.

2. The following may be taken as an example. Place in a test-tube 10 c.c. of solution of ammonium chloride, and add to it a few drops of solution of silver nitrate. A white substance insoluble in the liquid will appear, and on being shaken up will collect together in flakes and fall to the bottom, forming what is called a ' precipitate.'

The action which has taken place may be thus represented :—

Before Decomposition. Substances taken.	Composition.	After Decomposition. Substances obtained.
Silver nitrate, 170 parts.	Nitrogen, 14 parts Oxygen, 48 parts Silver, 108 parts	Ammonium nitrate, 80 parts,
Ammonium chloride, 53.5 parts.	Nitrogen, 14 parts Hydrogen, 4 parts Chlorine, 35.5 parts	Silver chloride, 143.5 parts.

Thus the compound radicles AMMONIUM and NITRATE combine, forming a salt (ammonium nitrate) soluble in water which therefore remains in the liquid; while the two simple radicles or elements SILVER and CHLORINE combine, forming an insoluble salt (silver chloride) which is precipitated as a white powder [1].

These radicles have, in many cases, not yet been isolated. No one, for example, has ever seen the ammonium radicle or the nitrate radicle. We can transfer them from one compound to another, but we have not been able to arrest them in their passage. They are like groups of figures in a money account, which have a definite value and may be transferred from the debtor to the creditor side without being actually exhibited in cash. The elements or 'simple radicles' may possibly be themselves compound, but even if their compound nature is discovered they will not cease to be radicles.

In fact, if we consider Chemistry as 'the study of the composition and properties of radicles and of the ways in which they act on each other to produce substances permanently differing from themselves,' we shall have a description of the science which is independent of any possible modification of the atomic theory, and of any discovery of the compound nature of the so-called elements.

EXERCISE 9.

Filtration and Washing of a precipitate.

Apparatus required.—Porcelain mortar; washing bottle with distilled water; funnel 8 cm. in diameter; filters 14 cm. in diameter; flask about 250 c.c. capacity; test-tubes and stand; glass rod; evaporating dish; solution of silver nitrate.

You obtained in the last Exercise (p. 74) a mixture of

---

[1] If this is left in the light you will observe that it becomes darker in colour. This is a good example of the action of another force, light, in analysis; it has separated chlorine from the silver, leaving the latter either as subchloride or as metallic silver, forming a dark powder.

The whole art of Photography depends on this or similar chemical actions of light.

mercury iodide and potassium chloride. You have next to learn how these may be separated from each other by taking advantage of the fact that one of them (mercury iodide) is insoluble in water, while the other (potassium chloride) is soluble. In consequence of this, when water is added to the mixture the potassium chloride will dissolve, and the solution may be separated from the precipitate of mercury iodide by decantation and subsequent filtration.

Add 50 c.c. of distilled water to the mixture in the mortar, and grind the whole together with the pestle: then pour the contents of the mortar into a beaker, rinsing out the last portions with a jet of water from the washing bottle. Leave it undisturbed for a few minutes that the precipitate may subside.

Meanwhile, fit a filter about 14 cm. in diameter, folded as already directed (p. 54) into a funnel about 7 or 8 cm. in diameter, taking care that the filter fits the funnel closely and does not project beyond its rim. Wet the filter with distilled water and allow it to drain for a few moments; then support the funnel in the neck of an empty flask, and pour the liquid from the beaker down the side of the filter, using a glass rod to direct the stream, and taking care not to disturb the precipitate more than you can help. When the greater part of the liquid has been poured off, fill up the beaker again with distilled water, stir up the precipitate with a glass rod[1], and leave it to subside. When the liquid above the precipitate is tolerably clear, pour it off again into the filter, still retaining the bulk of the precipitate in the beaker. Repeat this operation of filling the beaker with water, allowing the precipitate to subside, and then pouring off the clear liquid, two or three times. This process is called 'washing by decantation,' and is especially adapted for cases where we have a powdery, quickly-subsiding precipitate. Finally, transfer the whole of the precipitate to the filter, by stirring it up with a little water, pouring it quickly into the

---

[1] It will be found an advantage to fit upon the end of the glass rod a short ferule of india-rubber tubing, to avoid scratching or breaking the beaker while stirring.

filter and again rinsing the beaker with more water. What has
passed through the filter (technically called the 'filtrate') is, of
course, a solution of potassium chloride, and need not be
preserved.

It may happen that the first portions of the liquid which run
through the filter are turbid. If this is the case, they should be
poured back into the filter, another flask or beaker being placed
to catch the fluid. If however the filtrate is still turbid, there is
reason to suspect that there is a hole in the filter itself. In such a
case it will be best to return the precipitate to the flask by making
a large hole through the apex of the filter, by means of a glass rod
pressed vertically downwards, and washing down the precipitate by
a strong stream of water from the washing bottle. Another filter
should then be fitted to the funnel and the precipitate transferred
to it as before.

The washing of the precipitate must now be completed on
the filter, in order to free it from all traces of the solutions
from which it was formed. This is done by pouring over it a
gentle stream of water from the jet of the washing bottle, until
the funnel is nearly, but not quite, full; the level of the water
being on no account allowed to rise above the edge of the
filter. When the washing water has entirely run through the
filter, you may pour on a fresh supply, taking care to direct the
stream on the sides of the filter, so as to wash the precipitate
down towards the lowest point.

The washing of the precipitate can only be considered
complete when no potassium chloride can be detected in the
filtrate. The presence of a chloride in a solution may be
detected, as we have already seen (p. 66) by adding solution
of silver nitrate, which gives a white precipitate. The filtrate
should therefore be tested from time to time by allowing a little
of it to run into a test-tube and adding a drop of solution of
silver nitrate. The cloudiness produced will get less and less
on successive trials, and finally the filtrate will remain quite
clear on addition of the test. When this is the case the mercury
iodide may be considered sufficiently washed, and the funnel
may be covered with a piece of paper turned down at the sides

so as to form a cap, and put to dry (resting on its side in an evaporating dish) in a warm place. When dry, the mercury iodide may be detached from the filter, shaken into a short test-tube or small bottle, and kept (duly labelled) for use in a future exercise (see under MERCURY).

[Meanwhile it may be noticed that this substance when mixed with gum forms a splendid scarlet paint, the only drawback of which is its want of permanency. In illustration of this, the filter with the remains of the precipitate adhering to it, may be spread out flat on a piece of wire gauze and be heated gently over a lamp. It will be found to turn yellow, and at a slightly higher temperature, to volatilise. If the yellow substance is rubbed with a glass rod it will turn scarlet again.]

EXERCISE 10.

Use of the Pneumatic Trough.

Apparatus required.—Pneumatic trough; two cylindrical gas jars, 5 × 20 cm.; one ditto, 3 × 10 cm.; test-tubes; india rubber rings; corks; cork borers; pipette; jug of water; cloth.

Fill the trough with clean water up to the level of the over-flow-pipe, or about 2 cm. above the shelf, and place it before you on the table, the broad fixed ledge being farthest from you, and the movable shelf being placed near the left-hand end of the trough. Take one of the large cylindrical jars, and, holding it nearly horizontally in the right hand, plunge it into the water, the mouth of the jar being held a little higher than the closed end, in order that the air may more readily escape. When the jar is filled with water and wholly immersed in the trough, bring the closed end uppermost and raise the jar vertically until its mouth is on a level with the shelf, and then move it laterally until it rests over the hole in the shelf : taking care to keep the mouth always below the surface of the water. So long as this condition is fulfilled, you will find that the jar remains full of water, the column of water being retained in it by the pressure of the air on the surrounding water. Next, take a similar jar, holding it vertically, mouth downwards, the closed end being grasped by the right hand, and bring its mouth about 3 or 4 cm. below the water-level in the trough. It will now represent a jar full of gas, which is to be transferred or decanted into the other jar. This is effected by a manipulation quite analogous to that by which we pour water from one vessel to another, the only difference being due to the fact that instead of pouring water downwards through air, we have to pour air upwards through water. Hold the jar containing water steady with the left hand, bring the mouth of the other jar under the hole in the shelf, and gradually depress its closed end so that the air contained in it may ascend bubble by bubble through the hole into the jar placed over it. The level of the water in the latter will fall until the jar is completely filled with air. The reason for steadying it with the left hand will now be evident, since it will show a tendency to topple over owing to the upward pressure of the water; and it is best never to fill a jar completely with gas, but only so far that the level of the water inside and outside the jar may be the same. It can easily be filled completely, when required, from another jar.

Repeat the experiment, decanting air from the one jar into the other until you can do so without allowing a single bubble to escape. It is not necessary that the jar should rest on the shelf, but it may be held in the left hand, as in Fig. 49, p. 109, its mouth being retained about 2 cm. below the water-level, and the mouth of the jar of air brought close to its edge, and a very little to the right of it, since the bubbles of gas have a forward as well as an upward direction. You may, in the next place, fill the small jar with air and decant its contents into the larger one. Then try to decant the air back from the large jar into the small one. This will not be found so easy, since the bubbles of air from the mouth of the jar are so large that unless the jars are held steady and the decantation is very gradual a waste of gas is likely to occur, and must occur if the disproportion between the mouths of the jars is very great. You will know that none has been lost if the small jar is just filled with the air decanted from the larger one.

[When gas is to be transferred to a test-tube, it is better to decant it first into a jar of intermediate size, and then into the tube; or, the object may be effected in the following way. Fill the tube in the usual way with water, and insert into its mouth, still held downwards below the water-level, the up-turned beak of an inverted funnel about 7 cm. in diameter. Retain the funnel in its position by placing the little finger under the rim, the tube being held upright between the thumb and the other fingers. If, now, air be decanted from a larger jar, bubble by bubble, into the funnel, the bubbles will rise into the tube without any loss, the displaced water making its escape between the mouth of the tube and the funnel, which latter should not be held too closely in contact with the former.]

Additional Experiments.

A. Effect of change of temperature on the volume of a gas.

Take a test-tube about 2.5 cm. in diameter and 15 or 20 cm. in length; place round it, about the middle, a small india-rubber band; fill it with water, and place it inverted upon the shelf of the trough. Decant air into it from a small test-tube until, when your eye is brought on a level with the india-rubber ring, the lowest part of the curve which is formed by the surface of the liquid appears *just* to touch the upper edge of the ring[1]. Heat a little water in a test-tube nearly to boiling, and pour it over the tube containing the measured volume of air. The surface of the water in the tube will rapidly sink below the ring, showing that the air has expanded with the heat; but when a little water at the ordinary temperature is poured over the tube, the air will contract to its original volume. If you can obtain any ice or snow, you may place some in a flask and add some water; when the latter has been cooled down nearly to the freezing-point it should be poured over the tube containing the air. The surface of the liquid will now rise above the ring, the air contracting in volume as the temperature is lowered.

Try a similar experiment with coal gas, decanting the requisite quantity into the tube from an india-rubber tube connected with the gas supply.

The same experiment might be tried with any other gas, and you would find it to be a universal law that *the volume of a gas increases as its temperature is raised, and decreases as the temperature is lowered.*

Exact determinations of the amount of this variation in volume

---

[1] See p. 47, *note.*

were first made by Gay Lussac, who established the following law, applicable to all true gases, whatever their nature,—

FOR EVERY INCREASE IN TEMPERATURE OF 1°C A GAS EXPANDS $\frac{1}{273}$ (= 0.00366) OF THE VOLUME IT OCCUPIES AT 0°C.

B. Effect of change of pressure on the volume of a gas.

This is not quite so simply demonstrated at the pneumatic trough, since it is impossible to effect much variation in the length of the column of water which gives the pressure; and hence only a small, but still perceptible, difference in volume is obtained. The experiment may, however, be made in the following way. Fit a good cork to a test-tube about 1.5 cm. diameter: bore a hole in it (p. 37), and fit into the hole a straight piece of glass tubing about 20 cm. in length, and 6 or 7 mm. in diameter (the pipette you have made (p. 33) will do very well). Take out this tube, when fitted, and insert the cork tightly into the test-tube. Slip over the small tube a narrow india-rubber ring (cut from a piece of tubing), and sink it vertically in the trough almost to the bottom; then, holding the test-tube by the rim (lest the warmth of the hand should alter the volume of the enclosed air), fit the small tube into the cork, raise or lower the apparatus until the level of the water in the tube and in the trough is the same, and adjust the ring to this level. The air enclosed in the tube is now at the same pressure as the external air; both pressing on the surface of the water with the same force. Owing to the difference in diameter between the large and the small tube, a slight alteration in the volume of the air in the former will cause a considerable difference in the level of the water in the latter. Now slip under the extremity of the tube a gas jar, wholly immersed in the trough, and raise the whole out of the trough, the gas jar serving as a deep pneumatic trough. Set the jar on the table, and raise the tube until its extremity is only just below the water-level: you will observe that the water in the tube falls below its former level, showing that the volume of the air in the apparatus has increased. This enclosed air is obviously under less than the whole pressure of the external air, since the pressure of the latter is partly balanced by the column of water in the tube. Next, sink the tube in the jar until it touches the bottom: the water will rise above its original level in the tube, showing that the enclosed air has contracted in volume. This air is now exposed to more than the pressure of the external air, the latter being aided by the column of water in the jar which is above the level of the water in the tube[1]. We have thus an evidence that *the volume of a gas increases as the pressure to which it is exposed is lessened, and decreases as the pressure is augmented.* By measuring the pressures to which gases may be subjected, and the corresponding volumes, the following law, applicable to all gases, was established by Marriotte,—

THE VOLUME OF A GAS VARIES INVERSELY WITH THE PRESSURE UPON IT.

The practical lesson to be learnt from these experiments is that in measuring gases over the pneumatic trough especial care must be taken to deal with them under equal conditions of temperature and pressure (or to make allowance for any difference): otherwise the apparent volumes will not represent the real quantities we wish to take. The jars should not be held more than is necessary in the warm hands, and the volumes should be read with the water as nearly as possible at the same level inside and outside the jar.

EXERCISE 11.

The Use of the Mouth Blowpipe.

Apparatus required—Blowpipe; blowpipe lamp; piece of platinum wire; piece of platinum foil; two or three pieces of charcoal; small glass mortar; watch-glasses; crucible tongs; test-tube; knife; sodium diborate; sodium carbonate; potassium dichromate; potassium cyanide; strontium nitrate; solution of cobalt nitrate; manganese dioxide; tin peroxide; dilute hydrogen chloride.

1. General principles of its use.

The mouth blowpipe consists, in its simplest form, of a tapering tube, usually bent near its smaller end to a right

---

[1] Much greater variations in pressure can be obtained by connecting the end of the tube with a funnel by means of a long india-rubber tube; both this and the funnel being filled with water by immersing them in the trough (care should be taken to get rid of all air bubbles). If the funnel is then raised or lowered, the column of water in the tube is added to or subtracted from the pressure of the air, as above explained.

It may be convenient to note that a column of water 1 metre in height very nearly $= \frac{1}{10}$ of the average air-pressure.

angle, and terminating in a fine jet. But as the moisture of
the breath soon collects in the tube and interrupts the flow
of air, a much superior form of blowpipe is that which was
introduced by Dr. Black, and which is represented in Fig. 40.

Fig. 40.

All moisture is condensed in the wider part of the tube,
while the movable nozzle can be readily unscrewed and cleaned
out, if it should become stopped up by soot or oxide. Before
the instrument is used, the nozzle should be carefully examined
to see that the aperture is clear, round, and not very large,
otherwise the cone of flame will be ragged, irregular, and brush-
like.

The best fuel to be used with the blowpipe is undoubtedly
ordinary coal gas, where it can be obtained. The most efficient
form of burner is that represented by Fig. 41. It
consists merely of a short piece of brass tube about
1 cm. in diameter, flattened out at one extremity until
its edges form a narrow rectangular aperture about
0.5 mm. broad, inclined at an angle of 70° to the axis of the
tube. This tube, which need not be more than 3 cm. long,
may be screwed into the same iron foot which serves as the
base of the argand burner[1]. The orifice of the burner should
not be more than 10 cm. above the table, in order that the
greatest steadiness may be secured by resting the arms on the
table, while the right hand holds the blowpipe and the left hand

Fig. 41.

---

[1] If gas is not available, an oil lamp, a tallow lamp, or a spirit lamp fed
with a mixture of ten parts spirit of wine and one part turpentine, may be
used; or finally, but not preferably, a wax candle. The lamp should have
a flat wick, which just before use must be carefully trimmed smooth and
divided along the middle with the trimming scissors, so as to leave a furrow
along which, and about 2 mm. above it, the blast of air from the blowpipe
must be directed.

holds the support containing the substance to be examined.
The blast of air should be directed obliquely downwards,
parallel, in fact, with the orifice of the burner. The following
figure (Fig. 42) will serve to make the position clear.

Fig. 42.

The first thing to be learnt in the use of the blowpipe is the
method of keeping up a regular, continuous blast of air for
several minutes without interrupting the ordinary process of
respiration. This is by no means difficult. In the ordinary
double organ bellows, we can distinguish two essential parts;
(*a*) the lower compartment or 'feeder,' which by its alternate
expansion and contraction supplies air intermittently; and (*b*)
the upper compartment or 'reservoir,' which receives and stores
up the air thrown in by the feeder, and sends it out in a continuous stream at a uniform pressure to the pipes. When the

blowpipe is properly used, the lungs and the mouth are acting respectively as the feeder and the reservoir of the bellows; the muscles of the cheeks acting as the weights placed on the reservoir to secure a constant pressure on the gas; the tongue, slightly drawn back and applied to the roof of the mouth, representing the valve between the feeder and the reservoir; and the nostrils representing the aperture through which air enters the feeder.

These analogies being borne in mind, the following general directions will, it is thought, be sufficient to guide the student in learning the use of the blowpipe; a few minutes' practice being of more value than a prolonged description.

Begin by distending the cheeks with air from the lungs; keep them thus distended while you breathe freely through the nose. In doing this you will have unconsciously placed the tongue in the position it should occupy as a valve between the mouth and the lungs. Now insert the blowpipe between the lips, and expel the air from the mouth through the jet by compressing the muscles of the cheeks. When the stock of air in the mouth is nearly, but not quite, exhausted, introduce a fresh supply of air direct from the lungs, interrupting the ordinary respiration for a moment only, and slightly relaxing the muscles of the cheeks, so that they again become distended with air.

A difficulty will probably be found at first in keeping the pressure constant just at the moment at which air is thus thrown into the mouth from the lungs. A very little practice, however, will enable you almost unconsciously so to balance the action of the muscles of the chest and cheeks, that the latter yield exactly in proportion as the former impel air into the mouth, and thus no variation is perceptible in the blast of air from the jet. When, after one or two trials, you find that you can produce a fairly uniform stream of air for a minute or so, you may bring the jet into the lamp-flame (which should be about as large as the flame of an ordinary candle) in the position above described, and use the blast of air to deflect the flame. The appearance of the latter (see Fig. 43, p. 88) will sufficiently indicate the regularity and pressure of the stream of air. It should appear as a well-defined cone of blue light, burning noiselessly, surrounded by a faint nebulous yellowish envelope, which reaches for some distance beyond the apex of the cone. If it flickers, and burns with a roaring noise, either the jet is not introduced sufficiently far into the flame, or the aperture is too large, or the pressure of air too great. If it is irregular in outline, the aperture of the jet is not round, and must be cleaned out and rounded by introducing a large needle; or, if you are using a lamp with a wick, the latter may not be evenly trimmed, the projecting filament breaking the current of air. If the flame appears as a luminous tongue, either the aperture of the jet is too small, or the stream of air is not propelled with sufficient force; or, finally, the flame itself is too large.

The blowpipe-flame is used for three distinct objects.

1st, and always, to subject substances to a higher temperature than the lamp alone would give.

2nd. To promote the union of oxygen with substances capable of combining with it; in other words, to 'oxidise' them.

3rd. To abstract oxygen from substances which readily part with it; in other words, to 'reduce' them.

The possibility of employing the blowpipe for the last two purposes will be evident if we examine the structure of the lamp-flame before it is deflected by the blast of air. It will be readily seen to consist of three distinct parts, as shown in Fig. 43.

1. A dark central portion, *a*, composed. of gaseous compounds of hydrogen and carbon, unburnt as yet, since no oxygen can reach them, but raised to a high temperature by the surrounding portions of the flame.

2. A luminous zone, *b*, depositing soot on cold bodies placed in it, consisting of the gaseous hydrocarbons in the act of combining with the oxygen of the air; their hydrogen being wholly burnt to water, their carbon only partially burnt to carbon dioxide; the remainder of the carbon being set free and raised

to a white heat by the temperature of combustion, thus imparting to the flame all its luminosity. This carbon is itself con-
sumed as it reaches the exterior of the zone.

3. A scarcely visible, ill-defined external envelope, *c*, consisting of the gaseous or vaporised products of combustion, viz.
carbon dioxide and water, mixed with a great excess of air
strongly heated owing to its proximity to the zone of combustion.

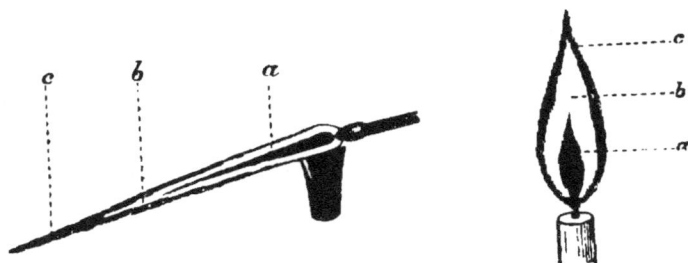

Fig. 43.

It is obvious that a substance placed in the centre of the
flame and gradually brought to the exterior will be successively
subjected to reduction, simple ignition, and oxidation.

1. The heated hydrocarbons which compose the central
portion will, if the substance contain oxygen loosely combined,
abstract that oxygen and be converted into carbon dioxide and
water.

2. As the substance approaches the exterior of the flame it
will arrive at an area of perfect combustion, where the hydrocarbons meet with just enough oxygen to burn them completely,
and where, consequently, the temperature is highest.

3. Finally, when it arrives at the extreme border of the
flame, it is in contact with excess of highly heated air, or
diluted oxygen, which will combine with it if combination is
possible under such conditions.

The cone of flame produced by driving a blast of air from
the blowpipe through the lamp-flame (see Fig. 43) is precisely
similar in its constitution; its powers being intensified, not

changed in kind, by a judicious adjustment of the relative proportions of gas and air.

We have, then, the following general rules for the use of the blowpipe,—

1. **To effect reduction;**—more gas and less air will be needed, the object being to burn the gas only partially, and thus to compel it to obtain the oxygen necessary for combustion from the substance submitted to its action. Admit more gas, therefore, until the flame is about 7 cm. high, and, holding the blowpipe-jet at the *border* of the visible flame, blow a gentle stream of air through it, so as to deflect the flame in the form of a long luminous tongue, within the tip of which (between *a* and *b* in the figure) the substance to be reduced must be held, wholly immersed in the flame, and thus exposed to the same conditions as the ore in the body of a smelting furnace, surrounded by ignited fuel and combustible gases.

2. **To effect oxidation;**—more air and less gas is needed, the object being not only to burn the latter completely, but also to provide, over and above, a supply of highly heated oxygen, and thus to expose the substance to conditions similar to those which exist on the hearth of a cupelling furnace. Diminish the supply of gas until the flame is only 5 or 6 cm. in height; introduce the blowpipe-jet about 1 mm. within the border of the flame, and blow more strongly so as to produce a well-defined blue cone, surrounded by a faintly luminous envelope, and hold the substance to be oxidised near the point marked *c* in the figure, about 1 cm. in front of the tip of the blue cone.

3. **To expose a substance to the highest attainable temperature,** it must be held at the point *b* in the figure, i. e. just at the tip of the blue cone mentioned above. Here the combustion is complete, and hence we have the maximum temperature which can be produced by the union of the gases [1].

---

[1] It must be understood that, where an oil or spirit lamp is used, the raising and lowering the wick will have the same effect as increasing or diminishing the supply of gas to the burner.

[The supports for substances to be exposed to the blowpipe-flame are principally of the following kinds.

1. Platinum foil or wire: the former in strips about 2 cm. broad and 5 cm. loug, held in a pair of crucible tongs or forceps; the latter, No. 26 brass wire-gauge, in pieces about 6 or 8 cm. long, terminating in a small ring about 2 mm. in diameter, formed by bending the end of the wire round any small cylindrical body; the tip of the blowpipe-jet answers very well.

Platinum foil is used as a support when substances are to be heated *per se*, to test their fusibility or volatility [1]. The loop of platinum wire serves to hold a bead of melted borax when we wish to examine the colour imparted to it by certain metals.

2. Charcoal, in the form of flat pieces of beech or elm charcoal, which should be carefully selected free from cracks and of close even texture. Good pieces may always be found among ordinary charcoal, sold for burning, and should be cut across the grain with a fine saw into pieces about 1.5 cm. thick. A cross section is

Fig. 44.

always to be preferred to a longitudinal one, since the former absorbs the flux more readily, and does not split when heated. With the point of a knife, or a spatula, cut a small conical hole in the charcoal, as shown in section in Fig. 44, so as to form a small crucible for containing the substance; taking care that the surface of the charcoal extends some little way beyond the hole, so that any incrustation may be retained for examination. The charcoal may be conveniently held in a pair of crucible tongs, their bent points being turned upwards, as in Fig. 42.

3. Pieces of hard, difficultly fusible glass tubing, about 6 cm. in length and 3 mm. in diameter internally, sealed at one end and expanded into a small bulb. The shape and method of making these tubes is given at p. 42.

4. Pieces of similar tubing about 8 or 9 cm. in length, open at both ends.

Such supports are extremely useful, when the effect of simply

---

[1] Pieces of broken porcelain, such as the fragments of an evaporating dish, will almost always serve in place of the platinum foil, when it is not essential that silica or alumina should be excluded. Platinum foil should be used only when nothing else will serve the purpose, since it is not easy to keep it clean, and it is soon acted on and pierced with small holes which render it useless.

heating a substance is to be tried, as a preliminary to its actual analysis; especially where there is reason to expect that any volatile products may be formed, since these will be retained in the cooler part of the tube for future examination. In general, however, the heat of a Bunsen's burner, or even a good spirit lamp, will be sufficient in such cases. Indeed, no glass tube will long withstand the intense heat of the blowpipe-flame without softening and enclosing the substance under examination.

The pieces of platinum wire for blowpipe use may be conveniently preserved in the following way:—

Fit a cork to a wide-mouthed 1 oz. bottle; then with a sharp cork-borer cut two or more holes in the cork, about 5 mm. in diameter. Push out of the borer the small cylinders of cork extracted from the holes, and make a pin-hole along the axis of each. Pass one end of the piece of platinum wire through the pin-hole and bend it into a hook on the other side of the cork to secure the wire in its place. Lastly, fill the bottle nearly to the neck

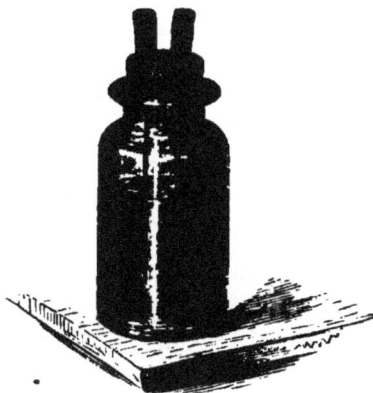

Fig. 45.

with dilute hydrogen chloride; cork it, and fit into each hole a cork cylinder with wire attached, so that the end of the wire may be immersed in the dilute acid. By this means the wires will be kept always clean and ready for use; while the cork cylinders will serve as convenient handles and prevent the wires from being lost. Fig. 45 represents a bottle thus fitted.]

2. Examples illustrating its use.

In the first place, light the lamp and see that it burns with a steady flame, about 6 cm. high, and that the wick, if it is an oil lamp, is properly trimmed and divided down the middle. If the flame wavers, owing to draughts in the room, a screen of some kind must be arranged to protect it. Next examine the blowpipe-jet, and clean it out, if necessary, with a needle, until it deflects the flame into a steady pointed cone. Arrange the lamp at such a height that the hand which holds the substance in the flame may rest steadily on the table, while the other hand holding the blowpipe may also be firmly supported either on some part of the stand of the lamp itself (Fig. 42, p. 85) or on a wooden block placed close to it. A little of each of the substances required should be placed in watch-glasses near at hand.

A. Simple ignition.

1. Make a ring, ⭘ (actual size), at the end of one of the bits of platinum wire by bending it round the tip of the blowpipe jet, moisten it with a drop of distilled water, dip it into the sodium carbonate, and hold it in the hottest part of the blowpipe-flame : notice the intense yellow colour imparted to the flame by the salt, which is highly characteristic of the metal sodium.

[The platinum wire must be rendered perfectly clean between each experiment by washing it with a few drops of dilute hydrogen chloride, and heating it strongly until it does not of itself impart any colour to the flame.]

2. Moisten the wire with strong hydrogen chloride, dip it into the powdered strontium nitrate, and hold it in the flame as before. The strontium present will impart to the flame a characteristic crimson colour.

3. Heat the loop at the end of the platinum wire, and dip it into powdered sodium diborate (borax) placed in a watch-glass. A portion will adhere to the wire, and must be again brought into the blowpipe-flame. It will at first swell up and give off water, but will finally fuse into a colourless transparent bead (of sodium metaborate), which undergoes no change on being further heated. When the bead is cold, moisten it slightly with solution of cobalt nitrate, and again heat it slowly before the blowpipe, holding it in the hottest part of the flame, i. e. at the apex of the blue cone. When all action appears to have ceased and the bead is as clear as at first, withdraw it from the flame and allow it to cool. It will now be seen to have acquired a deep blue colour, if the cobalt salt has been taken in the right

proportion. If the bead appears nearly black, too much of the cobalt nitrate has been added, and the wire after being again heated must be tapped on the edge of the table, so as to shake off the greater part of the still fluid bead, then dipped into the borax and again fused before the blowpipe. If the blue colour is very faint, a little more of the cobalt nitrate must be taken ; but the delicacy of the reaction is so great that it is hardly possible to take too little of the substance.

[To clean the wire, heat the bead strongly till it melts, and immediately shake it sharply off into the sink or upon a plate : melt some more borax upon the wire, and shake the bead off in the same way. On again making a bead, it will be usually found sufficiently colourless for other experiments.]

[Other examples of the use of the blowpipe in ignitions will be found under the heads of SILICATES, CHROMIUM, and ALUMINIUM.]

B. Oxidation and Reduction.

[In the preceding experiment it is immaterial into what part of the flame the substance is introduced, since the colour imparted to borax glass by cobalt is the same whether the oxidising or reducing flame be employed to melt the bead. In the following experiments, however, the distinction between the two flames must be carefully observed, and the bead should be kept steadily for at least half a minute in the one or the other flame, as directed.]

1. Form a borax bead as above, and, having added to it a trace of manganese dioxide, heat it in the oxidising flame, holding it 1 cm. at least in front of the visible flame. The bead will acquire an amethyst colour. Now heat it again, but this time in the reducing flame, holding it so that the luminous portion shall completely envelope it, and taking care that it does not, even for a moment, remain in the outer border of the flame. It will now be found that the amethyst colour has nearly or completely disappeared, and the bead is as colourless as at first. But if it be again held in the oxidising flame, the colour will return.

The reason is this,—Manganese unites with sodium borate to form two distinct compounds, one containing more oxygen than the other. That which contains most oxygen is deeply coloured : the other is colourless. By heating the substance in the oxidising flame, we determine the formation of the first, or highly oxidised salt; but when this is transferred to the reducing flame, oxygen is taken away from it and the colourless salt is formed.

2. Prepare a charcoal support, as described p. 90. Mix together in a mortar equal parts of tin dioxide and potassium cyanide, and transfer a small quantity of the mixture to the hole in the charcoal, taking care that none is spilled over the surface of the support. Bring the mixture into the reducing flame of the blowpipe, holding the charcoal slightly inclined towards the jet, so that the flame may play directly into the hole. The mass will readily melt, and bright globules of metallic tin will make their appearance, while the flux will be gradually absorbed by the charcoal. Maintain the heat steadily until the scattered particles of metal have run together into one globule, and the flux has disappeared ; then withdraw it quickly from the flame and let it cool. Try the malleability of the metal by detaching it with a knife from the charcoal, placing it in the mortar and pressing the pestle strongly down upon it. It will be found to spread out under the pestle into a flat plate, not crumbling to powder or even tearing at the edges.

3. Make a hole in another part of the charcoal support, and place in it the piece of tin which you have just obtained. Fuse it in the reducing flame, and notice that its surface can be kept quite bright so long as it is held in that flame. Remove it into the oxidising flame, lowering the gas flame and slightly increasing the blast of air. It will now become tarnished, a crust of white oxide being formed, which appears to grow out of the metal. Bring it again into the reducing flame, and a bright globule of metal will be again formed. The reduction should be assisted by the addition of a minute quantity of potassium cyanide.

[Other examples of oxidation and reduction will be found under the heads of COPPER, IRON (borax-beads), BISMUTH and CADMIUM (reductions on charcoal)].

SECTION II.

PREPARATION AND EXAMINATION OF THE PROPERTIES OF THE PRINCIPAL NON-METALLIC RADICLES AND THEIR COMPOUNDS.

[An asterisk is prefixed to those experiments which illustrate properties of the substance which are considered especially applicable for its detection in the course of qualitative analysis.]

---

1. OXYGEN and OXIDES.

Apparatus required—Pneumatic trough; argand burner; fish-tail burner, or spirit lamp; Bunsen's burner; retort stand; sand-bath; piece of wire gauze; corks; cork-borers; rat-tail file; three-square file; Florence flask; glass tubing, about 5 mm. in external diameter; porcelain mortar; two small porcelain dishes; scales and weights; sheets of writing-paper; deflagrating jar; deflagrating cup; one wide-mouthed stoppered bottle, holding about 700 c.c.; two ditto, holding about 200 c.c.; two cylindrical gas jars, 20 × 5 cm.; ground-glass disc, 8 cm. in diameter; taper on wire (Fig 9); cedar matches; piece of watch-spring; German tinder, or fusee; crucible tongs; jug of water; blotting-paper; potassium chlorate; manganese dioxide; clean sand; sulphur; box of test-papers.

Oxygen is generally obtained by the action of heat on some compound containing it. Thus when mercury oxide was heated (p. 69) it was found to give off oxygen; and this method is of interest as being the one used by Priestley in his discovery of oxygen. But practically another substance, potassium chlorate, is nearly always used in the preparation of the gas. This salt contains potassium chlorine and oxygen, and when it is heated the whole of the oxygen in it (about one-third of its weight) is evolved, and a compound of potassium and chlorine [1] (potassium chloride) remains. This may be tried on a small scale as follows.

[1] $2 \, K \, Cl \, O_3 = 2 \, K \, Cl + 3 \, O_2.$

Place a few crystals of potassium chlorate in a dry test-tube, supported in a slanting position in a Bunsen's holder (as in Fig. 39), and apply heat. The salt will fuse into a liquid, and at a temperature just above its melting-point, will give off a gas, with effervescence, which may be shown to be oxygen by its power of re-kindling a glowing match introduced into the tube. Continue heating the salt for half a minute, then remove the lamp, and when the fused mass in the tube is cool, warm it with a little water, and test the solution with a drop of silver nitrate. A white precipitate will be produced, which we have seen already (pp. 66, 75) to indicate the presence of a chloride.

[The reaction may be expressed as follows [1] :—

Substance taken.	Composition.	Substances obtained.
Potassium chlorate, 122.5 parts.	Potassium, 39 parts————————	Potassium chloride, 74.5 parts.
	Chlorine, 35.5 parts————————	
	Oxygen, 48 parts————————	Oxygen, 48 parts.]

A very pure gas may be obtained by this method, but the temperature at which the decomposition takes place is so high that ordinary glass flasks cannot support it without softening. It is found, however, that if the salt be mixed with manganese dioxide, the decomposition takes place at a much lower temperature, and with greater rapidity. The manganese dioxide does not, apparently, give up any of the oxygen which it contains : at any rate, it is found at the close of the action to be unaltered in composition, and may be employed repeatedly with fresh portions of potassium chlorate.

In the first place the materials for the preparation of the gas should be got ready, as it is important that they should have time to dry thoroughly. If this precaution be neglected

---

[1] The student is advised to write out all the chemical actions which he meets with, in some such form as the above. When he has studied the properties and combinations of two or three of the elements he will be better able to appreciate the grounds of the atomic theory, and the system of symbols by which all chemical changes may be concisely yet fully expressed in the form of equations.

the flask in which they are to be heated is not unlikely to crack during the experiment, owing to the moisture condensing in the neck and dropping down upon the hotter portions of the glass.

Weigh out on paper (as directed in Exercise 5, p. 53), 25 grms. of crystallised potassium chlorate; reduce the crystals to powder in a mortar; then place the salt in a porcelain dish on the sand-bath to dry.

Weigh out, similarly, 6 grms. of manganese dioxide, and place it also in a porcelain dish on the sand-bath near the potassium chlorate, stirring them both occasionally with a glass rod.

In the next place, you have to prepare an apparatus for generating and collecting the gas. Since heat is required to decompose the potassium chlorate, a flask supported in a re-tort-stand will serve to contain the materials; and since oxygen gas is scarcely soluble in water, it may be collected over the pneumatic trough filled with water; a bent glass tube will serve to convey the gas from the flask to the trough.

Take a clean Florence oil-flask, selecting one which has a smooth even mouth; roughen the sharp edges of the mouth with a file; or, if a Herapath's blowpipe is at hand, turn the edges outwards so as to form a spreading lip, as directed, p. 44. Cut off a piece of the glass tubing about 65 cm. in length, and round the edges in the usual way: then (using the fish-tail burner) bend the tube to an acute angle, making the middle of the bend not more than 8 cm. from one extremity; and lastly make a slight bend in the opposite direction as near the other extremity as possible, so as to give the tube the shape shown in Fig. 46.

Fig. 46.

The object of this second bend is to facilitate the escape of

H

the bubbles of gas by giving them a forward and upward direction. It will be found convenient to use a cork handle (p. 32), in case the end of the tube should become too hot to be held in the fingers. When the glass is quite soft, hold the tube out in front of you, so that the eye may be in the plane of the first bend ; it will then be easy to turn up the end which is nearest to you, so that both the bends may lie in the same plane but in opposite directions.

Next, choose a sound cork very slightly larger than the neck of the flask, squeeze it until it becomes soft and elastic, and bore a hole through it for the delivery tube, using a cork-borer which is rather smaller than the glass tubing which you have used.

The end *a* should now be fitted with gentle pressure and twisting motion into the cork, through which it should pass completely and project slightly at the opposite end. Do not attempt to use much force in pushing the tube into the cork, or you may break the tube and be cut with the splinters of glass ; but enlarge the hole with the rat-tail file, until the tube will enter it without much difficulty.

The potassium chlorate and manganese dioxide may now be taken from the sand-bath, and set aside to cool for a minute or two, during which time the pneumatic trough may be filled with water up to a level about 2 cm. above the shelf. It is well to have a jug of water at hand, in case more should be wanted. Take the deflagrating jar, slightly grease the stopper and fit it tightly in its place ; then immerse the jar sideways in the water, raising the open end a little, so as to allow all the air to escape. When the last bubble is gone depress the open end, and raise the jar by its neck (taking care to keep the lower end below the water-level), until it can be moved laterally to its place on the shelf. Fill the two cylindrical jars with water and place them inverted on the shelf. The large bottle should be filled with water from a jug, the (greased) stopper inserted and held in its place with one hand, while with the other the bottle is inverted, and its mouth brought below the level of the water in the trough, when the stopper may be withdrawn, and the bottle

moved to the shelf. It is better not to overcrowd the shelf with
bottles and jars, which, when filled with water and inverted, are
rather unsteady, and hence the other bottles should, after being
filled with water, be placed near the trough, to be brought to
the shelf when required.

Now mix the potassium chlorate and manganese dioxide in
the mortar with a few circular strokes of the pestle, add to the
mixture an equal bulk of fine dry sand[1], and grind the whole
until thoroughly mixed; then shake it out on a sheet of paper,
and transfer it to the flask. After wiping the neck of the latter,
fit the cork with the bent delivery tube into its place with careful pressure, and support the flask on a piece of wire-gauze
bent into a shallow cup, and resting on the largest ring of the
retort-stand. The smallest ring should be passed over the neck
of the flask to secure it in an upright position, and the latter
should be fixed at such a height that a lamp may be easily
placed beneath it, and that the end of the delivery tube may
pass under the shelf of the pneumatic trough, and be about 3
or 4 cm. below the surface of the water. The whole apparatus
will then be arranged as in Fig. 47, next page.

Before beginning to heat the mixture in the flask, you should
ascertain whether there is any leakage owing to the cork being
unsound or badly fitted. Place the warm hand on the flask for
a few seconds, and observe whether, owing to the expansion of
the air in the flask, the level of the water in the delivery tube is
depressed below the level of the water in the trough, and finally
a bubble or two of air escapes, and also whether on removal of
the hand the water only returns *by degrees*, as the air cools, to
its original level in the tube. If no depression of the level takes

---

[1] Unless this addition of sand is made, it often happens, especially if
there is rather more than the due proportion of manganese dioxide in the
mixture, that there is a sudden rush of gas at the last, a low incandescence
spreading through the half-fused mass. It is well to be prepared for this,
and to take away the lamp at the moment when you perceive any tendency
to a rush of gas. Of course the only possible risk would arise from the
delivery tube not being sufficiently large to carry off the rapid current of
gas, or becoming obstructed by particles of the mixture mechanically
carried over with the gas. When sand is added, however, as above
directed, the decomposition proceeds with perfect regularity from beginning
to end, and much less attention to the lamp is required.

place, there is a leak in the cork-joint, which must be stopped
before anything further is done. If the fault is not cured by
pressing the cork farther into the neck of the flask, it will
generally be best to take a new cork altogether.

When you have proved that the apparatus is air-tight, you
may proceed to heat the flask by a gas or spirit lamp. The
best form of gas lamp for the purpose is an argand, or other
form of ring-burner, since it distributes the heat over a larger
surface, and can be made to give a very small flame if required.

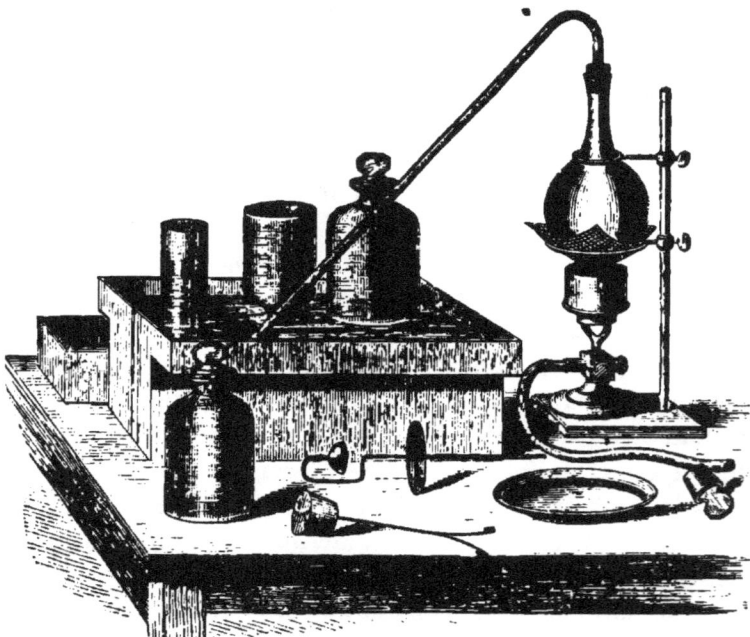

Fig. 47.

The first effect of the heat (which should be applied cautiously and gradually) will be to expand the air in the flask,
which will escape in bubbles through the water in the
trough.

In a short time the stream of bubbles, which had slackened,
will become more rapid, showing that gas is being evolved
from the mixture. As soon as this takes place, slide one of

the cylindrical jars along the shelf until its mouth is over the hole and the bubbles can rise freely into it. The heat must be carefully regulated, so as to keep up a moderately rapid stream of bubbles. It will be scarcely worth while to test the first jarful of gas, which will certainly be mixed with air. Fill the jar again with water, and place it as before to receive the gas. When it is full, slide it off the shelf with one hand, and with the other bring a ground-glass plate upon its mouth, while under water; raise the jar out of the water, still keeping the glass plate pressed against its mouth, and place it mouth upwards on the table. Now light a cedar match at the lamp, blow it out, and while its end is still glowing, plunge it into the jar of gas. If it is re-kindled, bursting sharply into flame, the gas is sufficiently pure for experiments. If this is not the case, return the jar at once to the trough, fill it again with gas, and test it in the same way.

When you have thus ascertained that the gas is pure, you may proceed to fill the jars and bottles with it, bringing each in succession over the hole in the shelf, and when it is full of gas sliding it (without raising its mouth above the level of the water) to one end of the shelf, and bringing another into its place. In order that the pneumatic trough may not be inconveniently crowded with bottles, you may remove them one at a time, when filled with gas, to the table, after inserting the stopper tightly, under water. When the deflagrating jar has been filled, slip under its mouth a shallow tray (a common plate or saucer will answer the purpose); then, keeping the tray horizontal and the mouth of the jar resting in it, raise both out of the water and place them on the table. The water remaining in the plate will thus act as a valve to prevent the gas in the jar from escaping.

As soon as the jars and bottles are filled with oxygen, take out the cork from the flask, and raise the delivery tube at once out of the water, withdraw the lamp, and set the flask aside to cool. The porous, half-melted mass which it contains may be readily washed out by a little warm water. It consists of a mixture of potassium chloride, manganese dioxide, and sand,

and will not be worth preserving. The flask will be found scarcely, if at all, injured, and after being cleaned and dried may be used again for the same purpose.

The properties of the gas should next be examined in the following manner.

1. **Its action on test-paper.**

Take strips of blue litmus and reddened litmus paper, and moisten them with distilled water: then introduce them for a moment into the large bottle of oxygen, replacing the stopper as soon as possible, as the bottle of gas is to be used for another experiment. You will observe that the colour of neither test-paper is changed[1]. Oxygen is therefore a 'neutral' substance.

*2. **Its relation to ordinary combustion.**

Place one of the cylindrical jars, its mouth upwards and closed with a glass plate, on the table before you, near the lighted lamp. Loosen the stopper with the left hand; take in the right hand the piece of wax taper attached to a bent wire, and light it at the lamp; then withdrawing the stopper introduce the taper into the bottle of gas. Notice that it is not extinguished but burns with a whiter, more intense flame.

Its action on a glowing match, causing it to burst sharply into brilliant combustion, has been observed already, and need not be tried again. The thick cedar matches are preferable for this test, as they retain a glowing end longest.

3. **Its union with non-metallic radicles, such as sulphur.**

Place the large bottle of gas on the table before you. Take the deflagrating cup and adjust its position by sliding the rod through the flange so that it may, when placed in the bottle, hang in the position shown in Fig. 10, p. 10, about 5 or 6 cm. from the bottom of the bottle[2]. Place in it a piece of sulphur, rather larger than a pea. Loosen the stopper of the bottle of gas with the left hand, hold the deflagrating cup for a few

---

[1] If the evolution of gas has been rapid, chlorine and chlorine oxides may be present in sufficient quantity to be recognised by their odour (pure oxygen being inodorous), and to redden and bleach the litmus-paper in a short time.

[2] It may be adjusted with sufficient accuracy by holding it against the side of the bottle.

moments in the flame of the lamp, until the sulphur melts and finally catches fire, and then immerse it in the bottle. The sulphur, which was burning in the air with a faint lambent blue flame, will immediately begin to burn much more brilliantly, while white clouds are formed in the bottle. When the combustion is over, take out the cup, and observe the pungent suffocating smell of the gaseous combination of sulphur and oxygen (sulphur dioxide) which has been produced.

Dip a piece of moistened blue litmus paper into the bottle and observe that it is strongly reddened. The sulphur dioxide has combined with the water to form an 'acid' substance (hydrogen sulphite, or 'sulphurous acid'). Thus, though oxygen is neutral itself, it forms compounds which when combined with water have an acid reaction.

Acids, however, do not all contain oxygen, as was formerly supposed. They are, in fact, 'hydrogen salts'; i.e. salts in which the electro-positive radicle (see p. 75) is hydrogen.

4. **Its combination with metals, such as iron.**

(*c*) Take a piece of thin watch-spring (readily obtained from any watch-maker) about 20 cm. in length; soften it by holding it in the lamp-flame until it becomes red-hot and loses its elasticity: when it is cool, straighten it and push one end into the flange of the deflagrating cup, securing it, if necessary, by a small wedge, such as the end of a match. Bend up the other end of the spring so as to form a small loop, thus ___⌒, in which place a bit of German tinder or (which will do quite as well) a bit of one of the common cigar-lights made of touch-paper, and sold in strips. Loosen the stopper of the deflagrating jar with the left-hand, while with the right hand you take up the flange with watch-spring attached. Light the tinder at the lamp, and immediately take out the stopper, and steadily lower the watch-spring into the gas, until the flange rests on the neck of the jar. The tinder will continue to burn, and the watch-spring, becoming red-hot, will also enter into brilliant combustion, sending out sparks and fusing into globules of iron oxide, which will fall off into the water in the tray. Be careful to take out the watch-spring as soon as the combustion comes near the neck of the jar, or the latter may be cracked by the heat.

[The beauty of the combustion is almost entirely due, not to the iron, but to the carbon present in the steel watch-spring; as may be shown by repeating the experiment, substituting for the spring a piece of pure iron wire ('binding wire') coiled into a spiral by being wound round a test-tube or glass rod about 1 cm. in diameter. In this case the metal will simply burn with a steady glow, emitting few or no sparks according to the purity of the iron.]

Reserve the remaining bottles of oxygen for use in future Exercises, placing a written label on each, to show what it contains.

2. HYDROGEN.

Apparatus required—Pneumatic trough ; two gas jars, 20 × 5 cm.; one ditto, 10 × 3 cm. ; flask, with flat bottom, holding about 250 c.c.; bent delivery tube, used in the last Exercise; thistle funnel; corks; cork-borers; rat-tail file ; retort-stand ; piece of wire-gauze ; taper on wire; glass disc, 8 cm. in diameter ; gas tray, 8 cm. in diameter ; Bunsen's burner; Bunsen's holder; drying tube; glass tubing, 6 or 7 mm. in diameter; india-rubber tubing, for connectors ; glass jet (p. 33); large beaker or funnel; glass tube about 30 cm. long and 2 cm. in diameter ; jug of water; cloth ; granulated zinc; strong (common) hydrogen sulphate ; distilled water; plaster of Paris.

Hydrogen is usually prepared by the action of the force of chemical affinity upon certain compounds containing it. An experiment has already been made (p. 71) which shows that when zinc is caused to act upon hydrogen chloride, the latter is decomposed, the zinc combines with the chlorine, while hydrogen gas is liberated. If hydrogen sulphate (common sulphuric acid), diluted with water, is acted on by zinc or iron, a precisely similar change takes place; and this is the usual process for preparing hydrogen.

The apparatus required is shown in Fig. 48.

Get ready the pneumatic trough as for the preceding exercise. Instead of bottles it is best to use plain cylindrical jars, ground at the mouth, about 5 cm. in diameter and 20 cm. high. Fill two of these with water and arrange them on the shelf of the trough. Now take the bent delivery tube, used in the preparation of oxygen; select a sound cork which will, after being squeezed, fit the neck of the flask; bore two holes in it (Fig. 23, p. 38), one large enough to admit the delivery tube, the other adapted to the tube funnel. If none of your cork-borers will

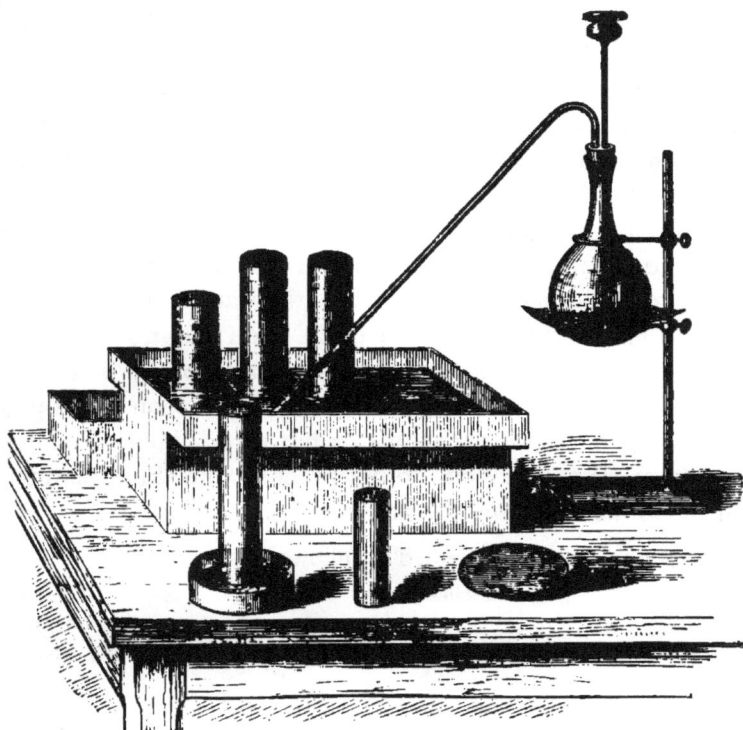

Fig. 48.

make holes of the precise size required, use the next smaller cork-borer, and enlarge the holes with the rat-tail file. Fit the short branch of the delivery tube into the cork until its end just appears on the opposite side; then with a screwing motion pass the tube of the funnel through the other hole in the cork until the lower extremity, when the cork is fitted into the neck of the flask, would reach nearly to the bottom. Place in the flask 30 grms. of granulated zinc, sliding the pieces down the neck, held slanting (not dropping them in, lest they should break the flask); support the flask in the retort-stand on wire-gauze [1], and fit the cork firmly into its place. The flask should be supported at such a height that the end of the delivery tube may be just below the hole in the shelf, and not unnecessarily deep in the water. Pour through the funnel enough distilled water to cover the zinc to the depth of 1 cm., and try whether the joints are tight by placing the warm hands on the flask for a few seconds. This will expand the air within, which will raise a column of water in the funnel tube, depressing, of course, the water in the delivery tube to an equal extent. If on withdrawing the hands this column remains steady for a moment, and only sinks gradually to its former level, the joints may be considered good. If however the column either does not rise at all, or sinks rapidly when the warmth is withdrawn, there is a leak somewhere in the cork. You may try to remedy this by pressing the cork further into the neck of the bottle [2]: if this does not succeed, drop a little sealing-wax on the top of the cork and spread it evenly with a hot wire. This will almost certainly prove a cure; but do not proceed with the experiment until the joints will stand the above test.

[The action may be expressed as follows [3]:—

Substances taken.	Composition.	Substances obtained.
Zinc, 65 parts.		Hydrogen, 2 parts.
Hydrogen sulphate, 98 parts.	{ Hydrogen, 2 parts / Sulphur, 32 parts / Oxygen, 64 parts }	Zinc sulphate, 161 parts.]

Now pour about 2 c.c. (a tea-spoonful) of strong common[4]

---

[1] Although the flask does not, in the present case, require to be heated, yet it is advisable to place under it a piece of wire-gauze which serves as an elastic cushion.

[2] A small leak may often be stopped, by simply moistening the cork and tubes with a few drops of water.

[3] Or by the following equation :—
$$Zn + H_2SO_4 = H_2 + ZnSO_4.$$

[4] The pure acid should not be used for this purpose: it should be reserved for analytical experiments.

hydrogen sulphate down the tube funnel, and shake the flask, so that the acid and water may mix. When this reaches the zinc an effervescence will commence, owing to the liberation of hydrogen. After a moment or two a little more acid may be poured through the funnel, so as to keep up a rapid stream of bubbles from the delivery tube; but especial care should be taken not to add too much at a time, lest the action should become too violent, partly from the undue strength of the acid, partly owing to the heat evolved from the mixture of the acid with the water. If the liquid in the flask should show a tendency to froth over, pour some distilled water down the funnel, to dilute and cool the acid. It is often the case that the action is slow at first, but it is better to give it a little time than to add acid recklessly[1]. Fill three jars with the mixture of air and hydrogen which first escapes from the flask, and reject their contents before proceeding to collect the gas for your experiments. This precaution is of more importance here than in the case of oxygen, because (as will appear) air forms with hydrogen a mixture which explodes upon contact with a light, and you may have an awkward, if not dangerous, mishap, if you unintentionally experiment with such a mixture instead of pure hydrogen. After sacrificing this quantity, however, you may safely collect some jars of the gas, precisely as directed in Exercise 1, sliding the jars over the hole in the shelf, and when each is full removing it to the side shelf of the trough. It is best not to fill the jar completely with gas (as it may topple over), but only so far that the water in the tray and in the jar may stand at the same level. Add from time to time a little more strong hydrogen sulphate, to keep up the stream of gas, but manage so that the action may be subsiding while the last jar is being filled, so as to prevent unnecessary waste of gas. When the jars are full, leave the flask and tube in their

---

[1] Pure zinc and some varieties of granulated zinc do not always act readily on the acid. If, after all, only a slight effervescence occurs, pour into the funnel one or two drops of solution of platinum perchloride and wash it into the flask with a little water. This will, owing to the galvanic action between the reduced platinum and the zinc, greatly increase the evolution of the gas.

position for the present, and examine the following properties
of the gas :—

1. **Its action on test-paper.**

This may be sufficiently tested by holding strips of blue
and red litmus paper over the bubbles as they rise from the
delivery tube. The gas will be found to be neutral, like
oxygen.

*2. **Its relation to ordinary combustion.**

Transfer one of the jars of gas (using a gas tray) from the
trough to the table. Take the taper attached to the bent wire
(Fig. 8, p. 9) and light it: then with the other hand raise the jar
of gas steadily from the water, keeping its mouth still downwards, and immediately pass the lighted taper up into it. The
gas will take fire with a slight noise where it is in contact with
the air, and burn with a pale almost invisible flame; but the
taper on being pushed further into the gas will be extinguished.
Now withdraw the taper and turn the mouth of the jar
upwards; the flame will pass quickly down the jar, and the gas
will be found to have disappeared entirely.

The jar may now be refilled with gas, for use in other
experiments.

3. **Its lightness as compared with common air.**

Transfer another of the jars of gas from the trough to the
table. Take a small gas jar (empty) and hold it inverted in
one hand, as shown in Fig. 49.

Raise the jar of gas out of its tray with the other hand, bring
its mouth close to the edge of the small empty jar, and steadily
depress the closed end, as in the figure, proceeding exactly as
if you were pouring a liquid upwards from it into the smaller
jar. Set down the small jar, which is in your left hand (mouth
downwards), in the tray of water; light the taper affixed to the
wire and bring it into the upturned mouth of the other jar.
The gas will be found to have escaped entirely, the taper
burning as in the outer air. Now raise the mouth of the small
jar above the level of the water in the tray, and pass up into it
the lighted taper. A slight explosion will take place, as the gas
catches fire, thus demonstrating that the hydrogen has really

ascended and displaced the air in the jar, precisely as it displaces the water in filling a jar at the pneumatic trough.

Fig. 49.

4. Its combination with oxygen, to form water.

(*a*) Fill with water both the jars used in the last experiment, and place them on the shelf of the pneumatic trough. Fill the smaller jar with hydrogen from one of the remaining jars of gas, and decant its contents into the larger jar. Fill it again with gas, and again decant its contents, as before. You have now two measures of hydrogen in the larger jar. Next, take one of the bottles of oxygen which you reserved (p. 104), invert it, and bring its mouth below the water-level in the trough; take out the stopper, and decant into the small jar sufficient oxygen to fill it. Decant this measure of oxygen into the jar containing the two measures of hydrogen, depress the jar in the trough, and shake it laterally to mix the gases (taking care that no air enters); then fill the small jar with the mixed gases, close its mouth with a glass plate, remove it from the trough and place it on the table, mouth upwards, retaining the glass plate in its position with the left hand. Light a match, withdraw the glass plate and apply the lighted match to the mouth of the jar. The explosion which follows, though loud, is quite unattended with danger if a small strong jar is used as above directed. Do not, on any account, apply a light to the mixed gases in the larger and thinner jar, since you may have a serious accident if the jar should break.

[You may, if time allows, try two or three similar experiments, varying the proportions of hydrogen and oxygen (taking, for instance, two measures of oxygen and one of hydrogen), and you will find that the explosion, which accompanies the chemical combination, is loudest when one measure of oxygen is mixed with two measures of hydrogen. This is found to be the exact proportion in which the two gases are combined in water [1]. And if the gases are mixed in any other proportion, the excess of one or the other remains uncombined.

You may, further, try a few similar experiments, using air instead of oxygen. You have noticed, in Experiment 2, that when a lighted taper is passed into a jar of hydrogen the gas takes fire at the mouth of the jar, where it is in contact with the air. This is due to its combination with the oxygen which is one of the constituents of air. Now, you will find that when air and hydrogen are mixed, and a lighted match applied to the mixture (this may be done in the larger jar), an explosion takes place, less violent than in the case of oxygen and hydrogen, but loudest when five measures of air are mixed with two measures of hydrogen [2]. But we have just seen that *two* measures of hydrogen unite with *one* measure of oxygen to form water. Hence five measures of air contain one measure of oxygen [3].]

That water is the product of the union of oxygen with hydrogen will be best shown by causing the combination to take place more slowly; for instance, by allowing a jet of hydrogen to burn in air.

(*b*) Raise the delivery tube out of the water, and support the end of it upon a wooden block placed on the table (the flask being still retained in the retort-stand). Attach to it one of

---

[1] $2H_2 + O_2 = 2H_2O$.

[2] If you have no jar large enough to contain seven measures of the gases, put an india-rubber ring round the small jar about 2 cm. from its mouth, and reckon the contents of the jar to this level as 1 measure.

[3] The composition of air will be further illustrated in the next Exercise.

the elbow-tubes which you have already made, stretching over the ends of both a short piece of india-rubber tubing, slightly moistened to make it slip over the glass more easily. To the other branch of the elbow-tube, which should point vertically upwards (Fig. 50), adapt in a similar way the glass jet you have already made (p. 33). Pour a little more acid on the

Fig. 50.

zinc, if the stream of gas has ceased; and since air may have entered the flask, it will be advisable to test the purity of the gas before lighting it at the jet, in order to avoid the chance of an explosion. To do this, hold over the jet·a small test-tube, bringing the jet nearly to the closed end of the tube. We have seen that hydrogen can be poured upwards, therefore the tube will soon fill with gas. After about ten seconds raise the tube slowly from the jet, close its mouth immediately with the thumb, remove it to some distance from the jet, and, still holding the mouth downwards, apply a lighted match to it. If the gas burns quickly, with a shrill noise, air is mixed with it; and it will not be safe to light it at the jet. Other trials should be made, and when the gas catches fire with only a slight noise, the flame passing slowly along the tube, it is pure and you may proceed with the experiments.

Pour a little more acid on the zinc in the flask, if necessary, and light the hydrogen issuing from the jet. When it burns steadily with a flame about 2 cm. high[1], hold over it a clean dry beaker or bottle, and notice that a dew is at once deposited on the glass, soon collecting into drops of a colourless liquid, which is pure water. It will be hardly worth your while to collect any quantity of water, but you may at any rate satisfy yourself that the product of the combustion is tasteless, and that when the outside of the glass is gently warmed by moving it to and fro over the lighted jet of gas the deposit volatilises without leaving any residue.

(c) When the current of gas has slackened, but is still burning at the jet, another experiment may be tried which shows in an interesting way that the flame, under certain conditions, although apparently steady, is really expanding and contracting; is, in fact, rather a series of flames quickly succeeding one another, than a continuous flame.

Hold a glass tube, about 30 cm. in length and 1.5 or 2 cm. in diameter, over the jet, and depress it gently until the flame is entirely within the tube. At a certain point the flame will become elongated and a musical note will be produced varying in pitch with the length of the tube, and also, for the same tube, with the size of the flame. This sound is due to the fact that the air in the tube is set in vibration by the extremely rapid succession of explosions caused by the combination of the hydrogen as it issues from the jet with the oxygen of the surrounding air[2].

5. Its diffusibility.

Put out the flame, and hold a piece of blotting-paper close

---

[1] The flame will shortly become yellow, owing to a trace of sodium, contained in the glass of the jet, becoming volatilised.

[2] Hydrogen and oxygen only combine when mixed in certain proportions. This mixture takes time to form, but when it is formed the combination is sudden. A distinct interval then occurs before the due mixture is again formed; this is exploded by the heat of the previous combination; another pause occurs, then another explosion, and so on.

above the jet from which the gas is still issuing. Bring a lighted match near the upper side of the blotting-paper, just over the jet. The gas will catch fire, showing that it has passed quickly through the pores of the paper, as through a sieve.

Additional Experiments.

6. Action of platinum in causing the combination of oxygen and hydrogen.

This will be more fully explained hereafter, but it may be tried in a simple way thus,—Take a small tuft of asbestos, and coil a platinum wire once or twice round it to serve as a handle: then moisten it with a drop of solution of platinum perchloride and hold it, first above and then in the flame of a Bunsen's burner until the salt is decomposed and a gray deposit of finely divided platinum left on the asbestos. Allow this to cool, and then hold it just above the jet, in the stream of gas. It will become red hot, showing that it is causing the combination of hydrogen with the oxygen of the air, and the gas will shortly catch fire.

7. Proof that water is formed when hydrogen and oxygen combine.

It must be observed that we have not yet rigorously proved that water is formed during the burning of hydrogen in air, since no pains has been taken to dry the gas, which of course contains moisture derived from the liquid in the flask. In order that the experiment may be decisive, the hydrogen must be passed over some drying material before it is burnt at the jet.

Take off the glass jet and india-rubber connector, and fit the end of the elbow tube into the cork at the end of a drying tube (Fig. 10 *c.* p. 12) filled with fragments of quicklime or of calcium chloride. Support the drying-tube in a vertical position by the Bunsen's holder, and fit the jet upon its upper end. After testing the purity of the gas, as already directed, you may light it at the jet, and hold a beaker over it, as in experiment 4 *b.* If moisture is now deposited on the beaker, it must have been formed during the combustion, since the gas has been dried, and the air contains too little moisture to form any deposit on the glass.

8. Lightness of hydrogen.

Make a solution of soap by dissolving in a small evaporating dish a bit of common soap as large as a pea in 5 or 6 c.c. of water. Remove the glass jet, and touch the end of the tube with a finger dipped in the soap solution. A bubble will soon be blown, which when it leaves the tube will ascend rapidly. One or two trials with soap solution of various strengths and more or less warm may be required to get a good result [1].

[If you possess one of the small fish-skin balloons sold by opticians, the lightness of hydrogen gas may be further demonstrated in the following way :—

In the first place ascertain that there are no holes in the balloon by expanding it with air and holding it between your eye and the light, turning it round to examine every part of it [2]. If any holes are visible, they may be mended by touching the margins with weak gum-water and covering them with small patches of gold-beaters' skin. A short piece of quill should be inserted in the neck of the balloon, and secured in its place by folds of thread. Take off the glass jet from the apparatus used in the preceding experiment, and, having carefully squeezed and sucked out as much as possible of the air from the balloon, stretch the india-rubber connector over its neck, and proceed to fill it with gas, pouring a little more acid on the zinc whenever the current of gas becomes slow. When the balloon is full, you may allow it to ascend in the room, after inserting a small plug of cork into the neck, and attaching a long piece of thread to the quill, in order to have the movements of the balloon under control.]

9. Diffusion of hydrogen through plaster of Paris.

Take a piece of glass tube about 25 cm. long, and 1 cm. internal diameter (a piece of combustion tube will do) and fit a

---

[1] A solution which keeps well and forms good bubbles may be made as follows.

Dissolve 5 grms. of pure sodium oleate (made by gently heating 5 grms. of sodium hydrate with 25 grms. of oleic acid) in 80 c.c. of distilled water : when cold add 5 c.c. of glycerine and mix thoroughly. The solution should be slightly shaken up before bubbles are blown with it.

[2] These balloons are very liable to be attacked by insects. They should be kept in a wide-mouthed bottle with a small bit of camphor.

cork to one end. Pour about 10 c.c. of water into a small porcelain dish, and add enough plaster of Paris to form a thick cream, stirring it thoroughly with a glass rod. Before it sets, plunge into it one end of the glass tube (the cork being withdrawn) until it rests on the bottom of the dish, and support it upright in a Bunsen's holder until the plaster has got moderately hard. Then carefully detach it from the surrounding plaster, and push the cork a little way into the tube so as to drive the plug of plaster before it. Take out the cork, and dry the plaster thoroughly by a gentle heat; leaving it for several hours in a drying cupboard, or before a fire. When it is quite dry, insert the cork and fill it with hydrogen by placing it over the end of the drying-tube, in the same manner as the test-tube was filled, p. 111. While it is filling, pour some water into a large beaker and add a drop of indigo sulphate to colour it. In about a minute, if there is a fair stream of gas, the tube will be sufficiently full. Its lower end should now be immediately dipped into the water in the beaker and the cork withdrawn from the upper end. The hydrogen will diffuse out through the plaster plug so much faster than the air diffuses inwards that the water will rise in the tube and nearly fill it.

After trying the above experiments, you may take the apparatus to pieces, wash the tubes, and set them aside to dry. The liquid which is in the flask is a solution of zinc sulphate, and may be filtered into an evaporating dish, and evaporated as directed in Sect. 1, Ex. 5, (p. 55) until it deposits crystals on cooling. The zinc sulphate thus obtained is very pure, and may be reserved for use in the laboratory.

3. NITROGEN and AIR.

Apparatus required—Pneumatic trough; porcelain dish 6 cm. in diameter; gas tray; deflagrating jar; two gas jars, 20 × 5 cm.; one gas jar, 10 × 3 cm.; small gas bottle; beaker; glass disc; taper on wire; jug of water; cloth; blotting-paper; crucible tongs; knife; test-tubes; piece of fine brass wire; graduated measure; porcelain mortar; india-rubber rings; watch glass; corks; cork borers; elbow tubes; phosphorus; tow; turpentine; test papers; calcium chloride; lime water; ether.

Nitrogen occurs in air, simply mixed with oxygen; and in

order to obtain a supply of it we have only to withdraw the oxygen by the action of some substance which has an affinity for it and not for nitrogen. Phosphorus is, on the whole, the best substance for the purpose, since it readily combines with oxygen and the compound formed (phosphorus pentoxide) is very soluble in water, and is therefore quickly withdrawn when the experiment is made over the pneumatic trough, leaving the nitrogen nearly pure.

Fill the pneumatic trough with water, and float a small porcelain dish on the water, retaining it over the movable shelf. Take a stick of phosphorus from the bottle with the crucible tongs[1], place it in a gas tray full of water, and cut off with a knife (still holding it under water) a piece about twice as large as a pea. Dry the piece of phosphorus thoroughly, by gently pressing, not rubbing, it between folds of blotting-paper; then place it at once, using the (dried) crucible tongs, in the floating dish. Take the stopper (which should be greased) out of the deflagrating jar, and, holding the jar in one hand over the phosphorus, light the latter by touching it with a burning match, and immediately lower the jar over the burning phosphorus until it rests upon the shelf of the trough, and insert the stopper. The level of the water in the jar will at first be depressed, owing to the expansion of the enclosed air by the heat of combustion, but it will soon rise above the level of the water in the trough, showing that one of the constituents of the air is being withdrawn. When the phosphorus ceases to burn, allow the jar to remain undisturbed for five minutes, until the white fumes (which consist of a compound of phosphorus with oxygen, phosphorus pentoxide) have been for the most part absorbed, and meanwhile fill all the gas jars with water and place them inverted on the shelf of the trough. Fill also the large beaker, which is to be used precisely as a gas jar, its large, spreading mouth rendering it easier to avoid loss in decanting gas from the deflagrating jar.

When the white fumes have nearly cleared away, depress the deflagrating jar in the trough and shake it laterally until the small dish is filled with water and sinks to the bottom, when it may be withdrawn. Decant some of the nitrogen first into the beaker (remembering not to fill it quite, lest it should topple over) and then from the beaker into the three gas jars. The properties of the gas may now be examined as follows.

1. Its action on test paper.

Raise the mouth of the smallest gas jar above the water of the trough (still keeping it inverted, since nitrogen is rather lighter than air), and pass up into it strips of moistened blue and red litmus paper. The gas will be found to be neutral, like hydrogen and oxygen.

2. Its relation to ordinary combustion.

(*a*) Slide one of the larger jars of gas off the shelf of the trough, cover its mouth with a glass plate, lift the jar out of the water, retaining the plate in its position with one hand, and place it on the table mouth upwards[1]. Light the piece of taper affixed to the wire, remove the glass plate from the mouth of the jar, and plunge the lighted taper into the gas. Observe that the taper is extinguished while the gas itself does not catch fire; thus showing it to differ from oxygen, in which the taper continued to burn, and from hydrogen, which extinguished the taper, but was itself inflamed.

(*b*) Twist a small piece of tow round the end of the copper wire in place of the taper, so as to form a ball about 1 cm. in diameter. Remove another jar of nitrogen from the pneumatic trough, and place it mouth upwards on the table, as in the last experiment. Pour a few drops of turpentine on the ball of tow, set it on fire, and immerse it in the jar of gas. The flame will be as instantly and completely extinguished as the small flame of the taper in the last experiment.

These are all the experiments which can readily be tried with nitrogen, which in the free state appears as an inactive gas, showing very slight chemical affinities.

[Reserve a jar or bottle full of nitrogen for use in the next series of experiments.]

---

[1] This position is most convenient, although the nitrogen will escape rather more quickly than from an inverted jar.

Composition of air.

We may obtain proofs of the composition of air in two ways :—

(*a*) By analysis ; i. e. by separating it into substances which we can prove to be oxygen and nitrogen, and observing the quantities of each which are obtained from a given amount of air.

(*b*) By synthesis ; i. e. by mixing oxygen and nitrogen in the proportions indicated by analysis and observing whether the resulting product has the same properties as air.

A.—Analysis of air.

For this purpose we may conveniently employ phosphorus : not, however, kindling it and thus causing it to withdraw the oxygen quickly, but allowing it to act at the ordinary temperature for several hours. The cylindrical measure will do pretty well instead of a regular gas measuring-tube, if one of the latter is not at hand.

Fill a small test-tube about half full of water, put into it enough phosphorus to form a stratum about 1 cm. deep, and place the tube in a beaker of nearly boiling water until the phosphorus is melted. Make a small loop at the end of a piece of fine brass wire about 30 cm. long, by coiling it round a glass rod or pencil, plunge the looped end of the wire into the liquid phosphorus, and cool the tube by holding it in cold water. When it is quite cold, pull out the wire with the lump of phosphorus attached to it (this may require the tube to be dipped into hot water again for a moment, to loosen the phosphorus from the sides) and put it into a dish of water until wanted.

Put the graduated cylindrical measure mouth downwards into a porcelain mortar nearly full of water, pushing up into it the end of a piece of india-rubber tube. Suck air out of the jar through the tube until the water level inside stands exactly at the mark 200 c.c.; then pinch the tube to prevent air entering or escaping, and withdraw it from the jar. Incline the jar a little and introduce the lump of phosphorus, pushing it up by means of the wire nearly to the closed end of the jar. Restore the jar to its vertical position and leave it undisturbed in a safe place for 12 or 14 hours [1].

[Meanwhile the other experiments on air may be proceeded with.]

At the end of this time, pull out the phosphorus attached to the wire, and read the volume of the remaining gas. To do this it will be best to transfer the jar to the pneumatic trough, and lower it in the water until the level inside and outside the tube is the same. You will find that out of the original 200 c.c. of air, only 160 c.c. (approximately) remain; and this residual gas may be shewn to be nitrogen by decanting some into a gas jar and passing up into the jar a lighted taper. And since the white fumes formed during the action of the phosphorus are known to consist solely of a compound of phosphorus and oxygen, the (200−160=) 40 c.c. of gas which have disappeared, may be assumed to be oxygen.

Then the percentage composition of air by volume can be found by the proportion sum

200 : 40 :: 100 : percentage of oxygen,

and 100−percentage of oxygen = volume of nitrogen and other gases in 100 parts of air.

B.—Synthesis of air.

Take one of the larger cylindrical gas jars, and graduate it into five parts in the following way. Try first whether the large jar will hold not less than six of the small jars full of water. If it holds less than this, slip an india-rubber ring over the small jar, about 1 cm. from the mouth, fill the jar with water up to the level of the ring and try whether about six times this reduced volume will fill the larger jar. One or two trials will be sufficient to ascertain the proper volume, which is to be the unit of measurement. Now fill the measure up to the proper point with water, pour the water carefully into the larger jar, and mark the level at which it stands by a small india-rubber ring slipped over the jar. Add another measure of water, and

---

[1] It should not be left for more than a day, otherwise the air which diffuses through the water will vitiate the results.

mark the level similarly with another ring.   Proceed thus until five measures of water have been poured into the larger tube. You have now a tube graduated with fair accuracy, which should be filled with water and placed inverted on the shelf of the pneumatic trough.   Decant into it sufficient nitrogen, from the bottle which was reserved, to fill it to the level of the fourth division : then add one measure of oxygen (from one of the bottles reserved in Exercise 1), cover the mouth of the jar with a glass plate, and invert it once or twice (retaining the glass plate firmly in its place with one hand) so as to mix the gases thoroughly.   Place the jar mouth upwards on the table and immerse in it a lighted taper.   The latter will not burn brilliantly as in oxygen (p. 102) or be extinguished as in nitrogen, but will continue to burn as it did in the external air [1]. Other experiments would show that the mixture possessed all the properties of common air, and thus we have a proof by synthesis that 5 volumes of air consist (approximately) of 4 volumes of nitrogen and 1 of oxygen.

Besides oxygen and nitrogen, perceptible quantities of carbon dioxide and water vapour are present in air.   Their presence may be shown as follows :—

1. **Carbon dioxide.**

Pour a little lime water into a watch glass, and leave it exposed to the air for ten minutes or more.   A white film will form on the surface, proving the presence of carbon dioxide in the air of the room.

2. **Water vapour.**

Pour a few drops of ether into a large dry test-tube and make it evaporate quickly by blowing air into it through an elbow tube (best with a pair of bellows, since the warmth of the breath interferes with the cooling effect [2]).

---

[1] The brilliancy may be slightly increased, since air is not *entirely* freed from oxygen by the action of phosphorus under the above conditions, and hence the mixture would contain rather more than the proper amount of oxygen.

[2] If ether is not at hand, a little pounded ice (or snow) and salt may be put into the tube. Or a freezing mixture may be made by mixing 8 grms. of finely powdered sodium sulphate with 5 c.c. of strong common hydrogen chloride.

The outside of the tube will become very cold, and cool the air near it; and a deposit of dew or even ice will appear on it, proving that water vapour is present in air, and only requires to be cooled down to a certain point in order to be condensed and show itself in a visible form.

The chief sources of both of these are very similar; viz. the combustion of candles, fires, &c., and the breathing of animals. These may be illustrated in the following way.

1. Carbon dioxide.

(*a*) Put a little lime water into a small gas bottle, and shake it up. It will not become cloudy, since the amount of carbon dioxide in the air contained in the bottle is extremely small. Now light the taper affixed to a wire, place it in the bottle, and allow it to burn there, laying the stopper loosely on the mouth of the bottle, to prevent entrance of air. When it goes out, remove it from the bottle, insert the stopper and shake up the lime water, which will become milky, proving that carbon dioxide has been formed; the carbon of the wax having combined with the oxygen of the air.

An experiment showing that carbon dioxide is produced in respiration has already been made, p. 61: it may be repeated in a more exact way, as follows.

Adapt a cork to a large test-tube (taking care to soften the cork thoroughly by squeezing it, or it may break the tube); bore two holes in the cork and fit into them elbow tubes as shown in Fig. 51. The long branch of one elbow tube should reach nearly to the bottom of the test-tube, the short branch of the other should only just pass through the cork. Fill the test-tube about half full of lime water, and suck air through the liquid for a few seconds, applying the

Fig. 51.

mouth to the extremity of the short elbow tube. No turbidity will be produced, the quantity of carbon dioxide in the air being very small. Now, having taken a deep breath, apply your

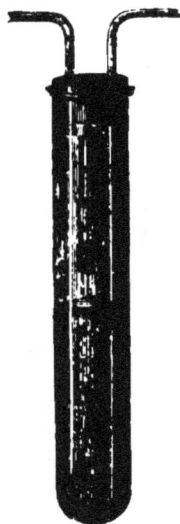

mouth to the other elbow tube (the long branch of which passes into the liquid), and blow air from the lungs through the lime water. The latter will almost immediately become turbid, showing that much carbon dioxide is present in air which is expired from the lungs [1].

2. Water vapour.

(a) Hold a beaker over the flame of a lamp or candle; its surface will become covered with moisture, owing to the combination of the hydrogen of the gas or wax with the oxygen of the air.

(b) Strew a little powdered calcium chloride upon a glass plate and breathe upon it. It will become liquid, showing that abundance of moisture is present in the breath.

COLLECTION OF GASES BY DISPLACEMENT.

This is a method of collection which depends upon the difference between the density of a gas and that of common air. It is especially useful in cases where the gas to be collected is so soluble in water that it would be impracticable to collect it over that fluid. In such cases a trough filled with mercury might be used, but the great expense of this would place it beyond the reach of most of those for whom these exercises are intended.

When we pour water into a vessel, the water, being heavier than the air contained in the vessel, descends to the bottom and displaces the air, which is driven out at the top. Precisely the same thing occurs when we lead a gas heavier than air into an open jar. The gas does not mix with the air at once, but collects at the bottom of the jar and gradually rises, driving the air before it, and finally overflowing the edge of the jar like other fluids. It is obvious that a gas lighter than air may be similarly collected by simply reversing the position of the receiving vessel, placing it mouth downwards, and leading the gas into its upper part; as has, in fact, been already done in filling a test-tube with hydrogen (p. 111).

[1] The reason why carbon dioxide does not accumulate in the air will be alluded to under the head of CARBON DIOXIDE.

Two points must, however, be attended to in order to obtain by this method jars of gas even approximately pure.

1. It is necessary to pass into the jar at least twice the volume of gas required to fill it; since diffusion must take place between the fluids with greater rapidity in proportion as their densities are less, and hence the boundary-line between the gas and air is never so sharply marked as in the case of water and air.

2. All currents of air in the room must be avoided as much as possible; the tendency of such currents being to stir up the gas, and produce a more rapid mixture with air than diffusion alone would cause.

COMPOUND OF NITROGEN AND HYDROGEN.

4. AMMONIA.

Apparatus required—Florence flask; corks; cork-borers; elbow tubes, one with long branch; drying tube, fig. 12, *c*; india-rubber connector; retort-stand; Bunsen's holder; wooden blocks; sand bath; argand burner; porcelain mortar; scales and weights; porcelain dish; test-tubes; taper on wire; gas jar, 20 x 5 cm.; one large and four small gas bottles; glass disc; pan of water; beaker; flask with flat bottom, holding about 250 c.c.; piece of platinum wire about 15 cm. long; card; glass rod; ammonium chloride (sal ammoniac); quicklime; small bits of flannel (or silk) and calico; box of test-papers; strong solution of ammonia; hydrogen chloride; hydrogen sulphate; indigo sulphate.

Nitrogen and hydrogen can only with great difficulty be caused to unite directly; but when organic substances, such as leather, wool, or gelatine, which contain both elements, are decomposed some of the nitrogen and hydrogen unite to form ammonia. The following experiment illustrates this :—

Place a small bit of flannel (or silk) in a moderate-sized test-tube, support it in a slanting position as in fig. 39, p. 69, and heat it rather strongly. The substance will become blackened or 'charred,' showing the presence of carbon: water will condense in the tube, proving that it contains hydrogen and oxygen;

while strong-smelling volatile products will be given off among which are ammonium carbonate and ammonia. The latter may be recognised (1) by its alkaline reaction on a strip of reddened litmus paper moistened and held in the tube: (2) by the white fumes (of ammonium chloride) formed when a glass rod dipped in strong hydrogen chloride is held in the issuing vapours.

If a piece of paper or calico (substances which contain no nitrogen) are heated in the same way, the vapours evolved will give no alkaline reaction or white fumes when tested as above.

Coal contains nitrogen, and the compounds of ammonia obtained from it in the manufacture of coal-gas are combined with hydrogen chloride (as illustrated in the second of the above tests) and form ammonium chloride ('sal ammoniac'), which is the usual source of ammonia, since it readily yields the gas when gently heated with quicklime.

The chemical change may be expressed as follows [1] :—

Weigh out 15 grms. of sal-ammoniac, and reduce it to powder in the mortar. The salt in its sublimed state is so tough as to render this rather difficult[2], and more may be effected by direct blows of the pestle, especially if aided by a thrust and slight twist of the arm at the moment of impact, than by the usual rubbing motion. Place the powdered salt in a basin and set it to dry on the sand-bath over the lamp, occasionally stirring it to prevent its caking together. Meanwhile reduce 20 grms. of quicklime to fine powder in the mortar[3]:

---

[1] Or by the equation,
$$2(H_4N)Cl + CaO = CaCl_2 + H_2O + 2H_3N.$$

[2] It should be bought in a state of powder if possible.

[3] If the quicklime is very hard and difficult to powder, it may be slacked as directed, p. 58, care being taken to use as little water as possible, so that the calcium hydrate may be perfectly dry.

then cover the mortar with a plate, to protect the lime from the moisture and carbon dioxide of the air.

An apparatus should now be fitted up as shown in Fig. 52.

Adapt a cork to a clean dry Florence flask, and fit into the cork an elbow tube each branch of which is about 8 cm. long. Fill the drying tube with fragments of quicklime about as large as split peas in the manner described at p. 13. Fit the end of the elbow tube into the perforated cork of the drying tube, and support the flask in the Bunsen's holder [1] at such a height that the lamp will go easily underneath it.

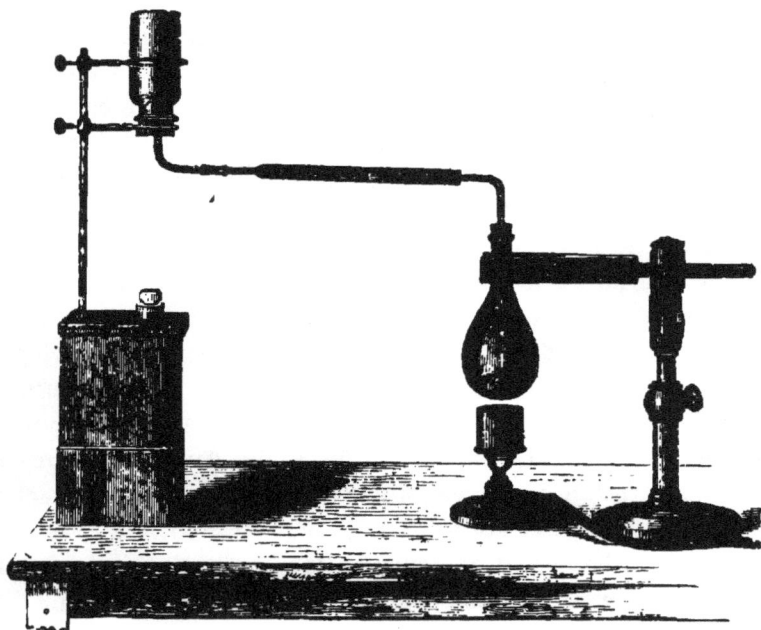

Fig. 52.

The ammonium chloride may now be taken from the sand-bath and set aside to cool; the cooling may be hastened by spreading it out on a sheet of paper.

[1] A notch should be cut across the cork at the end of each jaw of the holder (if this has not been done already) in order to grasp the neck of the flask more securely.

Next, adapt a large flat cork to the smallest ring of the retort-stand, and bore a hole in it, through which the long branch of the other elbow tube should be passed so that its end may be about 12 cm. above the cork. Clamp the retort-ring in the position shown in the figure ; one, at least, of the larger rings being fitted on above it, for the purpose of steadying the bottles while they are being filled. The retort-stand must now be supported on blocks at such a height that the drying tube may be horizontal when its end is connected with the elbow tube by a short bit of india-rubber tubing, as shown in the figure.

The ammonium chloride will by this time have cooled, and should be mixed quickly and thoroughly with the quicklime in the mortar. The mixture should be shaken out on a half-sheet of paper and transferred at once to the flask, in the manner shown in Fig. 34, p. 64 (the cork with elbow tube being left resting on the Bunsen's holder).

Having restored the flask to its place and fitted the cork tightly into its neck, place one of the smaller bottles inverted over the up-turned end of the elbow tube so that its mouth rests lightly upon the cork in the retort-ring (see the engraving). The flask may now be gently heated by the argand burner with a small flame. If the spirit lamp must be used, it should be moved to and fro under the flask, the flame being never allowed to rest in one place, otherwise the flask is apt to crack ; an accident which is not unlikely to happen in any case, as the substance to be heated is a solid of low conducting power, and not a liquid, which would distribute the heat by convection.

While the bottle is being filled you may grease its stopper and those of the other bottles, and also fit a good cork to the neck of the larger gas bottle for use in Expt. 5.

The gas comes off at a comparatively low temperature, and from its lightness collects in the highest part of the bottle, driving the air downwards before it, which will escape between the edge of the bottle and the cork. To ascertain when the bottle is full of gas, hold a piece of turmeric-paper (moistened by being breathed upon) near its neck and slightly above its mouth. If the gas is overflowing, it will quickly and decidedly redden the test-paper. Remember, however, that a very little ammonia is sufficient to act on the paper; and hence it is advisable to leave the bottle in its place for about half a minute longer after the above effect has been observed, to ensure its being really full of gas. The bottle should then be slowly raised until clear of the delivery tube, and the stopper inserted at once. Another bottle may then be placed in the same position and filled in like manner. When all the bottles have been filled with the gas, withdraw the lamp, disconnect the elbow tube from the drying tube, and place the flask (still fitted with the cork and elbow tube) in a draught-cupboard or in the open air.

The chief properties of ammonia may now be examined, as follows :—

*1. Its peculiar odour and its alkaline action on turmeric paper will have been noticed already.

2. Its lightness has been sufficiently proved by its collection by upward displacement.

3. Its rapid diffusibility.

Place an empty gas jar, mouth upwards, on the table, and drop into it strips of wetted reddened litmus and turmeric paper. Bring over the mouth of the jar a small bottle of ammonia inverted, remove the stopper and place the mouths of the jars in contact for a few seconds. The gas, in spite of its low density, will diffuse downwards through the air; the reddened litmus will become blue and the turmeric red; thus showing at once the alkaline reaction of ammonia and its quick rate of diffusion.

4. Its relation to ordinary combustion.

Support a small bottle of ammonia on the smallest ring of the retort stand (the cork being taken out) in the same position as when it was being filled. Withdraw the stopper and pass a lighted taper slowly up into the bottle.

The taper will be extinguished, but the gas will show a tendency to burn with a greenish flame at the mouth of the bottle.

[The feeble inflammability of ammonia in air may be shown by restoring to its place the flask, fitted with elbow tube, which was put aside just now, heating it so as to generate a little more ammonia, and bringing the flame of a Bunsen's burner close to the extremity of the tube from which the gas is issuing. A yellowish-green flame will be seen.]

5. Its solubility in water.

Fill the supplementary pan of the pneumatic trough (or a basin) with water. Place a dry glass plate on the table, bring over it a small bottle of ammonia, mouth downwards, remove the stopper, and bring the mouth of the bottle at once down upon the glass plate. Then, still keeping the bottle inverted and plate closely pressed against its mouth, plunge it below the surface of the water in the pan and withdraw the glass plate. The water will rise in the bottle especially if the latter is gently shaken, and will fill it entirely if the gas is unmixed with air. There will usually be a small residue of air, and you will thus gain an idea how far you have been successful in filling the bottle by displacement.

The water in the bottle will be found to have gained the smell and alkaline reaction of the gas, and to have a caustic taste. It has, in fact, combined with the ammonia, not merely dissolved it, and the liquid is a solution of ammonium hydrate (the common 'solution of ammonia' which you will often have occasion to use in testing).

[Another experiment illustrating the solubility of ammonia may be made, if time permits, as follows.—Take the cork which was recently fitted to the large gas bottle: bore a hole in the centre and fit into it a glass tube about 18 cm. long, ending in a jet (the pipette already made, p. 33, will do very well), so that the jet may project about 5 cm. within the bottle. Fill a beaker with water and add to it a few drops of solution of litmus together with one drop of dilute hydrogen chloride to make the litmus red. Twist a small piece of moistened tow round the tube just below the jet (this is for the purpose of beginning the absorption). Put the beaker on the iron base of the retort stand, and clamp the smallest retort ring at such a height that when the cork rests on it the end of the tube may reach nearly to the bottom of the beaker. Now invert the large bottle of ammonia, take out the stopper and quickly fit the cork into its place (still keeping the bottle inverted), and plunge the outer end of the tube into the beaker of coloured water. The pressure of the external air will force the water up the tube, as the ammonia is absorbed and a vacuum made in the bottle, and a fountain will be formed, the red litmus becoming blue under the action of the ammonia.]

*6. Its combination with other radicles.

Ammonia shows a great tendency to combine directly with salts of hydrogen (i. e. the substances called acids), associating their hydrogen more closely with itself to form the radicle 'AMMONIUM,' containing 14 parts by weight of nitrogen combined with 4 parts of hydrogen (instead of with 3 parts as in ammonia). Thus, when ammonia is brought into contact with hydrogen chloride the two gases combine to form the salt 'ammonium chloride'; the substance, in fact, from which you prepared the gas.

Dip a glass rod into strong hydrogen chloride and hold it just above the mouth of a bottle of ammonia, lifting the stopper for a moment only to let a little of the gas escape. White clouds of ammonium chloride will be formed when the gases mix.

Having cleaned the glass rod, dip it into strong hydrogen acetate (acetic acid) and allow a little ammonia to mix with its vapour in the same way. Similar white clouds of 'ammonium acetate' will be noticed.

This reaction is very useful as a test for ammonia; and hydrogen acetate is slightly preferable to hydrogen chloride for the purpose, since the latter itself gives slight white fumes in moist air, which might lead to a mistake.

Other experiments illustrating the formation of ammonium salts will be found under the heads of NITRATES, CHLORIDES, and AMMONIUM.

7. Its union with oxygen, under the influence of platinum.

Platinum, as noticed under HYDROGEN, has a peculiar power of causing chemical combination: and when a piece of it is heated and put into a mixture of ammonia and air, the oxygen of the air combines with both the nitrogen and the hydrogen of the ammonia (and not with the hydrogen alone, as in Expt. 3).

Form a piece of platinum wire, of the kind used for blowpipe experiments, about 15 cm. long, into a close spiral, by coiling it round a glass rod or pencil, and attach it to a strip of card sufficiently wide to fit rather tightly into the neck of a flask about 400 c.c. in capacity. The coil should hang down freely in the centre of the flask, about 1 cm. from the bottom, the card forming a diaphragm or partition in the neck, as shown in Fig. 53.

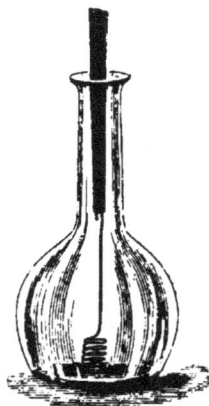

Withdraw the coil from the flask, pour a few drops of a strong solution of ammonia into the latter, and shake it so as to diffuse the ammonia gas through the air contained in the flask. Heat the platinum spiral to redness in a Bunsen's burner, and while it is still red-hot, plunge it into the flask. The coil will continue to glow for some time, and white fumes of ammonium nitrite will be almost immediately formed. The heated platinum wire has determined the union of the constituents of ammonia with the oxygen of the air. If, when the coil of wire has ceased to glow, the flask is gently heated by waving it for a few seconds over a lamp, more ammonia gas will be evolved from the liquid, and the platinum will again become red-hot. After the lapse of a minute or two, take out the coil of wire, add 2 or 3 c.c. of water, and shake it up in the flask; then pour the liquid (which will be slightly yellow, owing to the presence of nitrogen tetroxide) into a test-tube, add enough dilute hydrogen sulphate to render the solution strongly acid to test-paper (when the smell of nitrogen oxides will be perceived), then add one drop of solution of indigo sulphate, and heat the mixture. The blue colour of the indigo will disappear, proving, under the conditions of the experiment, that a nitrite or nitrate is present.

[Before putting away the drying tube, the quicklime in it should be examined and any of it which appears slaked by the moisture should be taken out and thrown away. A bit of glass rod should then be put into the hole in the cork, and the other end of the tube stopped with a plug of cork, to prevent entrance of moisture.]

COMPOUNDS OF NITROGEN AND OXYGEN.

Nitrogen forms five well-defined compounds with oxygen, a list of which will be found in Appendix D, and should be written out in the note-book. They or their compounds will be taken in descending order, beginning with compounds related to the highest oxide, nitrogen pentoxide.

5. NITRATES.

It has been seen in the last experiment (p. 130) that ammonia can be made to combine with oxygen to form a substance (a nitrite) which may be represented as containing a compound of nitrogen and oxygen. This oxidation of nitrogen goes on in nature on a large scale, during the decay or putrefaction of organic substances. The ammonia first formed is slowly oxidised to a nitrate; and thus soil is found to yield such salts as potassium nitrate ('nitre' or 'saltpetre'), and sodium nitrate. From either of these hydrogen nitrate ('nitric acid' or 'aqua fortis') can be prepared by heating it with hydrogen sulphate in the manner next to be described.

Preparation of Hydrogen Nitrate.

Apparatus required [1]—Stoppered (or plain) retort, about 2co c.c. in capacity; retort-stand; sand-bath with sand; argand, or spirit lamp; thistle funnel (Fig. 12, *a*); small flask; porcelain mortar; Bunsen's holder; funnel, 10 cm. in diameter; beaker; test-tube-stand; test-tubes in basket; watch-glass; washing bottle with distilled water; wooden blocks; blotting paper; lamp cotton, or tow; cloth; solutions of barium chloride, silver nitrate, litmus, indigo sulphate, ammonium hydrate, iron protosulphate; crystallised potassium nitrate; common hydrogen sulphate; bit of worsted or flannel; quill pen; piece of thin sheet lead; copper filings; charcoal.

[1] In future, lists of apparatus required will only be given where a more or less elaborate apparatus has to be set up, as for distillation, or preparation and collection of a gas.

Measure 20 c.c. of water into a beaker placed on a plate and add to it by degrees 20 c.c. of strong common hydrogen sulphate. This should be done carefully, as great heat is evolved by the union of the acid with the water (compare the action of quicklime on water, p. 58) and the beaker may be broken. Add only about 1 or 2 c.c. at a time, and stir with a glass rod.

While the mixture is cooling, weigh out 20 grms. of potassium nitrate, powder the salt roughly in a mortar, and transfer it to the retort, in the manner shown in Fig. 54.

Fig. 54.

Arrange the retort on the sand-bath, precisely as for the distillation of water, p. 63, putting a large test-tube temporarily in place of the flask as receiver, to collect the first portions of the distillate. Pour the diluted acid through the tubulure by means of a funnel, taking particular care that none of the liquid passes down the neck of the retort; then replace the glass stopper and proceed to heat the mixture.

Meanwhile, the condensing arrangement may be set up, in the way described at p. 64. Hydrogen nitrate will soon begin to distil over, not in the most concentrated form (since the hydrogen sulphate was slightly diluted) but sufficiently strong for experiments.

[The action which goes on in the retort is a 'double decomposition ' (see p. 74), the potassium and part of the hydrogen changing places, as shown in the diagram below [1],—

Potassium nitrate, 101 parts.
{ (Nitrogen, 14 parts) (Oxygen, 48 parts / Potassium, 39 parts) —62 } Hydrogen nitrate, 63 parts.

Hydrogen sulphate, 98 parts.
{ Hydrogen, 2 parts (Sulphur, 32 parts) (Oxygen, 64 parts / —39, 1 —96 } Potassium and hydrogen sulphate, 136 parts.

Thus only half the hydrogen in the hydrogen sulphate is replaced by potassium, in the process as ordinarily conducted; but at a higher temperature the whole of the hydrogen may be replaced and potassium sulphate formed.]

When 2 or 3 c.c. of acid have distilled over remove the test-tube to the test-tube stand and substitute for it a flask as a receiver. The heat should be regulated so as to keep the liquid in the retort gently boiling. It will be advisable to provide some means for cooling the receiver as well as the neck of the retort, to ensure the efficient condensation of the corrosive vapours of the acid. For this purpose lay over the flask as it rests in the mortar one or two folds of blotting-paper, and pour over it from time to time some cold water.

While the distillation is going on, you may examine the purity of the acid which came over first, and was collected in the tube. Add to it about 10 c.c. of water, and pour one half of the solution into another test-tube.

(*a*) Test the first half of the solution with a drop of solution of barium chloride; shake the mixture and hold it up to the light. If a turbidity is perceptible, hydrogen sulphate is present as an impurity[2].

(*b*) To the other half of the solution add a drop of solution

---

[1] Or by the equation,—

$$KNO_3 + H_2SO_4 = KHSO_4 + HNO_3.$$

[2] It must be borne in mind that barium nitrate is insoluble in strong hydrogen nitrate; and hence if the solution of hydrogen nitrate is not very dilute, a crystalline precipitate may be formed. This, however, is easily distinguishable in appearance from barium sulphate, and will readily dissolve when more water is added, and the liquid warmed.

of silver nitrate. If a precipitate is produced, chlorine is present as an impurity, derived from the potassium chloride which the nitre of commerce usually contains.

When crystals make their appearance in the liquid and on the sides of the retort, put out the lamp, and, as soon as the apparatus has cooled a little, withdraw the receiver and pour its contents into a clean test-tube or flask. The residue in the retort consists of potassium-and-hydrogen sulphate, and may when cool be washed out with a little warm water. The retort itself, after being thoroughly rinsed with distilled water, should be set aside to dry, supported with the tubulure downwards in a Bunsen's holder or otherwise.

Test two small portions of the acid which you have obtained (diluted with 10 c.c. of water, as above) with solution of barium chloride and with solution of silver nitrate respectively. If the distillation has been carefully conducted, traces only (at most) of a sulphate and chloride should be found to be present [1]. The liquid will perhaps be slightly yellow, owing to the presence of traces of lower nitrogen oxides (formed by the decomposition of the hydrogen nitrate which takes place if the temperature is somewhat too high), but this impurity is unimportant.

The following experiments may now be tried with the acid which you have obtained:—

*1. Its action on white wool, &c.

Put a bit of white worsted or flannel, and the feather of a quill pen into a small porcelain dish, add a few drops of the acid and warm gently. The substances will in a short time be stained yellow; but if the action is continued much longer they will be entirely destroyed. Treat a bit of calico or cotton wool in the same way: no yellow stain will be produced.

*2. Its action on litmus.

Put a little water into a test-tube, add enough solution of litmus to colour it strongly blue, and then add a drop of the acid. The blue colour will be at once changed to red, showing

---

[1] In order to obtain a perfectly pure acid it will generally be necessary to re-distil it after addition of a crystal or two of silver nitrate and of potassium dichromate.

that hydrogen nitrate belongs to the class of substances which have an 'acid reaction.'

3. Its combination with ammonia.

You have already seen (p. 129) that ammonia unites directly with hydrogen chloride to form the salt called ammonium chloride. It unites in a similar way with hydrogen nitrate, forming ammonium nitrate. These salts have neither the acid reaction of hydrogen nitrate nor the alkaline reaction of ammonia, but are neutral to test-paper.

Pour 5 c.c. of the liquid into a porcelain dish, and add about 3 c.c. of solution of ammonia (ammonium hydrate). Stir the mixture with a glass rod, and place the drop of it which adheres to the rod upon a strip of blue litmus-paper laid on a clean white plate, drawing a thin line across the strip. The liquid will redden the litmus-paper decidedly. Continue to add ammonia cautiously drop by drop, stirring the mixture, and placing a fresh drop from time to time on an untouched part of the strip of litmus-paper. You will find that a time will arrive when the liquid no longer alters the colour of the blue litmus-paper. When this is the case, place a drop of it on a piece of reddened litmus-paper, and observe whether the red colour is changed to blue. If so, pour a few drops of the hydrogen nitrate into a test-tube, dilute it with about ten times its volume of water, and add it by means of a pipette drop by drop to the solution in the dish, stirring thoroughly after each addition and constantly trying the action of the liquid on both blue and reddened litmus-paper. You will eventually obtain a solution which does not materially alter the colour of either test-paper, and is, in fact, neutral. When this point is reached, evaporate the solution to dryness on the sand-bath at a gentle heat[1], placing near it (for the sake of comparison) two watch-glasses, the one containing a little hydrogen nitrate, the other a little solution of ammonia. (These should not be put very near each other, or else the vapours will mix.) You will find that a white residue will be left in the porcelain dish, while there will

---

[1] The heat must not be high, otherwise the salt will fuse and be decomposed.

be scarcely a trace of a residue left in the watch-glasses. By
bringing hydrogen nitrate in contact with ammonia in a certain
definite proportion, you have obtained a substance which differs
from them in at least two respects, (*a*) in having no action on
vegetable colours, (*b*) in being a crystallisable solid at ordinary
temperatures. This 'salt' is, in fact, identical with the substance called ammonium nitrate, from which you will obtain
nitrogen protoxide.

4. Its action on metals, such as lead.

Put a bit of thin sheet lead (the lead foil used for lining tea
chests answers well) into a porcelain dish, pour upon it 8 or
10 c.c. of hydrogen nitrate, and heat it on a sand-bath (in a
draught cupboard, if possible, as the fumes are corrosive and
poisonous). Orange vapours of lower nitrogen oxides will be
given off and the metal will readily dissolve, forming lead
nitrate[1]. This salt will be left as a white residue, when the
solution is evaporated to complete dryness, and should be kept
for a future experiment, p. 139.

Properties of Nitrates.

[For the following experiments crystallised potassium nitrate and
a solution of the same salt (made by dissolving 0.5 grm. in 25 c.c. of
water) should be used[2]. In order to make the solution quickly,
place the proper quantity of the salt in a glass mortar, grind it
to powder, then add 2 or 3 c.c. of water, and grind the whole
together. Pour off the liquid into a large test-tube, add a little
fresh water to the residue in the mortar, and again grind the whole
together and decant the solution. In this way, by several successive
triturations with fresh quantities of water, the whole of the salt
will soon be obtained in solution, and the liquid should then be put
into a measure and made up to the required volume.]

---

[1] It is hardly necessary to remark that in most cases where a substance is
said to dissolve in an acid, the phenomenon really consists in the formation,
by the action of the acid, of a salt soluble in water.

[2] In cases where, as here, a solution of salt is prepared for the purpose
of being tested, the strength of the solution is such that 50 c.c. of it would
contain (approximately) a weight corresponding to that of a molecule of the
substance, expressed in centigrammes. See Appendix B.

*1. Their direct oxidising powers, e. g. on carbon.

Put a small crystal of potassium nitrate into an ignition tube, (p. 13) add a splinter of charcoal, and heat over a Bunsen's burner.

The salt fuses, and when the charcoal becomes red-hot a vivid deflagration will take place, owing to the decomposition of the nitrate and oxidation of the carbon; the oxygen uniting with the potassium and some of the carbon to form potassium carbonate, while nitrogen gas and carbon dioxide are liberated.

*2. Their action on indigo, destroying its blue colour.

Pour a little water into a test-tube, and add about half its volume of strong hydrogen sulphate, dropping in the acid little by little, lest the tube should be broken by the heat evolved. Next, add just sufficient solution of indigo sulphate to colour the liquid distinctly blue, and then put in a few drops of the solution of potassium nitrate and heat the mixture to boiling. The blue colour of the indigo will disappear entirely, a light brown colour only remaining.

*3. Their action on copper.

Pour into a test-tube a few drops of the solution of potassium nitrate; add an equal volume of strong hydrogen sulphate, and drop in a bit of copper wire (or, better, copper filings) and warm gently. Nitrogen dioxide will be given off, and may be recognised by the orange fumes which it forms with the oxygen of the air in the tube. These may be most distinctly seen by placing the tube on a piece of white paper and looking down into it.

*4. Their action on iron protosulphate.

Place a little solution of potassium nitrate in a tube, add an equal volume of hydrogen sulphate as in the last experiment, and cool the mixture by holding the test-tube in a stream of water. Then, holding the tube in a slanting position, pour as slowly as possible down the side (most conveniently from a pipette), a solution of iron protosulphate, so that the lighter solution may not mix with, but float upon, the heavier. A dark brown ring will eventually be formed at the line of junction of the two liquids, which will, when the tube is heated, disappear with effervescence due to the escape of nitrogen dioxide. The ring will be best seen by raising the test-tube to a level with the eye and holding a piece of white paper behind it.

5. Their reduction, with formation of ammonia.

Put 2 grms. of caustic potash (potassium hydrate) into a test-tube, add 5 c.c. of water, and shake until the potash is dissolved ; then drop in two or three small pieces of granulated zinc, add one drop of solution of platinum perchloride and warm gently. Hydrogen gas will be evolved [1], and may be recognised by pressing the thumb gently against the mouth of the tube for a minute, while the liquid is heated just to boiling, and then applying a light. Hold a piece of reddened litmus-paper at the mouth of the tube; no alkaline reaction will be observed. Now drop in a small crystal of potassium nitrate, keeping the mixture heated. Ammonia will very shortly be formed, and may be recognised by its odour and its action on a piece of reddened litmus-paper held at the mouth of the tube.

From the nitrates all the nitrogen oxides can be obtained. Thus hydrogen nitrate may by the action of phosphorus pentoxide be separated into nitrogen pentoxide and water; the process, however, is neither easy nor safe enough for the beginner to attempt. All the other oxides can be obtained by withdrawing the proper quantity of oxygen from a nitrate, and will be prepared and examined in due order.

[1] The zinc replaces hydrogen in acting upon potassium hydrate, precisely as it does in acting upon hydrogen sulphate,

$$Zn + 2\,KHO = Zn\,K_2O_2 + H_2.$$

The platinum reduced as a black powder upon the zinc promotes the action by forming with the zinc a galvanic couple.

6. NITROGEN TETROXIDE.

The nitrate usually employed to obtain this is lead nitrate, which when heated gives off a mixture of nitrogen tetroxide and oxygen.

[Thus:—

Lead
nitrate,
331 parts.
{
Lead, 207 parts ———————— 207 } Lead oxide, 223 parts.
Oxygen, 96 parts ———————— 16 }
Nitrogen, 28 parts ———————— 16 { Oxygen, 16 parts.
64 }
28 } Nitrogen tetroxide, 92 parts [1].]

Scrape together the residue of lead nitrate which you obtained by dissolving lead in hydrogen nitrate (p. 136), and heat it in the dish on a sandbath, stirring it constantly, until it is thoroughly dry. Then put it into a moderate-sized test-tube supported in the Bunsen's holder (as in Fig. 39, p. 69), and heat it pretty strongly over an argand burner. A deep orange vapour will be given off which is nitrogen tetroxide [2].

This substance combines readily with water, forming two acids, hydrogen nitrate and hydrogen nitrite. To prove this, moisten a piece of blue litmus paper with distilled water, and dip it into the tube; it will be strongly reddened.

Allow the tube to cool, closing its mouth loosely with a cork; observe that the vapour becomes much paler in colour as it cools.

7. NITROGEN TRIOXIDE and NITRITES.

This is obtained, though not in a pure condition, by the action of starch on hydrogen nitrate. The chemical change

---

[1] Or, in symbols:—

$$2\,Pb\,(NO_3)_2 = 2\,PbO + O_2 + 2\,N_2O_4.$$

[2] Besides this gas, oxygen is evolved, as explained above. Its presence may be shown by introducing a glowing match into the tube (nitrogen tetroxide itself having no tendency to support combustion).

is rather complicated, but the general nature of it is that the carbon and hydrogen of the starch take away some oxygen from the hydrogen nitrate, leaving enough combined with the nitrogen to form nitrogen trioxide.

Put a little starch (about as much as will lie on the end of a spatula) into a test-tube, add about 3 drops of water and mix it with the starch by a thorough shaking. Then add 3 c.c. of strong hydrogen nitrate, mix the whole thoroughly, and heat it cautiously, taking it away from the flame when an action begins. The liquid will turn yellow, and orange vapours of nitrogen trioxide, resembling those of the tetroxide, will be given off.

1. Nitrogen trioxide combines with water to form an acid, viz. hydrogen nitrite (nitrous acid); hence a piece of moistened blue litmus paper held in the tube will be strongly reddened.

2. Similarly, it combines with other metallic oxides or hydrates, forming the class of salts called NITRITES, which are chiefly remarkable for the readiness with which they not only give up their oxygen (and hence are called ' oxidising agents') but also combine with more oxygen (and hence appear as ' reducing agents').

In illustration of these facts, fold a small piece of blotting-paper into a strip narrow enough to go into the test-tube, lay it on a plate and moisten it with a few drops of solution of potassium hydrate; then introduce it into the tube filled with the vapours of nitrogen trioxide. The vapours will be quickly absorbed by the potassium hydrate, with formation of potassium nitrite [1].

Allow the paper to remain in the tube for half a minute, warming the mixture of acid and starch if the evolution of nitrogen trioxide has ceased; then take out the strip of paper, put it into a small beaker and add 10 c.c. of water to dissolve the potassium nitrite. The following experiments may be tried with the solution.

*1. Oxidising power of nitrites.

This property is best seen in their action upon hydrogen iodide, a compound of hydrogen and iodine; potassium nitrite

---

[1] $N_2O_3 + 2 KHO = 2 KNO_2 + H_2O.$

decomposes this, some of its oxygen combining with the hydrogen, while iodine is liberated.

Put two or three drops of solution of potassium iodide into a test-tube, add about 10 c.c. of water and five or six drops of dilute hydrogen sulphate. The liquid now contains hydrogen iodide, formed by double decomposition[1].

Add to this a drop of the solution of potassium nitrite; iodine will be at once liberated, colouring the liquid yellow [2].

[Compare this result with the action of nitrates; which, though they contain more oxygen than nitrates, do not give it up so readily.

Make a little more solution of hydrogen iodide, as above directed, and add a drop or two of solution of potassium nitrate. No immediate liberation of iodine will be observed. Hence this reaction may be used to distinguish between nitrites and nitrates, which resemble each other in many respects, e.g. in their action on indigo and iron protosulphate.]

*2. Reducing power of nitrites.

Pour the rest of the solution of potassium nitrite into a test-tube, and add a drop of solution of mercury protonitrate. A grey precipitate of metallic mercury will be produced; the metal having been 'reduced' (an old alchemical term) from its combination with the nitrate radicle, while the potassium nitrite is converted into potassium nitrate [3].

8. NITROGEN DIOXIDE.

Apparatus required—Pneumatic trough; flask, fitted with the delivery tube and funnel used in Ex. 2 ; glass measure; beaker; one large and three small gas bottles; two gas jars, 20 × 5 cm.; one ditto, 10 × 3 cm.; taper on wire; glass disc; test-tubes and stand; blue litmus-paper; india-rubber rings; copper clippings; hydrogen nitrate; solution of iron protosulphate; carbon disulphide; bottle of oxygen.

This oxide is formed when copper is made to act upon hydrogen nitrate, as in experiment 3, p. 137.

---

[1] Hydrogen sulphate and potassium iodide give potassium sulphate and hydrogen iodide. Compare the change which occurs in the preparation of hydrogen nitrate.
[2] The presence of free iodine may be confirmed by the characteristic reaction with starch (see under IODIDES)..
[3] $2\,Hg\,(NO_2) + K\,NO_2 + H_2O = 2\,Hg + 2\,HNO_3 + K\,NO_3$.

[The reaction is of the following nature :—

1. Copper displaces hydrogen in a portion of the acid, forming copper nitrate.

2. The displaced hydrogen, instead of being evolved as gas, acts upon another portion of hydrogen nitrate, combining with some of its oxygen, while the nitrogen and the rest of the oxygen are evolved as nitrogen dioxide [1].]

Measure out 50 c.c. of water and place it in a beaker large enough to hold more than twice the quantity; then measure out the same volume of strong (common) hydrogen nitrate and add it to the water in the beaker. Notice the heat evolved (due to the chemical combination of the water and the acid), which is not, however, so great as that produced by the union of hydrogen sulphate and water, p. 132.

While the liquid is cooling, place in the flask about 30 grms. (10 or 12 strips about 0.5 cm. broad and 5 or 6 cm. long) of sheet copper, and arrange the apparatus precisely as was done in preparing hydrogen, Fig. 48, p. 105. Pour a little water into the flask, but only *just* sufficient to cover the end of the funnel, and try whether the joints are tight as directed in p. 106. Fill the gas bottles and jars with water and place them inverted on the shelf of the trough, the stopper of each being placed upon its own bottle, to prevent confusion. When everything is ready pour the diluted acid into the flask. An effervescence will almost immediately commence, and the flask will be filled with orange vapours, which are not, however, nitrogen dioxide but one of the higher oxides formed by the union of the dioxide with the oxygen of the air in the flask. These will be soon absorbed by the water, so that the column of water in the delivery tube will rise a little and no permanent gas will come over at first. If there is little or no action after the lapse of a minute, the acid may have been too much diluted, and about 1 c.c. of strong acid may be poured into the flask. A colourless gas will shortly come over, with which, after rejecting two jars full, the bottles and jars may be filled for experiments. No heat is required at first to accelerate the action : indeed, it may

---

[1] This may be expressed by the following equation—

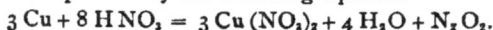

$$3\,Cu + 8\,HNO_3 = 3\,Cu(NO_3)_2 + 4\,H_2O + N_2O_2.$$

become too violent, and should then be moderated by surrounding the flask with a wet cloth or moistened blotting-paper. On the other hand, if the evolution of gas becomes slow before the receivers are all filled, 1 or 2 c.c. of strong acid should be added through the funnel. When sufficient gas has been collected, the retort-stand and flask should be taken away at once, and placed in the open air. The blue liquid is, of course, a solution of copper nitrate, and will hardly be worth preserving.

[Besides the process above given, the following is an excellent method of obtaining pure nitrogen dioxide. Measure 80 c.c. of water and place it in a beaker large enough to hold twice the quantity. Measure 30 c.c. of strong (common) hydrogen sulphate, and add it, little by little (see p. 132), to the water in the beaker, stirring the mixture with a glass rod. When all the acid has been added, place the beaker aside to cool. Meanwhile, mix in a mortar 6 grms. of potassium nitrate and 50 grms. of iron protosulphate. Arrange an apparatus as above directed, and place the mixture in the flask, adding about a teaspoonful of sand, to make the action regular. Pour on it the diluted acid through the tube-funnel, and apply a gentle heat to the flask, watching carefully for the first signs of evolution of gas, and taking away the lamp temporarily if a rush of gas commences. The mixture will become brown, and pure nitrogen dioxide will be evolved[1].]

The following experiments may now be tried with the gas you have collected :—

1. Its action on litmus paper, and on air.

Introduce a strip of blue litmus paper into one of the jars standing on the shelf of the trough, pushing it up with a glass rod, and leaving it adhering to the side of the jar (take care that no air enters). The colour of the litmus will remain unaltered. Now raise the mouth of the jar for a moment above the water, so as to allow a little air to enter. Orange vapours will be formed, and the litmus paper will be at once turned red. The explanation is, that nitrogen dioxide though a neutral substance itself, unites with the oxygen of the air producing one of the higher nitrogen oxides, which (as has been seen) combines with water to form an acid. The action will be more closely examined in the next experiment.

[1] $6\,FeSO_4 + 4\,H_2SO_4 + 2\,KNO_3 = 3\,Fe_2(SO_4)_3 + 4\,H_2O + K_2SO_4 + N_2O_2$.

*2. Its action on oxygen.

Refill with water the jar used in the last experiment, and decant into it one measure of oxygen (from one of the reserved bottles, p. 104) using the small cylindrical jar as a measure. Fill the small jar with nitrogen dioxide, and decant its contents, bubble by bubble, into the oxygen. Orange vapours of nitrogen tetroxide will be at once formed, and the bulk of the mixed gases, which will be at first slightly increased owing to the heat produced by their union, will soon diminish as the tetroxide dissolves in the water, until only about one-half of the original volume of oxygen remains. Add in the same way another measure of nitrogen dioxide; stopping, however, if the addition of a bubble no longer causes the formation of orange vapours. Eventually, if the oxygen and nitrogen dioxide were pure, no residue of permanent gas will be left. Hence we learn that two volumes (or 60 parts by weight) of nitrogen dioxide combine with one volume (or 32 parts by weight) of oxygen, to form nitrogen tetroxide [1].

[It was by the above method that Cavendish made his analyses of air, and the following experiment will serve to illustrate his mode of proceeding:—

Take one of the larger cylindrical gas jars, and graduate it into five equal parts in the way described in p. 119. Fill it to the fifth division with air, as there directed; then fill the small measure with nitrogen dioxide, and decant the gas gradually into the air enclosed in the graduated jar. Add one more measure of nitrogen dioxide in the same way; you will then have added two volumes of nitrogen dioxide to five volumes of air. Now agitate the graduated jar (taking care that its mouth is not raised above the surface of the water in the trough), in order that the higher nitrogen oxides may be entirely absorbed; leave it to cool undisturbed for a few minutes; and, finally, depress it in the trough until the water inside and outside stands at the same level, and note the volume of the residual gas. It will be approximately four-fifths of the original volume of air, and if no further red fumes are formed on addition of a little more

---

[1] $N_2O_2 + O_2 = N_2O_4$. It should be observed that the formulæ N_2O_2 and N_2O_4 represent weights of the gases which occupy twice the space of the usual unit of volume, viz. the volume occupied by the weight of hydrogen expressed by the formula H_2.

nitrogen dioxide, we may assume that the oxygen of the air is entirely removed, and consequently that five volumes of air contain approximately one volume of oxygen. The residual gas may be tested with a lighted taper and shown to be a non-supporter of combustion; further tests would absolutely prove it to be nitrogen.]

3. Its relation to combustion.

Insert the stopper into one of the bottles of gas, and place the latter mouth upwards on the table. Plunge into the gas the taper, lighted; it will be immediately extinguished, and the gas will not take fire [1].

4. Its tendency to give up oxygen.

It has been proved in experiments 1 and 2 that nitrogen dioxide has a great tendency to combine with more oxygen. But it can also be made to give up the oxygen it contains (although, as just now observed, a taper will not effect this) by the action of substances possessing a strong affinity for oxygen, such as carbon disulphide.

Pour into a small dry test-tube about 1 c.c. of carbon disulphide, and place it in the stand, within reach (taking care that no lighted lamp is near, since carbon disulphide is highly combustible). Place one of the larger cylindrical jars of nitrogen dioxide on the table, previously decanting into it, if necessary, a few bubbles of gas, so as to fill it completely. Slide the glass cover aside for a moment, and pour the carbon disulphide quickly into the jar (taking care to admit as little air as possible), shake it up once, to diffuse the vapour through the gas, replace the jar on the table, and apply a lighted match to its mouth. No explosion will follow, but a brilliant flash of light will pass down the jar. The oxygen of the nitrogen dioxide unites with both the carbon and the sulphur of the carbon disulphide, forming carbon dioxide and sulphur dioxide, while the nitrogen remains uncombined [2].

---

[1] The taper sometimes, if in full combustion and if the gas contains a little nitrogen protoxide (as it is liable to do when made by this method), continues to burn, and with increased brilliancy.

[2] A portion of the sulphur is also separated and appears as a yellow deposit on the sides of the jar.

5. Its absorption by iron protosulphate.

Place 4 or 5 c.c. of solution of iron protosulphate (green vitriol) in a test-tube, add 6 or 8 drops of dilute hydrogen sulphate, and pour it into a small jar of the gas. Close the mouth of the jar with a glass plate, and agitate the liquid. A dark-brown solution will be formed, which, when poured into a test-tube and boiled over a lamp, will give off the gas with effervescence, becoming nearly colourless again.

[Reserve one small bottle of nitrogen dioxide for use in the next exercise.]

9. NITROGEN PROTOXIDE.

Apparatus required—Pneumatic trough, filled with warm water; stoppered retort, with long beak; retort-stand; piece of wire-gauze; argand burner; scales and weights; deflagrating cup; two gas jars, 20 × 5 cm.; one ditto, 10 × 3 cm.; one gas bottle, holding 700 c.c.; three ditto, holding 150 c.c.; taper on wire; glass disc; cedar matches; sheet of writing-paper; ammonium nitrate, in crystals; sulphur; bottle of nitrogen dioxide.

This is the lowest in the series of nitrogen oxides, containing less oxygen than any other. It is usually prepared by heating ammonium nitrate, a salt of which the formation has been already illustrated in experiment 3, p. 135. The decomposition of this salt by heat is of the following nature. It contains hydrogen, nitrogen, and oxygen; the hydrogen unites with some of the oxygen to form water, while the nitrogen associated with the rest of the oxygen forms nitrogen protoxide.

[Thus,

Ammonium nitrate, 80 parts, { Nitrogen 28 parts ——— 28 } Nitrogen protoxide, 44 parts. { Oxygen 48 parts —— 16 } { Hydrogen 4 parts ——— 32 } Water, 36 parts[1].]

Weigh out 20 grms. of ammonium nitrate, and transfer it to the retort in the manner shown in fig. 54, p. 132. Support the latter on a piece of wire-gauze bent into the form of a basin, and placed on the largest ring of the retort-stand in such a

---

[1] Or, in symbols:—

$$(H_4 N) NO_3 = 2 H_2 O + N_2 O.$$

position that the beak may dip under the shelf of the pneumatic trough, and that there may be room for the argand burner under the wire-gauze. In order to keep the retort steady, the smallest ring may be brought down over the tubulure, and to do this it will be most convenient to arrange the retort-stand so that its stem is *behind* the retort, as shown in fig. 37, p. 63 [1].

Heat the retort, slowly at first, by a lamp, and while it is gradually becoming hot, fill the gas jars and bottles with water, and place them mouths downwards on the shelf of the trough. The salt will soon fuse into a clear liquid, and by a slight increase of the heat will begin to decompose with effervescence owing to the escape of nitrogen protoxide. Be careful not to use more heat than is necessary to keep the fluid steadily effervescing, and do not, on the contrary, withdraw the lamp entirely while the beak of the retort is under water; or the sudden condensation of the steam, which is one of the products of the decomposition, will be likely to draw up water from the trough into the hot retort, and crack it. When the evolution of gas begins, white fumes (consisting of particles of the salt tossed up as spray by the escaping gas) will appear in the retort. So long as these are formed only in moderate quantity no harm need be apprehended; but if they increase, and the effervescence becomes violent, the heat should be diminished (not withdrawn) at once.

When a steady stream of bubbles has escaped for about a minute, you may place one of the large cylindrical jars over the hole in the shelf, and allow it to fill with gas. When full, remove it to another part of the shelf and bring another jar into its place. While this latter is filling, you may test the gas in the first jar. Slide it off the shelf with one hand, cover its mouth with a glass plate held in the other hand, raise it out of

---

[1] It is a decided advantage to use a retort instead of a flask and delivery tube, in the preparation of this gas. If the latter is used, the water which is formed in the reaction collects in the neck, and drops down on the melted salt, producing a burst of vapour. The retort can be more uniformly heated, and the water which condenses in the neck flows down into the trough: moreover the beak of the retort is wider than the usual delivery tube, and hence the condensation of the water is more regular.

the trough and place it mouth upwards on the table. Remove the glass plate, and immediately introduce a cedar match which has been lighted and *just* blown out. If it is re-kindled, as in the case of oxygen, the gas is sufficiently pure; but it must be borne in mind that the end of the match should be glowing rather brightly, in fact only just blown out; a mere spark is not enough to begin the combustion, as with oxygen. Proceed with the filling of the other jars and bottles, carefully watching the decomposition of the salt.

When sufficient gas has been collected, take out the stopper of the retort, and set the latter aside to cool, still supported in the retort-stand.

The following properties of the gas may be examined.

1. Its solubility in water, odour, and taste.

Pour into a small bottle of the gas about 20 c.c. of distilled water, and replace the stopper, inserting a small strip of thin paper between it and the neck of the bottle. Invert the bottle and shake it briskly for a few seconds, then hold it up to the light. Observe that bubbles of air are entering at the side of the stopper, thus showing that there has been an absorption of the gas. If the piece of paper had not been inserted at the side of the stopper, the latter would have become so firmly fixed in its place by the pressure of the external air, that it would have been difficult to loosen it. Shake the bottle again for a few seconds, then pour out some of the solution into a beaker and taste it. It will have acquired a sweetish taste and smell.

2. Its weight, and its relation to combustion.

Put an empty cylindrical gas jar on the table, mouth upwards, and place in it the taper, lighted, leaning the wire against the side of the jar so that the taper may rest steadily at the bottom. Bring the mouth of one of the small bottles of gas close to the edge of the jar, withdraw the stopper, and gently pour the gas over the taper, precisely as if you were pouring out a liquid. The flame will become larger and more brilliant, but the gas will not itself ignite : thus proving it to belong to the same class as oxygen; viz. to be one of those gases which are not inflammable, but supporters of the combustion of a taper.

3. Its decomposition by sulphur.

Place the large bottle of gas on the table before you, and see that its stopper is sufficiently loose to be taken out when required. Put a small bit of sulphur into the deflagrating cup, heat it over the lamp until it enters into *full* combustion, then immerse it in the bottle.

The sulphur will burn as brilliantly as in oxygen[1], but the blue flame will have a yellowish border. Sulphur dioxide (as shown by its odour and action on test-paper) will be formed, as with oxygen, thus proving that the gas you are dealing with contains oxygen. You will probably observe brown vapours in the bottle : these consist of one of the higher nitrogen oxides.

4. Its decomposition by hydrogen.

Fill one of the larger gas jars with water and place it, inverted, on the shelf of the trough. Decant into it one measure of nitrogen protoxide, using the small thick jar as a measure. Add one measure of hydrogen (see note [2]) and mix the gases by shaking the jar laterally. Decant some of this mixture into the small jar, close its mouth with a glass plate, place it on the table (mouth upwards), and apply a light to the mixed gases. The detonation which takes place is as violent as when the combination of hydrogen with free oxygen was effected in a similar way.

In this reaction it is found that the hydrogen only combines

---

[1] The sulphur must be heated considerably above its melting-point, and must be in full combustion in the air; otherwise it is extinguished when plunged into the gas.

[2] A sufficient quantity of hydrogen for the above, and other similar experiments, may be made in the following way. Take a small gas jar; place in it one or two bits of granulated zinc, and fill it completely with dilute hydrogen sulphate (1 measure of acid to 6 of water). When a brisk action has commenced, place an inverted porcelain dish, 9 cm. in diameter, over the jar like a cap; then, keeping the dish pressed close to the mouth of the jar, invert the apparatus quickly. The zinc will fall to the lowest part of the jar, and the evolved hydrogen will gradually fill the jar, displacing the acid which collects in the dish and prevents the entrance of air. A minute bubble of air may possibly enter at the moment that the jar is inverted, if the operation is not skilfully performed, but so small a quantity will not materially vitiate the purity of the gas When the jar is full of gas, fill up the dish with water, raise the tube a little so that the remainder of the zinc may fall into the dish; then close the mouth of the jar with a glass plate, and transfer it to the pneumatic trough.

This was probably the earliest method devised for the collection of a gas in order to examine its properties.

with the oxygen of the nitrogen protoxide, and that a quantity of nitrogen remains, equal in volume to the gas originally taken[1]. Now, it has been proved (p. 110) that 1 measure of hydrogen combines with half a measure of oxygen: hence 1 measure of nitrogen protoxide yields one measure of nitrogen and half a measure of oxygen.

5. Its relation to respiration.

The most remarkable property of this gas, its property of producing a peculiar intoxication (whence its name of 'laughing gas') and finally insensibility when respired, is one which the student is advised not to try by himself; since particular precautions are necessary to obtain a perfectly pure gas for the purpose, and the effects of inhaling it vary with different individuals, and to some are decidedly injurious.

6. Test to distinguish it from oxygen.

We have seen that nitrogen protoxide in many respects resembles oxygen: it may, however, be readily distinguished from the latter gas by the fact that nitrogen dioxide has no action upon it.

Place two cylindrical gas jars, filled with water, on the shelf of the trough: decant into one some oxygen, into the other some nitrogen protoxide; then pass up into each a few bubbles of nitrogen dioxide (from the bottle reserved in the last exercise). Orange vapours will be formed in the jar containing oxygen, while no change will be observed in the other jar: nitrogen dioxide not having the power of decomposing nitrogen protoxide, although it will combine with free oxygen.

The series of nitrogen oxides which has just been examined affords excellent illustrations of the laws of chemical combinations; and the student is recommended at this point to gain a knowledge of these laws, and also of the atomic theory and the system of chemical symbols. A short account of these is given in Appendix D, but a general text-book should, of course, be consulted as well.

---

[1] $N_2O + H_2 = H_2O + N_2$.

10. CARBON.

The artificial forms of this element are obtained by heating various organic substances, which contain carbon as an essential constituent. For example, wood consists of carbon, hydrogen and oxygen, and when strongly heated in a close vessel is decomposed, the greater part of its carbon remaining as charcoal, while the rest, together with the hydrogen and oxygen, form a number of volatile compounds, among which are wood naphtha, hydrogen acetate (acetic acid) and water.

1. **Formation of charcoal.**

Place a little sawdust, or a few splinters of wood (a couple of matches cut up into small pieces, the end coated with phosphorus being, of course, rejected) in a small test-tube, and heat the tube rather strongly, supporting it horizontally in a Bunsen's holder, lest the condensed water should run down to the hot part and crack the tube. The wood will become black or 'charred' owing to the separation of carbon, while gases and vapours (such as wood naphtha) will be given off which will catch fire if a lighted match is held at the mouth of the tube. Drops of liquid will also condense in the tube, which will redden blue litmus paper; they consist mainly of water and hydrogen acetate.

2. **Formation of bone black (animal charcoal).**

Put some pieces of bone into a ladle, cover them completely with sand, and heat the whole to redness for about a quarter of an hour in an ordinary fire, while other experiments are proceeded with. Allow it to cool and then take out the charred pieces of bone, and grind them to fine powder in a mortar. This 'animal charcoal,' owing to its porosity and finely-divided state, has a strong power of absorbing some colouring matters and also many of the gases, as the following experiments will show [1].

---

[1] Animal charcoal as purchased at the shops, which has been recently ignited in a covered porcelain crucible, may be used in this and the following experiment, and will be found to act even more powerfully than the charcoal prepared as above directed.

(*a*) **Its combination with colouring matter.**

Put a little water into a test-tube, and add about 2 drops of solution of indigo sulphate. Add to this some bone black and shake the whole well together : then filter it into a clean test-tube, passing it again through the same filter if it does not run through clear. The filtrate will be nearly or quite colourless.

(*b*) **Its power of absorbing gases.**

Put about 2 c.c. of solution of hydrogen sulphide into a test-tube, add to it some bone black, and shake the mixture thoroughly for half a minute. You will find that the liquid has now lost all the offensive smell of the gas, which has been absorbed by the charcoal. The value of charcoal as a disinfectant is due to the property of absorbing gases which it possesses.

3. Formation of lamp-black.

1. Pour a few drops of turpentine into a small evaporating dish, place it in the centre of an ordinary china plate, putting in a small tuft of tow to serve as a wick. Take out the stopper of a deflagrating jar, and place it sideways on the neck, so as partially to close the opening. Apply a light to the turpentine, and place over it the jar, thus loosely stopped, and tilted slightly on one side, so as to admit a very limited supply of air for the combustion of the turpentine. Under these circumstances the latter will not be completely burnt, and a quantity of finely divided carbon (the ordinary lamp-black) will be deposited on the sides of the jar. Oxygen has less affinity for carbon than for hydrogen, and hence, while the whole of the latter is burnt, forming water, a part only of the former is converted into carbon dioxide, the rest appearing in the elemental condition. Compare the account of the Bunsen's burner, p. 4.

4. Direct combination of carbon with oxygen.

Fill a large gas bottle with oxygen gas [1].

---

[1] If a supply in a gas holder is not at hand sufficient gas for this experiment may be made in a tube-apparatus such as that which is described at p. 207 for making hydrogen iodide, a mixture of 6 grms. of potassium chlorate and 2 grms. of manganese dioxide being placed in the test-tube, and the gas collected by displacement.

Select a splinter of charcoal, about 3 cm. in length (charcoal made from the bark of a tree burns most brilliantly); lay it across the deflagrating cup, and confine it in its place by a piece of copper wire twisted round the stem of the cup and made to press upon the charcoal, as shown in fig. 55. Heat the projecting part of the charcoal until it glows, and then immerse it in the jar of gas. It will at once enter into brilliant combustion, as it combines with the oxygen to form carbon dioxide [1].

Fig. 55

To prove the formation of carbon dioxide, take out the charcoal when the combustion is over, and pour into the bottle 8 or 10 c.c. of lime water (from a test-tube and not direct from the bottle). This will, on being shaken up, become turbid ; a result which, as has been already seen, is characteristic of the presence of carbon dioxide.

5. Action of carbon in reducing metals from their oxides.

This depends upon the affinity of carbon for oxygen, illustrated in the last experiment, and is the principle of many of the smelting processes employed on a large scale.

Take about as much lead protoxide (litharge) as will lie on the end of a spatula, and powder it in a mortar. Make a small cup-shaped cavity in a piece of charcoal as directed in p. 90, fill it with the powdered substance, and direct the blowpipe flame (the mouth blowpipe being used, as in Exercise 11, p. 85) down upon it. The lead protoxide will soon melt and effervesce owing to an escape of gas, and finally nothing but a bright globule of metallic lead will be left on the charcoal.

The carbon has reduced the lead protoxide, combining with its oxygen to form carbon oxides which escape, while the lead is left in the metallic state [2].

[1] $C_2 + 2O_2 = 2CO_2$.
[2] $4PbO + C_2 = 4Pb + 2CO_2$.

COMPOUNDS OF CARBON AND OXYGEN.

11. CARBON DIOXIDE.

[Formula of molecule, CO_2,
Weight of molecule,   44 hydrogen-atoms.]

Apparatus required—Flask, fitted with cork and tube-funnel as in
Ex. 2; elbow tube, with branches 8 cm. in length; ditto, with branches
respectively 8 cm. and 18 cm. in length; corks; cork-borers; rat-tail
file; retort-stand; wire-gauze; wooden blocks; card; one gas bottle,
holding 700 c.c.; four ditto, holding 150 c.c.; gas jar, 20 × 5 cm.; glass
disc; test-tubes; porcelain dish; Bunsen's burner; taper on wire;
deflagrating cup; marble; strong common hydrogen chloride; lime
water; litmus-paper; magnesium ribbon.

This gas can, as has been seen, p. 152, be formed by direct
synthesis of carbon and oxygen: but it is usually prepared by
an analytical process, i. e. by the decomposition of marble or
chalk (calcium carbonate) by hydrogen chloride.

To illustrate this, put a fragment of marble into a small test-
tube, and pour upon it 2 or 3 c.c. of dilute hydrogen chloride.
An effervescence will at once begin, carbon dioxide being given
off, while the marble gradually disappears, calcium chloride
being formed which dissolves in the liquid[1]. Hydrogen sulphate cannot be so conveniently used, since it forms a nearly
insoluble compound with calcium, which covers the marble and
hinders further action. Thus, if a few drops of dilute hydrogen
sulphate are added to the mixture in the tube, a white precipitate of calcium sulphate appears, and the action soon ceases
almost entirely.

Owing to its high density (about $1\frac{1}{2}$ times that of air) it can
readily be collected by downward displacement, the tube con-

---

[1] Calcium   Hydrogen   Carbon   Calcium   Water.
  carbonate.   chloride.   dioxide.   chloride.

$$CaCO_3 + 2HCl = CO_2 + CaCl_2 + H_2O.$$

veying the gas being led to the bottom of the bottle which is to be filled. This is preferable to collecting it over the pneumatic trough, since the gas is, as will be seen, rather soluble in water, although it may be collected over warm water, like nitrogen protoxide.

Fit up an apparatus as shown in fig. 56, the flask, cork, and

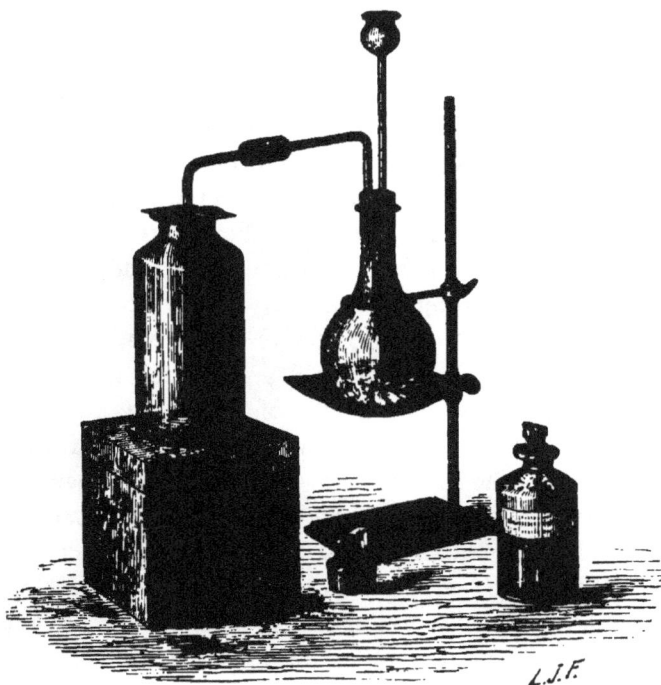

Fig. 56.

tube-funnel being the same as used in preparing hydrogen; but another form of delivery tube is to be substituted for the bent tube then employed. It is made up of two elbow tubes united by a cork joint; one of the elbow tubes having two equal branches about 8 or 9 cm. in length, the other having one branch about 8 cm., the other about 18 cm. in length. The cork joint is made by boring a central hole through a small cork (as free from fissures as possible), and pushing one branch of each tube into the hole until they meet in the middle of the cork. This forms an excellent joint, and has one special advantage over an india-rubber connector, viz. that it is stiff, and yet allows the tube to be turned round in any direction, retaining it in whatever position it is placed[1].

Place in the flask 20 grms. of marble, previously broken with a hammer into lumps rather larger than peas. Secure the flask in the retort-stand at such a height that the open extremity of the delivery tube may be raised about 14 cm. above the table. Pass the end of the delivery tube into one of the bottles, supporting the latter on blocks, so that the tube may reach almost to the bottom. Over the mouth of the bottle lay a card in which a slit has been cut, half way across it, wide enough to admit the delivery tube. Pour through the funnel sufficient water to cover the marble, and then add, a few drops at a time, some strong common hydrogen chloride, until a brisk effervescence is set up, due to the evolution of carbon dioxide. After the lapse of half a minute, light the taper, remove the card from the mouth of the bottle, and pass the taper steadily down into it. It will probably be extinguished before it reaches the bottom, showing that the gas has partially filled the bottle. It may be withdrawn, re-lighted, and after a few seconds plunged again into the bottle, to ascertain the point to which the gas has now risen. When the taper is quickly and decidedly extinguished when held at, but not within, the mouth of the bottle, the latter may be considered to be full of gas. It should then be withdrawn, by first taking away the wooden blocks, then lowering the bottle slowly until clear of the delivery tube, and lastly inserting the stopper previously greased. Another bottle may now be substituted and filled in like manner; and so on until all are filled, more acid being added from time to time through the funnel. Reserve the large bottle for the last experiment, and examine the properties of the gas as follows:—

[1] A similar but neater joint may be made by cutting off a piece of glass tubing about 3 cm. in length and 8 or 9 mm. in diameter internally, and (after fusing the sharp edges in the lamp) fitting into it the two elbow tubes by means of rings of india-rubber tubing.

1. Its high density, and relation to ordinary combustion.

Take one of the larger gas jars, place in it the lighted taper, resting the wire against the side of the jar so as to support the taper steadily at the bottom. Bring the mouth of one of the small bottles of gas close to the edge of the jar, withdraw the stopper, and gently pour the gas over the taper, precisely as if you were pouring out a liquid[1]. The taper will be immediately extinguished.

2. Its diffusion into other gases.

Carbon dioxide, although it has just been proved to be much heavier than air, will diffuse upwards into air, but the rate at which it does so is comparatively slow.

Place one of the small bottles of the gas on the table, and replace its stopper by a glass plate. Invert over it a similar, but empty, gas bottle; its stopper being removed and the mouths of the two bottles brought into contact. After the lapse of half a minute, remove the upper bottle, replace its stopper, pour into it a little lime-water and shake it up. The cloudiness produced proves that carbon dioxide had really ascended, in spite of its density, into the upper bottle.

[Fill the partially-emptied bottle again with gas, for other experiments.]

3. Its solubility in water.

Pour into a bottle of the gas sufficient distilled water to fill it one-third, and replace the stopper, inserting a small strip of thin paper between it and the neck of the bottle. Invert the bottle and shake it violently for a few seconds, then hold it up to the light. Observe that bubbles of air are entering at the side of the stopper, thus showing that there has been an absorption of the gas. Shake the bottle again for a few seconds, then pour out a little of the solution into a beaker and taste it. The water will have acquired a distinct acid taste. Pour some more of the solution into a test-tube, and heat it over a lamp. Bubbles of gas will be given off as the temperature rises, and if

---

[1] Do not pour the gas directly over the centre of the jar, but at its edge, since the gas receives a forward as well as a downward impetus while the bottle is being inverted.

the liquid is boiled for a minute or two, the whole of the gas will escape.

4. Its action, in presence of water, on litmus.

Pour a few drops of solution of litmus into the solution of carbon dioxide obtained in the last experiment. It will be at once turned red, although the colour will not be so bright a red as that produced by the action of hydrogen nitrate (p. 134). It is probable that carbon dioxide (like sulphur dioxide, p. 103), forms a definite combination with water, hydrogen carbonate; but this has not been obtained in a pure condition, since it is easily decomposed by heat. To prove this, pour some of the red liquid into an evaporating dish and boil it for a minute or two. Carbon dioxide will escape, and the blue colour of the litmus will be restored [1].

5. Its action on calcium hydrate.

Fill a small test-tube about half full of lime water, and pour the liquid into another bottle of the gas. Replace the stopper, inserting a slip of paper as in experiment 3, and shake it slightly at first, afterwards vigorously. The liquid will at first become milky, owing to the formation of insoluble calcium carbonate, but will afterwards, if the gas is in excess, become clear again, because a soluble calcium-and-hydrogen carbonate is formed [2]. (If the liquid still remains turbid after agitation, pour it into another bottle of gas, and agitate it again).

If, now, some of the clear solution be poured into a test-tube and heated to boiling over the lamp, the calcium-and-hydrogen carbonate will be decomposed; carbon dioxide will be given off, and the insoluble calcium carbonate re-precipitated [3].

This experiment illustrates the cause of the hard crust of calcium carbonate formed in kettles and boilers in which hard spring water (which usually contains carbonates, see p. 67), has been boiled.

---

[1] If the colour remains red, it shows that a little vapour of hydrogen chloride has passed over from the flask. To obtain a pure gas it should be passed through a wash-bottle (see p. 164) containing a little distilled water (in which hydrogen chloride is extremely soluble) before being collected.

[2] 1st action,—$CaH_2O_2 + CO_2 = CaCO_3 + H_2O.$
2nd action,—$CaCO_3 + H_2O + CO_2 = CaH_2(CO_3)_2.$

[3] $CaH_2(CO_3)_2 = CaCO_3 + H_2O + CO_2.$

6. Its action on potassium hydrate (caustic potash).

Put a small bit of solid potassium hydrate, about half the size of a pea, upon a watch-glass. Fill a small test-tube with carbon dioxide by displacement, withdrawing it from the delivery tube slowly, and closing its mouth at once with the thumb. Introduce 3 or 4 drops of water from the jet of the washing bottle, and then put in the fragment of potash, closing the mouth again tightly at once. Invert the tube, and shake it so as to moisten the sides with the solution of potassium hydrate which is rapidly formed. You will soon feel the thumb strongly pressed inwards owing to the absorption of the gas. In about half a minute, place the mouth of the tube below the surface of some water in a jug or pan, and withdraw the thumb gradually. The water will rush in violently, and nearly or quite fill the tube, showing that the absorption of the carbon dioxide by potash has been rapid and complete. Hence a solution of potash can be effectively used to withdraw carbon dioxide from a mixture of gases (as will be seen in Exercise 13). The action is quite analogous to that on calcium hydrate, potassium carbonate being formed [1].

7. Its decomposition by magnesium.

Affix a piece of magnesium ribbon, about 16 cm. in length, to the flange of the deflagrating cup (the cup itself with its wire being removed) or to a cork, in the manner described in p. 103, so that its extremity may reach nearly to the bottom of the large bottle of carbon dioxide. Hold the end of the ribbon in the lamp-flame until it begins to burn with a brilliant white light, then plunge it slowly into the bottle of gas. It will continue to burn brilliantly, forming white flakes of magnesium oxide interspersed with black particles which consist of carbon. In order to separate the latter, pour into the bottle, when the combustion is over, a little distilled water, shake it up and pour it, with the suspended particles, into an evaporating dish. Add 8 or 10 drops of strong hydrogen chloride, and heat the liquid over a lamp. The magnesium

[1] $2KHO + CO_2 = K_2CO_3 + H_2O.$

oxide will disappear, while the black flakes of carbon will remain floating undissolved in the clear liquid. This experiment proves that the gas which you are examining contains carbon, and also that some substances will burn in a gas in which a taper will not burn. Combustion is purely a relative term, and may be defined as the chemical action of one substance upon another accompanied with the evolution of light and heat.

[That these black particles are really carbon may be readily proved by collecting them on a small filter, washing them with several changes of water (until the washings give no precipitate when tested with silver nitrate) and then drying them thoroughly, the filter being put in a porcelain dish on a sand-bath. When they are quite dry, detach some of them from the filter, and put them into a small dry test-tube. Put a cork *loosely* into the tube, and heat it rather strongly over a lamp. At a red heat the particles will glow as they combine with the oxygen of the air in the tube to form carbon dioxide. To prove that this gas is really formed, set aside the tube in some sand to cool, and then pour into it a little lime-water, which will, on being shaken up, become cloudy.]

Additional Experiments.

8. Its decomposition by iron.

When carbon dioxide is passed over heated iron, the latter combines with only one half of the oxygen, and carbon prot-oxide remains. This reaction is important as showing that the molecule of carbon dioxide contains two atoms at least of oxygen (since the quantity of oxygen in it can be divided and taken out in two instalments), and it may be tried on a small scale as follows :—

Powder 0.5 grm. of marble or chalk, and mix it thoroughly with 2 grms. of fine iron filings. Fill the bulb of an ignition tube with the mixture, and attach a glass jet to the open end of the tube by an india-rubber connector. Heat the bulb strongly by a gas blowpipe, and when it is red-hot, hold a lighted match to the jet. The gas given off will be found to burn with a blue flame, and cannot, therefore, be carbon dioxide.

What has happened is this,—the calcium carbonate has been decomposed by the heat, carbon dioxide being given off. This has given up half its oxygen to the red-hot iron, being thus reduced to carbon protoxide, which, as will be seen in Exercise 13, burns in air with a characteristic blue flame.

9. Its decomposition by plants in presence of light.

This action, which it is best to try on a bright, sunny day, supplies us with a direct proof that carbon dioxide contains oxygen, and also explains why it does not accumulate in the air, although it is being constantly produced by the processes of respiration and combustion (p. 121).

Fill a large gas bottle with common fresh water and pass into it some carbon dioxide from an apparatus fitted up in the usual way (fig. 56, p. 155)[1]. When the gas has bubbled through the water for 8 or 10 minutes, withdraw the elbow-tube and put into the water a bunch of freshly-gathered parsley or water-cress, or any vigorously growing leaves, taking care that no air-bubbles remain entangled among the leaves. Fill up the bottle to overflowing with water, place a large porcelain dish or a common plate over the mouth, and invert the bottle quickly, holding the plate firmly against its mouth so that no air may enter. Pour a little water into the dish to ensure the exclusion of air, and expose the whole to bright daylight (sunlight if possible) for 6 or 8 hours. Bubbles of gas will be observed to collect on the under side of the leaves, and rise to the top of the bottle. The growing plant has inhaled the carbon dioxide (as animals inhale oxygen) and has appropriated its carbon, while the oxygen is set free as gas. If, however, the bottle is kept in a dimly-lighted room, or if the day is very dull, only a slight action will take place; showing that light is required for the decomposition.

If 8 or 10 c.c. of gas are obtained, it may be proved to be oxygen by immersing the bottle in a deep pan, or the pneumatic trough, shaking it to detach all bubbles from the

---

[1] Or some of the contents of a bottle of soda water (which is simply a strong solution of carbon dioxide) may be added to the water in the bottle.

leaves, then decanting the gas into a test-tube (using a funnel as explained, p. 81), and testing it with a glowing match, which will be re-lighted.

12. CARBONATES.

Formula of
Molecule.

[Typical examples,—Calcium carbonate, $Ca\,CO_3$.
Sodium carbonate, $Na_2\,CO_3$.]

These are salts which contain the carbonate radicle, a group consisting of 1 atom of carbon and 3 atoms of oxygen. Their most characteristic property is, that they give off carbon dioxide when acted on by almost any acid. For example,—

*Put a small fragment of crystallised sodium carbonate (common 'washing soda' will do) into a test-tube, and add 1 or 2 c.c. of dilute hydrogen chloride. Effervescence will take place, and the gas given off may be proved to be carbon dioxide in the following ways,—

(a) Dip the long branch of an elbow tube in lime water, and introduce it into the test-tube (if possible without touching the sides). If suction is now applied to the outer end of the elbow tube, the drop of lime water, retained in the other branch by capillary action, will be drawn up and wet the sides of the tube, being thus freely exposed to the action of the gas. If the tube is now withdrawn and held against any dark background, the drop as it subsides again to the end of the tube will be seen to be turbid [1].

(b) Take the cork fitted with tubes out of the washing bottle (fig. 11, p. 11), and adapt it (the tubes still remaining in it) to a test-tube large enough to fit it [2], adjusting the long branch of the jet-tube so that it may reach nearly to the bottom

---

[1] A drop of lime water at the end of a glass rod will show turbidity in the same way, when lowered carefully and steadily into the test-tube. In either way very small traces of carbon dioxide may be detected.

[2] If you have no test-tube of the proper size, a cork must be fitted to a test-tube and holes bored in it to receive the tubes of the washing bottle as above; or ordinary elbow tubes may be used, as in fig. 51, p. 121.

of the test-tube. Put 4 or 5 c.c. of lime-water into the test-tube, then fit the cork again into its place, and insert the jet of the tube into the test-tube containing carbon dioxide. Apply your mouth to the other tube and suck a little of the gas through the lime water, which will become turbid.

[The tubes and cork should, of course, be rinsed with distilled water before being replaced in the washing bottle.]

A similar experiment should be tried with at least one other carbonate, such as chalk, oolite, egg-shell, or ammonium carbonate (smelling salts). It will be found that in every case carbon dioxide will be given off.

* The carbonates of potassium, sodium, and ammonium have a strong alkaline action on test paper. This should be proved by putting a drop of solution of sodium carbonate or ammonium carbonate upon strips of reddened litmus and turmeric paper.

13. CARBON PROTOXIDE.

[Formula of molecule, CO.
Weight of molecule, 28 hydrogen-atoms.]

Apparatus required—Pneumatic trough; flask, with elbow tubes and tube funnel, used in Exercise 11; wide-mouthed bottle holding about 150 c.c.; corks; cork-borers; rat-tail file; bent delivery tube used in Ex. 1; wooden blocks; retort-stand; wire-gauze; argand burner; india-rubber connector; two jars, 20 × 5 cm.; one ditto, 10 × 3 cm.; three small gas bottles; glass disc; taper on wire; crystallised hydrogen oxalate; strong common hydrogen sulphate; solution of potassium hydrate (caustic potash); solution of calcium hydrate (lime water); potassium ferrocyanide; small bottle of oxygen.

One reaction in which this gas is formed has been already tried, p. 160; it is, however, usually prepared by the decomposition of hydrogen oxalate (oxalic acid). This substance contains the elements of water, carbon dioxide, and carbon protoxide, as seen by the formula of its molecule, $H_2C_2O_4$ (which $= H_2O + CO_2 + CO$). When it is heated with strong hydrogen sulphate, the latter combines with the elements of water, and the two carbon oxides are liberated in a state of mixture. The carbon dioxide may· be withdrawn by allowing the mixed gases to pass through a solution of potassium hydrate, which (as has been seen, p. 159) unites readily with carbon dioxide, but has, under ordinary conditions, no action on carbon protoxide. A bottle containing this solution must therefore be interposed between the generating flask and the pneumatic trough.

Adapt a cork to a bottle with a rather wide mouth, of about

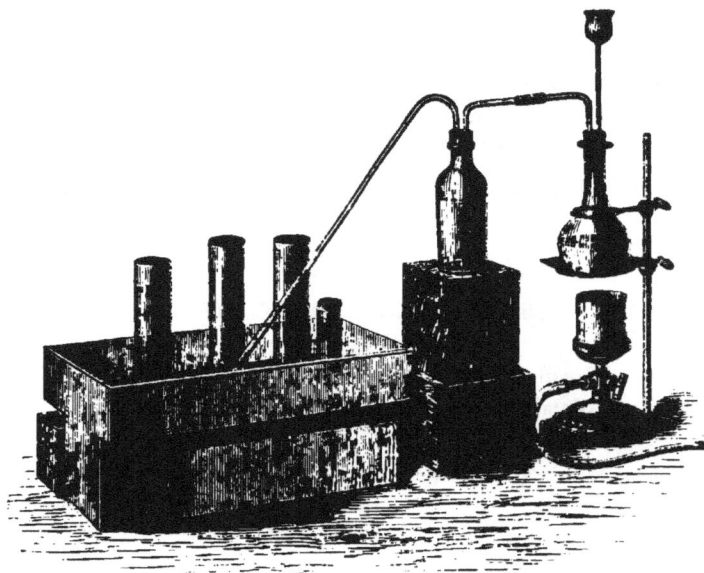

Fig. 57.

150 c.c. capacity. Bore two holes in the cork: into the one fit the bent delivery tube used in the preparation of hydrogen ; into the other fit the longer branch of the elbow tube used in Exercise 11, passing the tube through the cork to such a distance that its extremity may reach almost to the bottom of the bottle. Before the cork is finally fitted into its place, fill the bottle to about three-fourths of its height with a solution of caustic potash or soda, made by dissolving 20 grms. of the substance in 100 c.c. of water. Support it on blocks at such

a height that the delivery tube may dip under the shelf of the pneumatic trough. In the next place, take the flask, fitted with the short elbow tube and funnel, used in Exercise 11, place in it 20 grms. of crystallised hydrogen oxalate, and fix it in the retort stand, with a piece of wire-gauze under its bottom, at such a height as to admit of the argand-burner being placed under it, and to allow the elbow tube to be connected with the tube of the wash-bottle by a short piece of india-rubber tubing.

The whole apparatus will then appear thus, Fig. 57.

Pour into the flask about 60 c.c. of strong hydrogen sul-phate. As there are several chances of leakage, it will be necessary to test the joints as directed in p. 99, and not to proceed with the experiment until they are all made tight. The flask may then be gradually heated with the lamp, especial care being taken to avoid any chance of cracking it, and thus spilling the strong acid to the damage of table, clothes, and hands. While the flask is being heated, the jars and bottles may be filled with water and inverted on the shelf of the trough.

When the mixture in the flask begins to effervesce, the heat should be moderated, so that the stream of gas may pass through the solution in the wash-bottle in bubbles not suc-ceeding each other too quickly to be counted.

After a sufficient quantity (about two jars full) of the gas has been rejected, the jars and bottles may be filled for experi-ments. As soon as this is done, the india-rubber connector between the flask and the wash-bottle should be taken off, and the flask removed at once into the open air, as the gas is poisonous and should not unnecessarily be allowed to escape into the room.

[The best method of procuring carbon protoxide is undoubtedly that which was discovered by Fownes, depending on the decompo-sition of potassium ferrocyanide when heated with hydrogen sul-phate [1].

---

[1] $K_4 Fe(CN)_6 + 6 H_2 SO_4 + 6 H_2 O$
$= 3 (H_4 N)_2 SO_4 + 2 K_2 SO_4 + Fe SO_4 + 6 CO.$

The apparatus may be the same as that described above. The wash-bottle, although not absolutely necessary, is useful to absorb traces of hydrogen carbonate, cyanide, and sulphite, which come over.

Place in the flask about 60 c.c. of strong hydrogen sulphate. Weigh out 15 grms. of potassium ferrocyanide, reduce the salt to powder, and add it little by little to the acid in the flask, shaking between each addition, in order to prevent the salt caking into lumps. Add also about a teaspoonful of fine dry sand, to promote the regular evolution of gas. Now fit in the cork and proceed to heat the mixture, watching carefully for the first evolution of gas. No gas comes off until the temperature is pretty high, and then there is a sudden rush, which must be moderated by withdrawing the lamp at once. The temperature should not be allowed to rise higher than is necessary to obtain a slow stream of the gas. It is better not to pour water into the flask for the purpose of washing out the residue, until as much of the latter as possible has been shaken out; otherwise the sudden heat produced by the addition of water to the strong acid is likely to crack the flask.]

The following properties of the gas should now be examined, and the many points of contrast between the two carbon oxides should be carefully noted.

1. **Its insolubility in water, and neutral action on litmus.**

Pour some water into a bottle of the gas, put a slip of paper between the neck and the stopper, invert the bottle, and shake it. No entrance of air in bubbles will be observed. Pour a few drops of solution of litmus into the water in the bottle: there will be no change of colour [1].

*2. **Its relation to combustion.**

Plunge a lighted taper into a jar of the gas, placed mouth upwards on the table. The taper will be extinguished, and the gas will burn with a characteristic blue flame. In burning it unites with the oxygen of the air to form carbon dioxide, as will be presently proved.

3. **Its direct union with oxygen.**

Decant into one of the gas jars two measures of carbon

---

[1] If the gas has been insufficiently washed, the colour of the litmus may be slightly changed owing to the carbon dioxide present.

protoxide (the measure being the small gas jar). Add one measure of oxygen, transfer the jar to the table, and apply a light. The gases will unite with a slight explosion, forming carbon dioxide [1].

4. Its want of action on lime-water.

To try this satisfactorily, all traces of carbon dioxide must first be got rid of. For this purpose, pour into a bottle of the gas about 10 c.c. of solution of potassium hydrate, and shake it up thoroughly for half a minute ; then decant some of the purified gas into the small gas jar, filling it completely, and place it, covered with a glass plate, on the table. Pour into it quickly a little lime water (remembering that carbon protoxide is rather lighter than air, and hence the cover should not be removed more than necessary) and shake it up. No material turbidity should be produced [2], otherwise the gas is not pure and must be again shaken up with potash, and another portion decanted and tested in the same way.

Now apply a light to the gas and allow it to burn in the jar. Replace the glass plate as soon as the combustion is over, and shake up the lime-water in the gas. It will now become very turbid, proving that carbon dioxide has been produced in the combustion.

COMPOUNDS OF CARBON AND HYDROGEN.

One, out of the very large number of these compounds, is selected for preparation and examination; partly as being a typical example of an important class of hydro-carbons connected with the alcohol series, partly as being the most valuable constituent of coal gas.

When coal is decomposed by a strong heat some of the carbon and hydrogen in it combine to form ethylene (olefiant gas) and methane (marsh gas). These, mixed with a large

[1] $2 CO + O_2 = 2 CO_2$.
[2] It must be remembered that common water usually gives a cloudiness when mixed with lime water (p. 67), and a slight action of this kind due to the water used in the trough must be allowed for.

quantity of hydrogen and some carbon dioxide (which is separated by means of lime) and carbon protoxide, form ordinary coal gas, which owes all its illuminating power to the hydrocarbons present in it. The decomposition of coal may be tried on a small scale as follows :—

Powder a little coal very finely, and fill the bulb of an ignition tube with it. Attach a glass jet to the ignition tube by an india-rubber connector, and support it horizontally in a Bunsen's holder; then heat the bulb pretty strongly. Water and a black, tarry liquid will condense in the tube, and gases will come off which burn with an intensely white flame when a lighted match is held to the jet.

14. ETHYLENE (Olefiant Gas).

[Formula of molecule, $H_4 C_2$.
Weight of molecule, 28 hydrogen-atoms.]

Apparatus required—Pneumatic trough; flask, fitted with delivery tube and funnel, used in Ex. 2; beaker; glass measure; one small and two large gas jars; three small gas bottles; taper on wire; glass disc; elbow tube; alcohol; strong common hydrogen sulphate; solution of calcium hydrate; bottle of chlorine; bottle of oxygen.

This gas is obtained in a pure condition by heating a mixture of alcohol and hydrogen sulphate. The simplest (although probably not a complete) view of the action is, to regard alcohol as containing the elements of ethylene and water; the latter combining with the hydrogen sulphate, while ethylene is liberated[1].

Measure 20 c.c. of alcohol (rectified or methylated spirits) into a beaker or flask capable of holding at least 100 c.c. Place the vessel in a basin of cold water, or hold it in the pneumatic trough, and add to it by degrees 40 c.c. of strong hydrogen sulphate. The mixture will grow dark, and probably boil at first, showing that chemical combination is going on

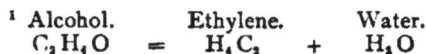

[1] Alcohol.   Ethylene.   Water.

$$C_2 H_6 O \;=\; H_4 C_2 \;+\; H_2 O$$

between the alcohol and the acid, and the temperature should be kept down as much as possible by agitating the flask in the surrounding water and adding the acid slowly. While the liquid is cooling you may arrange an apparatus similar to that used for preparing hydrogen, fig. 48, p. 105, putting about a teaspoonful of sand in the flask in order to lessen the tendency of the liquid to froth up [1]. Pour the mixture of acid and alcohol into the flask, shake it until thoroughly mixed with the sand, and heat it cautiously to the boiling point (about 150°); being careful so to regulate the heat that the contents of the flask may not froth over into the delivery tube. Reject the two first jars full of the gas, and then proceed to fill the jars and two small gas bottles, for examination of the properties of the gas.

1. **Its peculiar fragrant odour** will have been already noticed.

2. **Its insolubility in water** is proved by the fact that it can be collected without loss over the pneumatic trough.

3. **Its union with oxygen.**

(*a*) Place a jar of the gas mouth upwards on the table, and plunge a lighted taper into it. The taper will be extinguished, but the gas will burn with a bright white flame. By pouring some water from a jug into the jar, the gas will be forced out, and the flame seen to better effect.

(*b*) Attach to the end of the delivery tube an elbow tube, as in expts. with hydrogen, fig. 50, p. 111 (omitting the jet), and heat the mixture in the flask until more gas is evolved. Test carefully the purity of the gas by filling a test-tube with it, as directed, p. 111; and when you are satisfied that the gas is pure, light it at the end of the tube. Hold over the flame (for a few seconds only) a small empty gas bottle and observe that moisture is deposited in it. Pour a little lime-water into the bottle, and shake it up ; it will become cloudy. You have thus proved that water and carbon dioxide are produced during the

---

[1] If a very pure gas is required a wash-bottle containing caustic soda must be interposed between the flask and the delivery tube, as in preparing carbon protoxide, fig. 57, p. 164, in order to retain any sulphur dioxide and carbon dioxide which may come over.

combustion of ethylene, and therefore that the gas must contain hydrogen and carbon.

Depress a porcelain dish or white plate into the flame; it will at once be covered with a dense black soot. The reason of this will be clear from what has been learnt already respecting the principle of the Bunsen's burner (p. 4) and the blowpipe flame (p. 87). The gas issuing from the jet cannot at once obtain sufficient oxygen to combine with all the carbon and all the hydrogen which it contains; hence some of the carbon, having less affinity than hydrogen for oxygen, is separated and heated white-hot in the flame, imparting to the latter all its brilliant whiteness. This carbon is cooled down and deposited upon the surface of porcelain.

[The flask may now be set aside, to be washed out when the mixture is quite cold.]

(*c*) Since it can be proved that nothing but water and carbon dioxide are formed by the action of oxygen upon ethylene, it is easy to construct an equation expressing the change [1], and to calculate from it the volume of oxygen required to combine with a given volume of ethylene. We thus find that three measures of oxygen are required for a single measure of ethylene.

Fill one of the larger gas jars with water, and decant into it one measure of ethylene and three measures of oxygen gas, using the small strong jar as the measure. Fill this last jar with the mixture, and apply a light to it. The gases will unite with an explosion which is even more violent than that produced by the union of oxygen and hydrogen under the same conditions; and the experiment should only be made with small quantities of the gases in a thick jar. After the experiment, allow the remainder of the mixed gases to escape into the air, to avoid any chance of an accident.

[If no chlorine is at hand, the two following experiments may be deferred until the exercise on CHLORINE, a small bottle of ethylene being reserved for the purpose.]

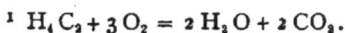

[1] $H_4 C_2 + 3 O_2 = 2 H_2 O + 2 CO_2$.

4. Its direct union with chlorine.

Decant into a gas jar one measure of chlorine [1], and add an equal volume of ethylene; then leave the mixture in a good light for a minute or two. Combination will readily take place, 1 molecule of ethylene uniting with 1 molecule of chlorine to form ethylene dichloride [2], a liquid which collects in oil-like drops on the sides of the jar and on the water as the latter rises in the jar.

It was owing to the oily appearance of this liquid that ethylene originally received the name of 'olefiant' gas.

5. Its decomposition by chlorine.

At ordinary temperatures chlorine simply unites directly with ethylene, as above shown. But at high temperatures its action is entirely different : its affinity for hydrogen is so great that it decomposes the gas, uniting with the hydrogen and leaving the carbon wholly uncombined [3].

Decant into a gas jar two measures of chlorine and add one measure of ethylene : then, without delay, cover the mouth with a glass plate, place the jar on the table mouth upwards, and apply a light. The action will be quick but not explosive, accompanied by a reddish flame, and formation of a dense cloud of soot, consisting of the unconsumed carbon.

It will be seen that the action is very analogous to that of oxygen already noticed, but in this case the whole of the carbon remains unacted on.

COMPOUNDS OF CARBON, HYDROGEN, AND OXYGEN.

These constitute the majority of what are called 'organic' substances: a term which may be explained to mean, substances derivable from plants or animals. Only a few typical

---

[1] This should be done quickly, to avoid loss of chlorine, owing to its solubility in water.

[2] Ethylene. Chlorine. Ethylene dichloride.

$$H_4C_2 \quad + \quad Cl_2 \quad = \quad H_4C_2Cl_2$$

[3] Ethylene. Chlorine. Carbon. Hydrogen Chloride.

$$H_4C_2 \quad + \quad 2Cl_2 \quad = \quad C_2 \quad + \quad 4HCl$$

examples of radicles belonging to this class will be examined here; those being selected which the student is most likely to meet with in the ordinary course of elementary work.

15. ACETATES.

Formula of
Molecule.

[Typical examples,—Hydrogen acetate,   H $(C_2 H_3 O_2)$.
Lead acetate,   Pb $(C_2 H_3 O_2)_2$.]

These are principally formed in the last stage of the fermentation of beer, wine, &c.; the alcohol undergoing a further change, owing to the action of the oxygen of the air, and becoming sour from the formation of hydrogen acetate, which is common vinegar or acetic acid.

Another source of them has been already illustrated, p. 151, viz. the decomposition of woody fibre by heat, when acid vapours of hydrogen acetate were found among the products.

*1. They become charred when strongly heated.

Put a small bit of lead acetate into an ignition-tube and heat it to redness. It will turn black owing to separation of carbon, while strongly-smelling vapours of acetone and other products will be given off.

*2. They yield hydrogen acetate when heated with hydrogen sulphate.

Place a small quantity of lead acetate in a test-tube, pour on it a few drops of strong hydrogen sulphate, and heat it gently. The substance will not blacken (as a tartrate does), but pungent-smelling vapours of hydrogen acetate will be given off, which will redden a piece of litmus-paper held within the mouth of the tube.

[If time permits, a little hydrogen acetate may be prepared from lead acetate in an apparatus similar to that which was used for the distillation of hydrogen nitrate (p. 132), 20 grms. of lead acetate (or, better, sodium acetate) being taken instead of potassium nitrate.]

*3. They yield ethyl acetate when heated with alcohol and hydrogen sulphate.

Repeat the above experiment, adding a few drops of pure alcohol (not methylated) before the addition of the hydrogen sulphate. Vapours of ethyl acetate (acetic ether) will be given off, which possess a characteristic fragrant smell, not unlike that of fresh apples (best observed after closing the mouth of the tube with the finger, and slightly shaking the warm liquid).

[A comparative experiment should be made by heating a little of the alcohol alone with hydrogen sulphate in order to discriminate between the slight smell of the ethylene thus formed and the much stronger and sweeter smell of ethyl acetate.]

*4. Their neutral solutions give a dark red colour when tested with iron perchloride.

Place a little hydrogen acetate in a watch-glass and add solution of ammonia drop by drop until the liquid is just neutral to test-paper. If an excess of ammonia has been unintentionally added, it may be driven off by evaporating the solution for a few minutes on the sand-bath. Now place the watch-glass on a piece of white paper, and add one drop of solution of iron perchloride. No precipitate will be formed, but the solution will become dark red (owing to the formation of iron acetate), and when boiled in a test-tube it will give a brown-red precipitate of basic iron acetate.

16. TARTRATES.

[Typical examples,—   Formula of Molecule.

Hydrogen tartrate $H_2(C_4H_4O_6)$.

Sodium and hydrogen tartrate, $NaH(C_4H_4O_6)$.]

These occur in grape juice, the crust deposited from wine consisting principally of calcium tartrate.

*1. They become charred when strongly heated.

Heat a little solid sodium-and-hydrogen tartrate in a test-tube nearly to redness, holding the tube with the open end slightly inclined downwards, lest any moisture should run back to the heated part and crack the tube. The substance will become charred, and give off vapours possessing a characteristic smell, like burnt sugar. The residue, besides carbon, also contains sodium carbonate as may be proved by pouring on it, when cold, a few drops of dilute hydrogen chloride. Carbon dioxide will be evolved, and may be tested for by holding within the tube a glass tube dipped in lime water (p. 162).

2. They are decomposed and turn black when heated with strong hydrogen sulphate.

Place a small quantity of solid sodium-and-hydrogen-tartrate in a test-tube, add a little strong hydrogen sulphate, and heat cautiously. The mixture will blacken, owing to the separation of carbon, and give off carbon dioxide and other gases with effervescence.

[For the following experiments a solution of sodium-and-hydrogen-tartrate (1 grm. of the salt dissolved in 25 c.c. of water) should be used.]

*3. Their solutions give a white crystalline precipitate when tested with a potassium salt.

Add one drop (not more) of solution of potassium hydrate to a portion of the solution ; stir the mixture with a glass rod, and shake it for a few seconds. A crystalline precipitate of potassium-and-hydrogen tartrate will be gradually deposited, the crystals forming first along the lines where the glass rod had rubbed against the surface of the tube [1]. If a few more drops of potassium hydrate are added, the precipitate will re-dissolve.

*4. Their solutions give a white precipitate with calcium salts.

To another portion of the solution of sodium and hydrogen tartrate add at least three or four times its volume of lime water. A white precipitate of calcium tartrate will be produced, which will dissolve on addition of a few drops of solution of ammonium chloride.

*5. They readily reduce silver salts.

Make another portion of the solution slightly alkaline by

---

[1] This should be especially noticed as an effective method of promoting the formation of a crystalline precipitate. The contact of the rod with the glass alters the surface in some way so as to form nuclei to which crystals will attach themselves.

adding solution of potassium hydrate drop by drop and testing after each addition with reddened litmus paper. Then add a drop of solution of silver nitrate, which will produce a brown precipitate of silver oxide. Lastly heat the mixture to boiling, when the precipitate will turn black, owing to its reduction to metallic silver, the oxygen uniting with the carbon and hydrogen of the tartrate [1].

17. OXALATES.

Formula of
Molecule.

[Typical examples,—Hydrogen oxalate, $H_2C_2O_4$.

Ammonium oxalate, $(H_4N)_2C_2O_4$.]

These occur in the sap of many plants, especially wood-sorrel (*oxalis acetosella*). Hydrogen oxalate (oxalic acid) is generally prepared by heating sugar (or starch) with hydrogen nitrate, as the following experiment will show.

1. Weigh out 4 grms. of loaf-sugar, reduce it to powder, in a mortar, place it in a large test-tube, and pour upon it 30 c.c. of strong common hydrogen nitrate; then heat the tube cautiously on a sand-bath (in a draught cupboard, if possible). Carbon dioxide and nitrogen oxides will be copiously evolved, and hydrogen oxalate will be formed in the liquid. When the action has moderated, pour the solution into a porcelain dish, add 10 or 12 c.c. more of the acid, and evaporate it down until a drop taken out with a glass rod and placed on a watch-glass deposits crystals on cooling. The liquid may then be left to cool and crystallise, while other experiments are proceeded with. Transparent prisms of hydrogen oxalate will thus be obtained, which, after the liquid has been poured off, should be placed in a filter and left to drain, and finally washed with a little water and dried by pressing them between folds of blotting-paper.

They should be proved to be hydrogen oxalate by tests 2 and 3.

[1] For a modification of this experiment see under SILVER.

*1. They are decomposed by hydrogen sulphate, with evolution of carbon oxides.

Pour 5 or 6 c.c. of strong hydrogen sulphate on a few crystals of hydrogen oxalate in a test-tube, and heat the mixture. Carbon dioxide and carbon protoxide will be given off with effervescence. (Compare the method of making carbon protoxide, p. 163.) The latter gas will burn with its characteristic blue flame if a lighted match be held to the mouth of the tube ; the presence of the former gas may be shown by holding a glass tube dipped in lime water within the test-tube (p. 162).

[For the following experiments a solution of ammonium oxalate may be used, containing 1 grm. of the salt in 30 c.c. of water.]

*2. They are readily decomposed by oxidising agents, e. g. manganese dioxide, with formation of carbon dioxide.

Place a little manganese dioxide in a test-tube, add enough water to cover it, and then a few drops of strong hydrogen sulphate. If there should be any effervescence (manganese dioxide sometimes contains carbonates) wait until this has ceased, and then add a few drops of a solution of ammonium oxalate. Carbon dioxide will now be given off, and may be tested for with lime water, as above.

*3. Their neutral solutions give a white precipitate with calcium salts, insoluble in hydrogen acetate.

Add 5 or 6 c.c. of solution of calcium sulphate to some of the solution of ammonium oxalate, made alkaline with ammonia. A white precipitate of calcium oxalate will be formed.

(*a*) Pour off about half of the liquid, containing the precipitate in suspension, into another test-tube, add a little hydrogen acetate, and warm slightly. The precipitate will remain undissolved.

(*b*) To the remainder of the liquid add some dilute hydrogen chloride, and warm. The precipitate will dissolve, but without effervescence (differing in this respect from calcium carbonate).

4. They do not blacken when heated, but are (with one or two exceptions) decomposed, yielding a carbonate.

(*a*) Put a little potassium oxalate, or calcium oxalate [1] into a

---

[1] This may be prepared, if none is at hand, by heating about 6 or 8 c.c.

test-tube and heat nearly to redness. Only a slight darkening or charring will be noticed, while carbon protoxide will be given off and may be tested for with a lighted match[1]. When the residue in the tube is cool, pour on it a little dilute hydrogen chloride; carbon dioxide will be given off with effervescence, showing that a carbonate has been formed.

(*b*) Place a few crystals of hydrogen oxalate in a small test-tube, and heat them slowly. The salt will melt and give off its water of crystallisation, which should be prevented from condensing by warming the whole of the tube (held in a paper holder, p. 60, note). Finally, on slightly increasing the heat, the hydrogen oxalate will sublime unaltered, condensing in long prismatic crystals.

COMPOUND OF CARBON AND NITROGEN.

18. CYANOGEN.

[Formula of molecule, $(CN)_2$ or Cy_2
Weight of molecule, 52 hydrogen-atoms.]

[This gas and its compounds are extremely poisonous, and great care is necessary in experimenting on it and them. The vapours of cyanogen and hydrogen cyanide (prussic acid) should not be inhaled more than is necessary for recognising their presence; not a particle of a cyanide should be allowed to get into a scratch or cut; and the room in which the experiments are made should be well ventilated.]

This substance (which is of interest as being the first compound which was recognised as a radicle, from its property of uniting with other substances like an element) is usually ob-

---

of solution of ammonium oxalate nearly to boiling, and adding about the same quantity of solution of calcium chloride (also heated). Boil the liquid for a few moments, to promote the separation of the precipitate; then filter it, and wash and dry the precipitated calcium oxalate as already directed, p. 77. If the precipitate passes through the filter, as it is rather apt to do unless separated from boiling solutions, a filter of Swedish paper must be used.

[1] $Ca C_2 O_4 = Ca CO_3 + CO$.

tained by the decomposition of substances containing carbon and nitrogen associated with other elements. Thus, if ammonium oxalate (which contains, as has been seen, carbon, nitrogen, hydrogen, and oxygen) is strongly heated, the carbon and nitrogen unite to form cyanogen [1]; and if sodium carbonate is present, the cyanogen combines with sodium forming sodium cyanide. This may be shown as follows,—

Weigh out 2 grms. of ammonium oxalate and 1 grm. of anhydrous sodium carbonate, and mix them intimately in a mortar : heat the mixture in a small test-tube, supported horizontally, that the water which is given off may not run down to the hot part of the tube and crack it. Besides water, white fumes of a substance called oxamide will be given off, and a peculiar pungent sweetish smell (somewhat like essence of bitter almonds) will be noticed [2], which is that of cyanogen : the greater part of this, however, will remain combined with sodium in the residue. Continue heating the substance carefully until nothing but a blackish residue remains ; then leave the tube to cool while other experiments are proceeded with. When cool, warm the residue with a little water, filter the solution from the particles of carbon, and prove the presence of a cyanide in it by applying the iron protosulphate test, as directed in Expt. 3 under CYANIDES.

2. Cyanogen is obtained in a pure state by heating certain cyanides, preferably mercury cyanide. This may be used for the following experiment, if it is at hand ; if not, a mixture of potassium ferrocyanide and mercury perchloride (which yields mercury cyanide by double decomposition) may be employed as described on the next page [3].

---

[1] The action may, neglecting the formation of oxamide and other products, be expressed thus—
$$(H_4N)_2C_2O_4 = 4H_2O + (CN)_2.$$
[2] This will be most safely recognised by waving with the hand a little of the vapour towards your face.
[3] Silver cyanide also yields cyanogen readily. A little of it may be made by dissolving 1·5 grm. of silver nitrate and 0·5 grm. of potassium cyanide in separate portions of water and mixing the solutions, when a white precipitate of silver cyanide will fall. This should be filtered, washed, and thoroughly dried : then it may be heated in a tube as above.

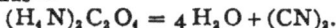

Reduce a little potassium ferrocyanide (yellow 'prussiate of potash') to fine powder, and heat it in a porcelain dish on a sand-bath, with constant stirring, until all the water of crystallisation has been driven off. This may be known by holding a cold watch glass just above the salt; if moisture is deposited, the dehydration is not complete.

Weigh out 1 grm. of the dried salt, and 1 grm. of mercury perchloride ('corrosive sublimate'); mix them thoroughly in a mortar, and put the mixture into a small test-tube, to the mouth of which has been fitted a cork with a bit of glass tubing, about 6 or 8 cm. long, passing through it; support the test-tube slanting in a Bunsen's holder and heat it moderately over an Argand burner. The mercury cyanide is decomposed by heat (compare the decomposition of mercury oxide, p. 69) into mercury, which condenses in bright drops on the side of the tube, and cyanogen which escapes as gas [1].

After the action has begun, and the peculiar, penetrating smell of cyanogen has been noticed, apply a light to the end of the tube. The cyanogen will take fire, burning with a fine purple flame. Hold a small gas bottle over the flame for a few seconds; no moisture will be deposited, but on pouring a little lime-water into the bottle and shaking it up, the cloudiness will prove that carbon dioxide is produced during the combustion, nitrogen being the only other product [2].

[As soon as the above experiment has been tried, remove the tube at once to a draught cupboard, or to the open air.]

19. CYANIDES.

Formula of
Molecule.

[Typical examples,—Potassium cyanide, KCN.

Hydrogen cyanide, HCN.

For the experiments a dilute solution of potassium cyanide should be used, made by dissolving half a gramme of the salt in 100 c.c. of water.]

---

[1] $Hg\,Cy_2 = Hg + Cy_2$.
A portion of the cyanogen remains behind in a different condition, as a black solid, called paracyanogen.
[2] $(CN)_2 + 2\,O_2 = 2\,CO_2 + N_2$.

*1. They evolve hydrogen cyanide when acted on by an acid.

Add a few drops of dilute hydrogen sulphate to some solution of potassium cyanide, and warm the mixture. A smell like that of cyanogen itself will be noticed, resembling that of peach kernels or essence of bitter almonds, both of which, in fact often contain prussic acid. Solid potassium cyanide has a similar smell, being easily decomposed, even by the carbon dioxide in the air.

*2. Their solutions give a white precipitate when tested with silver nitrate.

Add one drop of solution of silver nitrate to some solution of potassium cyanide. A white, curdy precipitate (of silver cyanide) will be produced which will redissolve on agitation. Add more silver nitrate until a permanent precipitate is produced : then pour off half of the liquid (containing some of the precipitate in suspension) into another tube.

(*a*) To this portion add some solution of ammonia and warm. The precipitate will readily dissolve, but will be thrown down again on adding sufficient dilute hydrogen nitrate to neutralise the excess of ammonia.

(*b*) Allow the other portion of the precipitate to subside, pour off the liquid, retaining the precipitate in the tube, and boil the latter with a little strong hydrogen nitrate. The silver cyanide will be slowly decomposed and dissolved by the acid. You may, without waiting until it has all dissolved [1], prove the presence of silver in the solution by pouring it into another tube, and adding a drop or two of dilute hydrogen chloride, when a white precipitate of silver chloride will be formed.

3. Formation of ferrocyanides.

Formula of
Molecule.

[Typical examples,—Potassium ferrocyanide, $(K_4 Fe Cy_6)$.

Iron ferrocyanide, $(Fe_4 (Fe Cy_6)_3)$.]

Iron protocyanide has a great tendency to associate itself with more cyanogen to form a radicle in which the iron is no

---

[1] Potassium cyanide often contains a chloride, and in this case a residue of silver chloride would remain, undecomposable by hydrogen nitrate.

longer present as a basic or electro-positive radicle (p. 75), but forms part of the electro-negative radicle called the 'ferro-cyanide' radicle. This is always formed when a protosalt of iron is added to a solution of a cyanide : and if a persalt of iron is also present, a deep blue substance, iron ferrocyanide (the ordinary 'prussian blue' paint), is produced. These reactions furnish an extremely delicate test for the presence of a cyanide, which should be tried in the following way,—

* Add to some solution of potassium cyanide two or three drops of solution of potassium hydrate, and then about the same quantity of solution of iron protosulphate, and shake the mixture thoroughly, warming it gently. A dull greenish precipitate consisting of a mixture of iron protohydrate and perhydrate will be formed. Now add an excess of dilute hydrogen chloride (sufficient to make the liquid strongly acid); a deep blue precipitate will be formed.

[The delicacy of the test may be shown by repeating the experiment, using only one drop of the solution of potassium cyanide diluted with 8 or 10 c.c. of water. Even in this case the liquid will be coloured distinctly blue.]

The changes which occur are of the following nature ;—On addition of the iron protosulphate a portion of iron cyanide is formed; this unites with the rest of the potassium cyanide to form potassium ferrocyanide [1]. At the same time, a mixture of iron protohydrate and perhydrate (the latter formed by the action of the oxygen of the air) is precipitated by the action of the potassium hydrate on the excess of iron sulphate. When hydrogen chloride is added, it dissolves these hydrates, and the iron perchloride formed acts upon the potassium ferrocyanide to form the deep blue precipitate of iron ferrocyanide.

4. **Formation of cyanates.**

[Typical example,—Potassium cyanate (KCNO).]

In the cyanate radicle, cyanogen is associated with oxygen, and this combination is readily effected by fusing a cyanide

---

[1] The presence of this in the alkaline liquid may be shown by filtering it: the filtrate will be yellow, and on acidifying it and adding a drop of solution of iron perchloride a blue precipitate will be formed.

with any substance capable of giving up oxygen. In fact,
potassium cyanide is a very powerful reducing agent, and is
used for that purpose in blow-pipe experiments.

* Mix together 2 grms. of potassium cyanide and 4 grms.
of lead protoxide (litharge), and heat them strongly in an iron
spoon over the Herapath's blow-pipe, stirring the mixture with
a piece of iron wire. The potassium cyanide combines with
the oxygen of the lead oxide forming potassium cyanate, while
the lead separates in the metallic form, appearing as a large
bright globule under the layer of fused salt.

Pour the contents of the spoon upon a clean plate of iron
(such as the sand-bath). When cold, the mass may be warmed
with water in a beaker, when the globule of lead will be left undissolved. If dilute hydrogen chloride be added to the solution
of potassium cyanate thus obtained, the salt is decomposed;
carbon dioxide escapes with effervescence, while ammonium
chloride is formed in the solution[1]. The presence of an
ammonium salt may be shown by gently warming a little of
the solution with some quicklime in a beaker, when free ammonia will be given off and may be recognised by its smell and
its action on reddened litmus-paper.

It will be remembered that a cyanide was obtained (p. 178)
from an ammonium salt (ammonium oxalate): we have in this
experiment effected the reverse change and obtained an ammonium salt from a cyanide.

5. Formation of sulphocyanates.

[Typical example,—Potassium sulphocyanate, (KCNS).]

These are very analogous to the cyanates (as seen by the
formula), but contain sulphur in place of oxygen. They are
formed whenever a cyanide is heated with sulphur, or a substance which yields it. Iron persulphocyanate is of an intense
red colour, and its formation furnishes a test for a cyanide even
more delicate than the one last described. Thus,—

* Pour one or two drops of solution of potassium cyanide
into a porcelain dish, add a drop of solution of yellow am-

[1] $KCNO + 2 HCl + H_2O = KCl + H_4NCl + CO_2.$

monium sulphide, and evaporate the mixture to dryness. The potassium cyanide will unite with the sulphur present in the ammonium sulphide, forming potassium sulphocyanate.

Add to the residue a few drops of dilute hydrogen chloride and then one drop of solution of iron perchloride. Iron persulphocyanate will be formed, which will colour the liquid deep red.

It will be advisable at this stage of chemical work, to apply the knowledge already gained of the properties of some of the non-metallic radicles to their detection in a substance of unknown composition. This process, which is called 'Analysis' in a narrow, technical sense, and the principles of which will be more fully explained in the introduction to Part II, consists in the application of tests in a certain definite order to the substance under examination, in order to find out which of the known radicles it agrees with in properties.

A short course of experiments, applicable to the detection of the radicles already treated of, will be found in Appendix C; and the student is advised to take at least 4 or 5 solutions of which nothing is known but that they contain some one of these radicles (associated with a metal, such as potassium, which will not interfere with the tests) and to find out which radicle is present by the methods there given.

---

20. CHLORINE.

[Formula of molecule, Cl_2.
Weight of molecule, 71 hydrogen-atoms.]

[We have in this case to deal with a gas which is soluble to a considerable extent in water, though much less so than ammonia or hydrogen chloride. Its solubility rapidly diminishes as the temperature of the water is raised, and hence it may be collected without much loss over warm water, the ordinary pneumatic trough being employed. But its density is so high that it can be most readily collected by downward displacement, like carbon dioxide.

Its action on the lungs is so extremely irritating and injurious, that it should never be prepared in a room which cannot be thoroughly ventilated. Every laboratory has, or should have, a cupboard communicating with a flue, for dealing with noxious gases; and, in default of this, a shed out of doors is the best place for such experiments. If, however, the chlorine must be prepared in a room, observe, (1) to collect it over warm water, not by displacement; (2) to allow *none* to escape into the air unnecessarily[1]; (3) as soon as the experiments are performed, to throw away the water in the trough, and wash it and the bottles at once with clean water.

A little ammonia sprinkled on a warm plate is the best means of getting rid of the gas. If the lungs are affected, pour a little pure alcohol into a test-tube, warm it, and inhale the vapour.]

The chief source of this element is common salt, which contains it united with sodium. In order to obtain chlorine from salt it is not sufficient to add an acid, since the chlorine unites at once with the hydrogen of the acid, forming hydrogen chloride. But if we add to this latter a substance which readily gives up oxygen, e.g. manganese dioxide, the oxygen unites with the hydrogen, and chlorine is set free. To illustrate this,—

Put a little sodium chloride into a test tube and pour on it some strong hydrogen sulphate. A brisk action will begin, and a colourless, pungent-smelling gas will be given off, which reddens a strip of moist litmus paper held in the tube. This gas is hydrogen chloride, and its properties will be more fully considered in the next exercise.

Now add a little manganese dioxide, and warm slightly. A greenish-yellow gas will soon fill the tube, which has a most irritating and suffocating smell, and which bleaches a strip of moist litmus paper. This gas is chlorine[2].

[1] The first portions of the gas, which come over mixed with air, should be collected in a large gas bottle; into this, when full, some solution of caustic soda should be poured, and shaken up in the gas until the smell of chlorine has disappeared. Or, the bottle may be taken out of doors and left open for some minutes.

[2] The following equations express the action :—

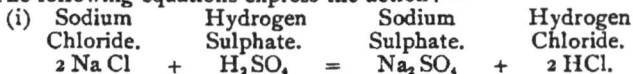

(i) Sodium Chloride.	Hydrogen Sulphate.	Sodium Sulphate.	Hydrogen Chloride.
$2NaCl$ +	H_2SO_4 =	Na_2SO_4 +	$2HCl$.

Preparation of Chlorine.

Apparatus required.—Flask; elbow tubes and cork-joint used in Ex. 11; argand burner; wooden blocks; card; one large and six small gas bottles; taper on wire; deflagrating cup; crucible tongs; porcelain dish; one large and one small cylindrical gas jar; glass disc; supplementary pan of the pneumatic trough, filled with warm water; sodium chloride (common salt); manganese dioxide; strong (common) hydrogen sulphate; blue litmus-paper; solution of indigo; phosphorus; brass filings, or metallic antimony; Dutch gold leaf; turpentine.

Measure 40 c.c. of water into a beaker holding 100 c.c., and add to it by degrees 40 c.c. of strong common hydrogen sulphate. While the mixture is cooling, weigh out 20 grms. of common salt (sodium chloride) and an equal quantity of manganese dioxide, and mix them together.

Arrange an apparatus like that used for carbon dioxide (fig. 56, p. 155) [1], put the mixture into the flask, and add the diluted acid through the funnel: then shake the flask until the contents are thoroughly mixed. Place the whole apparatus in the draught cupboard, and apply a very gentle heat. Collect one large and six small bottles of the gas (remembering to keep the door of the draught cupboard closed as much as possible, so that none of the gas may escape into the room). The yellowish-green colour of the gas will sufficiently show when the bottles are full, especially if a sheet of white paper is held behind each bottle while the gas is passing into it.

When the bottles are all filled, pass the gas into about 30 c.c. of solution of potassium hydrate placed in a large test tube (noting its ready and complete absorption by the potash [2]), until the liquid acquires a yellow colour and the gas is no longer

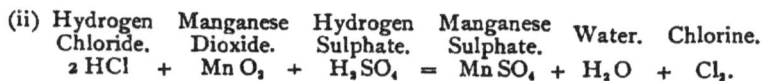

(ii) Hydrogen   Manganese   Hydrogen   Manganese   Water.   Chlorine.
Chloride.   Dioxide.   Sulphate.   Sulphate.

$$2\,HCl \;+\; MnO_2 \;+\; H_2SO_4 \;=\; MnSO_4 \;+\; H_2O \;+\; Cl_2.$$

[1] If the gas is required to be quite pure, a small wash-bottle should be interposed between the flask and the delivery tube, partly filled with plain water, to retain any hydrogen chloride which may come over.

[2] The action will be explained in Exercise 22.

absorbed. It should then be reserved in a closely corked tube or bottle for use in Exercise 22.

If the gas is still coming off (the end of the action may be known by the mixture in the flask becoming reddish brown) pass it into a bottle or flask filled with water, to form the solution of chlorine which is used as a test in analytical work.

Put away one of the small bottles of chlorine in a dark cupboard for use in Exercise 22.

[Another good method of preparing chlorine is to heat a mixture of 60 c.c. of strong hydrogen chloride with 20 grms. of manganese dioxide. In this case, however, half the chlorine is retained in combination with manganese[1].]

The following properties of the gas should be examined, all experiments being made in the draught cupboard, if possible.

*1. Its suffocating smell and its colour** will have been noticed already.

2. Its weight, nearly 2½ times that of air.**

This has been proved by the mode of collecting it, and will be further illustrated in Experiment 7.

3. Its solubility in water.**

Pour about 20 c.c. of water into a small bottle of chlorine; insert a bit of paper between the stopper and the neck, invert the bottle, and shake it for a few seconds. The water will readily dissolve the gas, as will be proved by its acquiring the colour and smell of chlorine.

*4. Its bleaching action on indigo, litmus, and other colouring matters.**

(*a*) Add a little of the solution of chlorine made in the last experiment to some dilute solution of indigo sulphate (about 3 drops of the strong solution in 10 c.c. of water). The blue colour will be at once destroyed.

(*b*) Put into the bottle containing the solution of chlorine, a strip of blue litmus paper, a flower, a bit of coloured calico or cloth, and a bit of paper having some printed words upon it

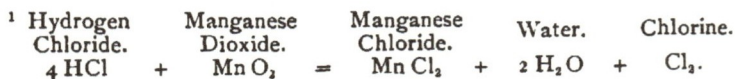

[1] Hydrogen Chloride.		Manganese Dioxide.		Manganese Chloride.		Water.		Chlorine.
4 HCl	+	Mn O$_2$	=	Mn Cl$_2$	+	2 H$_2$O	+	Cl$_2$.

and also some words written with common ink. The colour will be more or less quickly discharged in all cases, except that of the printer's ink: this consists of carbon, for which chlorine has very slight affinity.

5. Its relation to ordinary combustion.

Immerse a lighted taper in a small bottle of chlorine. It will not be extinguished, but will burn with a dull smoky flame, depositing abundance of soot; and the gas itself will not catch fire.

Observe that you have here combustion taking place in a gas which contains no oxygen. In fact, although oxygen usually takes part in combustion, it is not (as was formerly thought) necessary. The phenomena of combustion are simply due (as already mentioned, p. 160) to the light and heat given off during violent chemical action. In the present case the chlorine combines with the hydrogen of the candle, while the carbon for which it has only a slight affinity, remains quite unaltered (see also Expt. 9).

6. Its combination with non-metals, such as phosphorus.

Cut from a stick of phosphorus a small piece about half as large as a pea, observing the precautions given at p. 72. Dry it thoroughly on blotting-paper, pressing, but not rubbing it; place it in a deflagrating cup, and immerse it in the large bottle of chlorine. It will catch fire spontaneously, burning with a greenish flame, and a yellowish crystalline deposit of phosphorus pentachloride will be formed in the bottle.

7. Its combination with metals.

Place two or three pieces of Dutch gold-leaf (an alloy of zinc and copper) in a cylindrical gas jar, and invert over them a bottle of chlorine resting its neck on the mouth of the jar. The heavy gas will quickly descend, and when it reaches the leaves they will catch fire, burning with a dull flame, and forming zinc and copper chlorides.

[If Dutch metal is not at hand, the following experiment may be substituted.

Pour into a bottle of chlorine just enough water to cover the bottom[1]. Place some metallic antimony reduced to fine powder in a mortar, in a paper gutter, and shake the substance little by little into the bottle. The metal will catch fire and burn brilliantly when it comes into contact with the gas, forming antimony perchloride.]

8. Its combination with hydrogen.

(*a*) Suddenly, under the action of heat.

Fill the supplementary pan of the pneumatic trough with warm water. Fill the small stout gas jar with hydrogen gas (from a gas-holder or by the method given at p. 149), and decant its contents into a larger gas jar, placed in the pan of warm water. Next, fill the small jar with chlorine from one of the small bottles of the gas, and add it to the hydrogen in the larger jar[2]. You have now a mixture of equal volumes of hydrogen and chlorine. Agitate the jar for a second or two, in order to mix the gases intimately, keeping its mouth below the water-level[3]; then transfer some of the mixture to the small jar, remove it to the table and apply a light to its mouth. The gases will unite with explosion, forming hydrogen chloride[4].

[Another interesting method of causing the sudden combination of hydrogen and chlorine is the action of a bright light. This, although not always successful, since it requires the gases to be mixed in *exactly* the right proportions, may be attempted as follows.

Have ready at hand a cork rather larger than the mouth of the small gas jar, with one of its ends greased, and also a bit of magnesium ribbon about 12 or 14 c.m. long. Decant into a jar, as above, equal volumes (carefully measured) of hydrogen and chlorine, mix them quickly but thoroughly by agitation; then decant some of the mixture into the small jar, close its mouth with the cork (used as a glass disc, *i. e.* laid on the mouth), and place it mouth upwards on the table. Light the magnesium ribbon, held in the crucible tongs, at a Bunsen's burner, and bring it at once close to the side of the small jar (or take the latter out into full sunshine). An explosion

---

[1] This is to prevent fracture of the bottle by the burning particles of metal.

[2] If this is done neatly and rapidly, the loss of chlorine by solution in the water will be very small.

[3] The gases differ so much in density that their mixture by diffusion alone would be comparatively slow.

[4] $H_2 + Cl_2 = 2 HCl$.

will probably take place, and in any case the union will be quickly effected, white fumes being produced, due to the formation of hydrogen chloride.]

(*b*) Slowly, under the action of light.

Fill the small jar with the remainder of the mixed gases, close its mouth with a glass plate (greased), remove it from the trough and allow it to remain for at least half an hour in a good diffused light, not in direct sunshine, for fear of an explosion. If you then examine the contents of the jar, you will find that the colour and characteristic odour of chlorine have disappeared, and that the gas fumes in the air, reddens and does not bleach litmus-paper, and possesses the pungent odour of hydrogen chloride.

9. Its action on compounds of hydrogen and carbon.

This usually results in their decomposition, the chlorine uniting with their hydrogen, while carbon is set free. (Sometimes, however, especially under the influence of light, the chlorine takes the place of the hydrogen, atom for atom, in combination with the carbon; a process called 'substitution'). Examples of this have been already given in Expt. 5, and also under ETHYLENE, p. 171 (the experiments there mentioned should be tried now, if not already performed). One further experiment in illustration may be tried, if any chlorine remains unused.

Pour a few drops of turpentine upon a strip of blotting paper, and plunge it into a bottle of chlorine, leaving the mouth of the bottle open. Dense white fumes will be produced, owing to the formation of hydrogen chloride, and carbon will be deposited : the action being generally so violent as to set the turpentine on fire.

COMPOUNDS OF CHLORINE WITH OTHER RADICLES.

21. CHLORIDES.

[Typical examples,—Hydrogen chloride, (HCl).
                 Sodium chloride,   (Na Cl).
                 Ammonium chloride, (H_4NCl).]

Preparation of Hydrogen Chloride.

Apparatus required.—Flask, elbow-tubes, &c., as used in the preceding exercise; argand burner; one large and three small gas bottles; taper on wire; two glass discs; porcelain dish; beaker; sand-bath; test-tube; sodium chloride; common hydrogen sulphate; distilled water; blue litmus-paper; turmeric-paper.

The formation of this substance by the action of hydrogen sulphate on common salt has already been noticed and explained under CHLORINE. The action of the strong acid upon ordinary crystallised salt is so violent that it is necessary to melt the salt previously into compact lumps, so as to expose less surface to the acid[1]. If, however, the acid is slightly diluted, the ordinary fine-grained table salt can be used. The gas will not be quite free from moisture, but this is unimportant for most experiments. In preparing and experimenting on it the same precautions should be used as in the case of chlorine, on account of its irritating and corrosive nature.

Measure out 20 c.c. of distilled water into a beaker and add to it gradually 30 c.c. of strong common hydrogen sulphate[2]. Set this aside to cool.

Fit up an apparatus similar to that which was used for pre-

---

[1] This may be done in an iron ladle over a good fire. The salt melts at a red heat, and the liquid should be poured out on a clean piece of iron or slate, and when cold, broken up into lumps about the size of almonds, which should be kept in a well-stoppered bottle.

[2] The liquid will probably become turbid, owing to the presence of lead as an impurity in the hydrogen sulphate, but this may be neglected.

paring carbon dioxide (fig. 56, p. 155), placing a piece of wire-gauze under the bottom of the flask, and fixing it at such a height that the lamp will pass easily under it[1]. Place in the flask 20 grms. of sodium chloride, and having set a bottle under the delivery tube, pour into the flask the dilute hydrogen sulphate you have just prepared. Apply a gentle heat and fill three small bottles with the gas by displacement, using a lighted taper to ascertain when they are full, precisely as directed in the case of carbon dioxide. Fill also one of the larger cylindrical gas jars, greasing its mouth and closing it with a glass disc, when full.

The following properties of the gas may be examined.

1. **Its pungent, penetrating smell, and the white fumes which it forms in moist air**[2] will have been already noticed.

2. **Its high density, 1¼ times that of air.**

3. **Its acid reaction on litmus.**

4. **Its relation to combustion.**

These three properties may be tried in one experiment, as follows.

Place a piece of blue litmus-paper at the bottom of a cylindrical gas jar, and support in the jar the taper, lighted, resting the wire against the side of the jar. Pour some hydrogen chloride from one of the bottles into the jar. The taper will be extinguished, and the litmus-paper will be strongly reddened.

5. **Its solubility in water.**

Fill the supplementary pan of the pneumatic trough (or a deep basin) with water. Take the gas jar which was filled with hydrogen chloride, plunge its mouth (still closed with the glass disc) deeply into the water, and slowly withdraw the disc. The water will rush violently up into the jar, and will fill it entirely if the gas is pure and unmixed with air, showing that hydrogen chloride is extremely soluble in water. Add a few drops of solution of litmus to the liquid; it will at once be reddened.

---

[1] If the gas is required to be quite pure, it should be passed through a small wash-bottle containing strong hydrogen sulphate. This dries the gas, and at the same time frees it from any hydrogen sulphate which may be carried over as spray.

[2] These are due to its combination with the water-vapour in the air to form a product less volatile than itself, which is therefore condensed into fine liquid drops forming a mist.

6. Its direct combination with ammonia.

This has been already noticed under AMMONIA : it may now be tried more fully in the following way.

Place a bottle of the gas on the table, and replace its stopper by a glass plate. Take a similar bottle of ammonia [1], invert it, and replace its stopper quickly by a glass plate. Place the mouths of the bottles close together, and remove both plates, thus allowing the gases to mix. Dense white clouds will be produced in both bottles (showing that the gases quickly diffuse though they differ greatly in density), and still more if the position of the bottles is reversed, their mouths being still held together, so that the bottle of hydrogen chloride is uppermost, causing the heavier gas to descend into the lighter. Enough heat is evolved by the union of the gases to make the bottles sensibly warm ; and a snow-white powder is deposited in flakes on the sides of the bottles. The two gases, in combining, form a single substance, ammonium chloride, or 'sal ammoniac.' [2]

When the action is over examine the properties of this salt as follows. Pour 10 c.c. of distilled water into one of the bottles, shake it up until the white powder is dissolved, then pour the liquid into the other bottle and dissolve the deposit in it also. Place the liquid in a porcelain dish and evaporate it to dryness, using a gentle heat.

(*a*) Scrape together some of the white residue with a spatula, place it in a test-tube, and heat it over the lamp. It will volatilise entirely, forming a white ring in the cooler part of the tube. This distillation of a solid substance is called 'sublimation.' Observe that the substance does not fuse before volatilising, but passes at once from the solid to the gaseous state. This is an exception to the general rule that solids become liquids before they volatilise.

---

[1] If no bottles of ammonia are at hand, pour a little of the *strongest* solution of ammonia into a gas bottle, and shake it up, taking care that the stopper is not blown out. Loosen the stopper for a moment to let air escape, and then shake up the liquid again. In this way enough ammonia gas will be given off to fill the bottle. Lastly, invert the bottle, and loosen the stopper so as to let the liquid escape into a basin or the sink: you may then use the bottle of gas in the above experiment.

[2] $H_3N + HCl = H_4NCl.$

(*b*) Dissolve the remainder of the residue in a little water and test the solution with blue litmus and turmeric-paper. No alteration of colour will be produced in either case [1], while we have seen that hydrogen chloride reddened litmus-paper, and that ammonia reddened turmeric-paper. By the combination of the two gases a neutral salt has been formed.

7. Its decomposition by oxidising agents.

In this case chlorine is evolved, as already noticed, p. 184, and this at the moment of its liberation acts upon substances with even more than its usual intensity. Hence the solution of hydrogen chloride while undergoing this decomposition acts as a very powerful solvent.

Mix about 3 c.c. of strong hydrogen chloride with one-third its volume of strong hydrogen nitrate in a test-tube, and warm gently. The fluid will become yellow, and will give off an orange gas, which is a mixture of chlorine and nitrosyl chloride [2]. Put into the liquid a bit of gold leaf wrapped round a glass rod : it will be immediately dissolved. This mixture of acids is called 'Aqua regia,' (an old alchemical name,) and is much used for dissolving gold, platinum, and some other substances, which are insoluble in either hydrogen chloride or hydrogen nitrate alone.

Tests for Chlorides.

[A solution of sodium chloride containing 0.5 grm. of the salt in 40 c.c. of water may be used.]

*1. They give a white precipitate soluble in ammonia, when mixed with silver nitrate.

Put a few drops of the solution of sodium chloride into a test-tube, add 6 or 8 c.c. of water and then a few drops of solution of silver nitrate. A white precipitate of silver chloride will

---

[1] It may be found *slightly* acid, since the salt is dissociated to a small extent by long boiling with water, traces of ammonia being given off: the residue, therefore, would contain a corresponding amount of free acid.

[2] $3\,HCl + HNO_3 = 2\,H_2O + NOCl + Cl_2$.

be formed, which, on addition of a little hydrogen nitrate and after thoroughly shaking the mixture, will (if the silver nitrate is in excess) readily separate in flocks, leaving the liquid almost clear. Pour off the solution, shake up the precipitate with some fresh distilled water, and again pour off the fluid, after allowing the precipitate to subside. Silver chloride is one of the easiest substances to wash thus, by decantation. Divide the precipitate into two parts by shaking it up with a little water and, before it has had time to settle, pouring off one-half of the liquid into another test-tube.

(*a*) To one portion, after pouring off the water, add some solution of ammonium hydrate, and warm it. The precipitate will readily dissolve, but will be thrown down again on adding dilute hydrogen nitrate to the solution.

(*b*) To the other portion, after pouring off the water, add a little strong hydrogen nitrate and boil for half a minute. It will remain quite unacted on; a property which distinguishes it from the somewhat similar precipitate formed by a cyanide (p. 180).

2. They give off chlorine when heated with manganese dioxide and hydrogen sulphate.

Pour a few drops of solution of sodium chloride upon a little manganese dioxide in a test-tube and add 2 or 3 c.c. of strong hydrogen sulphate. On warming it gently chlorine will be evolved, which may be recognised by its smell, its colour (best seen by looking across the tube placed against a sheet of white paper), and its action on a slip of wetted litmus-paper held in the tube, which will be first reddened, then bleached.

———

Chlorine forms a series of compounds with oxygen very analogous to the nitrogen oxides. They are, however, so unstable and explosive that few experiments can easily be made with them. But several important series of salts are known, which contain radicles consisting of chlorine associated with oxygen. The principal of these are the hypochlorites and chlorates.

22. HYPOCHLORITES.

[Typical examples,—Potassium hypochlorite, (KCl O).
Calcium hypochlorite, (Ca Cl$_2$ O$_2$).]

These are formed by the action of chlorine upon metal-hydrates (such as potassium hydrate). To illustrate this,—

Pour 10 c.c. of solution of potassium hydrate into the bottle of chlorine which was reserved in Ex. 20, put a strip of paper between the stopper and neck of the bottle, and shake the bottle for a minute or so. The chlorine will entirely disappear (but the liquid will not become yellow, as was the case when chlorine dissolved in water, p. 186): it acts upon the potassium hydrate to form two salts, potassium chloride and potassium hypochlorite [1].

Pour the solution into a test-tube, noticing its peculiar sweetish smell, very different to that of chlorine, and examine with it the following properties of hypochlorites.

1. **They give up oxygen readily.**

(*a*) Add a few drops of the solution to a little solution of lead acetate in a test-tube, and warm the mixture. The white precipitate at first formed, containing lead protoxide, soon turns brown; the lead combining with more oxygen to form lead dioxide.

(*b*) Add a few drops of dilute hydrogen sulphate to a dilute solution of potassium iodide, and then add one drop of the solution of potassium hypochlorite. The liquid will turn yellow, owing to the liberation of iodine. In this case, hydrogen iodide was formed in the liquid, on adding the hydrogen sulphate ; the potassium hypochlorite gave up oxygen to this, forming water and iodine [2].

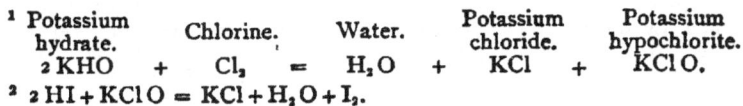

---

[1]

Potassium hydrate.		Chlorine.		Water.		Potassium chloride.		Potassium hypochlorite.
2 KHO	+	Cl$_2$	=	H$_2$ O	+	KCl	+	KCl O.

[2] 2 HI + KCl O = KCl + H$_2$ O + I$_2$.

*2. They yield chlorine when acted on by hydrogen chloride.

2. Place a little ordinary bleaching powder at the bottom of a large test-tube, and pour on it a few drops of strong hydrogen chloride. A violent action will take place, and a greenish yellow gas will fill the tube, which may be proved to be chlorine by its odour and bleaching action on litmus-paper [1].

In the case of ordinary bleaching powder, any acid will produce the same effect as hydrogen chloride; since the substance contains the elements of calcium chloride as well as calcium hypochlorite; and the calcium chloride yields hydrogen chloride when decomposed by an acid.

3. They bleach organic colouring matters, but only when a free acid is present.

(*a*) Dip a piece of blue litmus-paper into the solution of potassium hypochlorite. Its colour will not be altered. Lay it on a plate and pour on it a drop or two of dilute hydrogen chloride; it will be immediately bleached. The reason is that hydrogen hypochlorite is formed by the action of the acid, and this is much less stable than the potassium hypochlorite, and gives up its oxygen to the colouring matter forming a colourless compound.

(*b*) Shake up a little bleaching powder with 10 c.c. of water, filter it into a small beaker, and add to the liquid a drop or two of solution of blue litmus. The colour of the litmus will remain unchanged. Now blow air from the mouth into the solution through an elbow tube. The blue colour will soon disappear; the carbon dioxide of the breath having decomposed the calcium hypochlorite in the same way as the hydrogen sulphate in the previous experiment.

(*c*) Add a few drops of solution of potassium hypochlorite to a dilute solution of indigo (which always contains free acid). The blue colour will be at once discharged.

---

[1] The reaction is a remarkable one. Hydrogen hypochlorite is first formed by double decomposition; and then a molecule of chlorine is formed by the union of 1 atom derived from the hydrogen hypochlorite and 1 atom from the molecule of hydrogen chloride. Thus,—

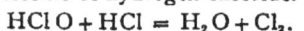

$$HClO + HCl = H_2O + Cl_2.$$

[(*d*) It is clear from the above experiments that white patterns may be formed on coloured cloth by putting an acid on certain parts of it, and then dipping it into a solution of a hypochlorite, which will only discharge the colour where it meets with the acid. This is the principle of the 'discharge' process of dyeing, and may be illustrated as follows.

Powder about a gramme of hydrogen tartrate (tartaric acid[1]), add to it in the mortar about 1 c.c. of ordinary thick solution of gum arabic (this prevents the solution from spreading over the cloth), and grind the whole together. Take a small piece of coloured cloth or calico (the thin cotton cloth dyed with madder answers well), lay it flat on a double fold of blotting paper, and draw on it letters with a glass rod dipped in the acid solution. Dry this partially before a fire or by hanging it at some distance above a lamp-flame, and meanwhile make a strong solution of bleaching powder by grinding about 20 grms. of it with enough water to make a thin cream, then rinsing it into a flask, adding about 80 c.c. of water, and shaking it up thoroughly. Lastly, it must be filtered from the residue, mainly of calcium hydrate, which is sure to remain. Warm the filtered solution gently, and pour it upon the piece of coloured cloth laid flat on a plate, taking care to soak the whole of the cloth in the liquid, and seeing that it is kept flat and that no folds overlie each other. The acid will at once decompose the calcium hypochlorite : chlorine will be liberated, as above explained, and will discharge the colour in those parts where the acid was placed. As soon as the design appears in white, remove the cloth and rinse it well in plain water.]

23. CHLORATES.

[Typical example,—Potassium chlorate, $(KClO_3)$.]

These are formed when solutions of hypochlorites are heated to boiling ; a part of the salt giving up oxygen to the rest, and being itself reduced to a chloride.

To illustrate this, take the solution of potassium hypochlorite which was obtained in Ex. 20 by saturating potassium hydrate with chlorine, and evaporate it down in a dish until a drop

---

[1] Hydrogen oxalate will do if hydrogen tartrate is not at hand.

placed on a watch glass deposits crystals on cooling; then leave it to crystallise. In this reaction two molecules of the potassium hypochlorite give up all their oxygen to a third molecule which is thus converted into potassium chlorate [1]. Thus the liquid contains a mixture of potassium chlorate and potassium chloride; and the former being much less soluble in water than the latter, crystallises out in the form of flat rhombic plates as the solution cools. These may be drained from the liquid, washed with a little water, and left to dry in a funnel. The liquid drained from the crystals will contain much potassium chloride, as may be proved by testing it with silver nitrate. The crystals should be proved to be a chlorate by tests 2 and 4. Meanwhile the properties of the chlorates may be examined, another sample of potassium chlorate being used.

[For some of the following experiments a solution of potassium chlorate will be required, containing 1 grm. of the salt dissolved in 40 c.c. of water.]

*1. They give up oxygen readily.

This has been already proved in the first experiment made in the preparation of oxygen, p. 96.

Put a small crystal of potassium chlorate into an ignition-tube, add a splinter of charcoal, and heat nearly to redness. The same deflagration will be observed as took place in the similar experiment with nitrates, p. 137, the potassium chlorate giving up all its oxygen to the carbon, with formation of potassium chloride and carbon dioxide.

*2. They give off chlorine tetroxide when acted on by hydrogen sulphate.

This experiment requires caution, but is quite safe if made with quantities not larger than those mentioned below.

Put 2 c.c. of strong hydrogen sulphate into a test-tube, add half a gramme (not more) of potassium chlorate, coarsely powdered, and place the tube in a beaker containing warm water (not hotter than 40° C., i.e. not too hot to be borne by the hand). The mixture will become deep orange and a

---

[1] $3 \, KClO = 2 \, KCl + KClO_3$.

greenish yellow gas (chlorine tetroxide) will gradually fill the tube, having a very characteristic smell, somewhat resembling (but easily distinguishable from) that of chlorine [1].

Take a piece of wire bent at right angles (the wire to which the taper is affixed, fig. 8, p. 9, the taper being removed, will do very well), heat one end of it in a lamp-flame, and plunge it while hot into the gas in the test-tube. The gas will decompose with slight explosion, and its greenish yellow colour will almost instantly disappear, owing to the formation of chlorine, the colour of which is less intense, and oxygen.

When the experiment is over, pour the mixture in the tube into a jug of water at once, and throw it away: the tube may then be safely washed out with water.

[The instability of chlorine tetroxide may also be shown by its violent action on phosphorus.

Nearly fill a large wine-glass or beaker with water, and drop into it a few crystals of potassium chlorate, which will sink to the bottom without much loss, as the salt is not very soluble in water. Drop upon the crystals a bit of phosphorus about as large as a pea, and support a tube funnel in the glass in such a position that the extremity of the tube may just touch the crystals at the bottom of the glass. Now pour down the funnel 2 or 3 c.c. of strong hydrogen sulphate; chlorine tetroxide will be evolved when the acid comes in contact with the potassium chlorate, and the phosphorus will take fire, as in chlorine gas, and burn underneath the water.]

*3. They bleach indigo when heated with it.

Add one drop of solution of indigo sulphate to some solution of potassium chlorate, and boil the mixture. The blue colour of the indigo will be destroyed; as in the case of nitrates, p. 137.

---

[1] The following equations express the action.

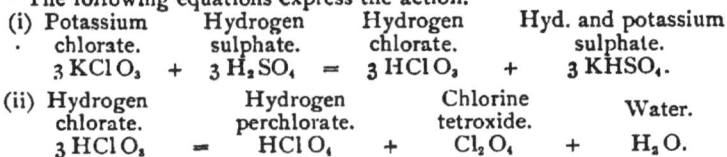

(i) Potassium chlorate. | Hydrogen sulphate. | Hydrogen chlorate. | Hyd. and potassium sulphate.

$$3\,KClO_3 + 3\,H_2SO_4 = 3\,HClO_3 + 3\,KHSO_4.$$

(ii) Hydrogen chlorate. | Hydrogen perchlorate. | Chlorine tetroxide. | Water.

$$3\,HClO_3 = HClO_4 + Cl_2O_4 + H_2O.$$

It will be seen that the chlorate is split up into a higher and a lower oxidised compound of chlorine.

*4. They bleach indigo, even in the cold, when hydrogen sulphite is added.

This action depends upon the reduction of chlorates by hydrogen sulphite, which has a great tendency to absorb oxygen (as will be explained under SULPHITES); lower and less stable compounds of chlorine are formed, which readily give up chlorine, decomposing the indigo.

Add a drop of solution of indigo sulphate to some solution of potassium chlorate. Put into another test-tube some solution of hydrogen sulphite (or, if this is not at hand, sodium sulphite to which a few drops of hydrogen sulphate have been added; hydrogen sulphite is thus formed by double decomposition) and add to it also a drop of indigo sulphate. In neither case will the blue colour be discharged. Now mix the contents of the two tubes; the blue colour of both the solutions will disappear, for the reason above explained.

This reaction serves to distinguish a chlorate from a nitrate, since the latter will not under the same conditions bleach indigo at once. The experiment should be repeated, using a solution of potassium nitrate instead of potassium chlorate, to prove this fact.

5. They give no precipitate when tested with silver nitrate.

Add a drop of solution of silver nitrate to some of the solution of potassium chlorate. No precipitate will be produced if the chlorate is pure, since silver chlorate (as indeed every other chlorate) is soluble in water [1].

Observe that although chlorine is present in the chlorate, it is present in such a condition as to be unable to combine *per se* with silver to form a chloride.

---

[1] Commercial potassium chlorate generally contains a trace of chloride, sufficient to give a turbidity with silver nitrate, but it may be readily purified by dissolving in as little hot water as possible (about 20 grms. in 60 c.c.) and re-crystallising.

24. BROMINE.

[Formula of molecule, Br_2.
Weight of molecule, 80 hydrogen-atoms.]

This element is obtained from sodium bromide or potassium bromide by a reaction precisely analogous to that by which chlorine was obtained from sodium chloride (p. 184). Its vapour is even more offensive in smell and poisonous than that of chlorine, and all possible precautions should be taken not to inhale any of it: all experiments being done in a draught-cupboard.

Mix intimately in a mortar about half a gramme of potassium bromide with twice as much manganese dioxide. Put the mixture into a dry test-tube, add about 3 c.c. of strong hydrogen sulphate, and heat the tube very gently on a sand-bath, supporting it upright by passing over it the smallest retort ring, and placing behind it a sheet of white paper. Deep red vapours of bromine will soon fill the tube, and will condense near the top into an intensely-coloured liquid. Some of its properties may be examined as follows.

1. Its bleaching action on organic colours.

Dip a wetted piece of blue litmus paper into the vapour. It will be at once bleached.

2. Its weight, and solubility in water.

Put about 5 or 6 c.c. of water into a test-tube, and pour on it some bromine by inclining the tube containing the vapour (taking care that none of the mixture in the tube is poured out as well). The vapour will readily be transferred in this way, owing to its weight. Now shake up the water with the bromine vapour, closing the mouth of the tube with the finger. An orange-coloured solution of bromine will be readily formed. Pour one or two drops of this into a dilute solution of indigo sulphate; the blue colour will be discharged, as in the case of chlorine.

3. Its action on potassium hydrate.

Add a few drops of solution of potassium hydrate to the

solution of bromine just obtained.   The orange colour will
disappear, the bromine uniting with the potassium and oxygen
to form a mixture of potassium bromide and potassium bromate [1].   This result is analogous to that which takes place
when potassium hypochlorite is heated (p. 198), but in the
case of bromine the potassium hypobromite, if formed at all,
is decomposed even in the cold.

25. BROMIDES.

[Typical example,—Potassium bromide (K Br.).
A solution of this salt containing 1 grm. of it dissolved in 40 c.c.
of water may be used.]

***1.   They are decomposed by hydrogen sulphate, yielding
hydrogen bromide mixed with vapours of bromine.**

Place a few crystals of potassium bromide in a test-tube
and add a few drops of hydrogen sulphate.   A strong action
takes place, and a gas giving white fumes in moist air is given
off which is hydrogen bromide ; but this is itself partly decomposed by hydrogen sulphate, orange vapours of bromine being
set free.   Hence hydrogen bromide cannot be obtained pure by
the same method as hydrogen chloride : it is usually prepared
by a process of which the principle will be explained under
IODIDES.

***2.   They give a yellowish-white precipitate, difficultly
soluble in ammonia, when tested with silver nitrate.**

Add a drop of solution of silver nitrate to solution of
potassium bromide.   A yellowish white precipitate of silver
bromide will be formed.   Divide the liquid in which the precipitate is suspended into two parts; add to the one some
hydrogen nitrate ; the precipitate will not dissolve.   To the
other portion add some ammonia, and warm ; the precipitate
will dissolve by degrees, but not so readily as silver chloride.
The fact of its solubility may be proved, without waiting until
all the precipitate has disappeared, by pouring off a little of
the clear liquid into another tube and adding to it some dilute

[1] $6 \, KHO + 3 \, Br_2 = 5 \, KBr + KBr O_3 + 3 \, H_2 O.$

hydrogen nitrate : the portion of silver bromide which had dissolved will be again precipitated.

*3. They are decomposed by chlorine, with liberation of bromine.

Add a few drops of solution of chlorine to some solution of potassium bromide in a test-tube. The bromine will be displaced by the chlorine from its combination with potassium, and will dissolve in the excess of potassium bromide, colouring the solution yellow.

Now add a small quantity of carbon disulphide [1], shake the mixture thoroughly, and allow it to stand for a few moments in order that the scattered particles of the carbon disulphide may collect at the bottom of the tube in one globule, which will be found to have acquired an orange tint, while the fluid above it will be colourless. The bromine has been withdrawn entirely from the solution by the carbon disulphide, in which it is very soluble. Pour off the upper stratum of fluid, fill up the test-tube with water, and again pour it off; then add some solution of potassium hydrate, shake it up, and allow it to stand as before. The globule of carbon disulphide will have lost its colour, the bromine having acted upon the potassium hydrate, to form potassium bromide and bromate, as just now explained.

26. IODINE.

[Formula of molecule, I_2.
Weight of molecule, 254 hydrogen-atoms.]

Iodine has many striking analogies to chlorine and bromine, but it is a solid at ordinary temperatures, and its vapour when formed is not so injurious and unpleasant as the other two elements. Hence it will be unnecessary to take the same precautions in experimenting upon it. It is obtained from potassium iodide by a reaction similar to that by which chlorine and bromine are prepared.

---

[1] If carbon disulphide is not at hand, ether will serve the purpose, as it will form a coloured stratum on the top of the aqueous solution.

Weigh out 3 grms. of potassium iodide and the same quantity of manganese dioxide, mix them intimately in a mortar and
put the mixture into a porcelain dish about 8 or 9 cm. in
diameter; then add 5 c.c. of strong hydrogen sulphate, stir the
whole thoroughly together, and cover the dish with an inverted
funnel just large enough to fit within the rim.   On warming
the dish on a sandbath, deep violet vapours of iodine will be
formed, and will condense in the upper part of the funnel in
glittering rhombic plates, the crystalline form of which may
easily be made out with a magnifying glass [1].   In a few minutes
sufficient iodine will be obtained for use in the following experiments on its properties.

[It should be borne in mind that iodine stains the skin yellow,
and therefore it should be handled as little as possible. A glass rod
or platinum (not aluminium or bone) spatula should be used.
Stains may be removed by dilute solution of potassium hydrate.]

***1. It melts and volatilises readily, forming a violet vapour,
much heavier than air.**

Put a few crystals of iodine into a large dry test-tube, and
heat them gently.   They melt into an almost black liquid, and
on further heating turn into a splendid violet vapour.   Warm
the upper part of the tube, to prevent the condensation of the
vapour, and when the tube is full of it pour some out on
a white plate.   The heavy vapour will pour out almost like
a liquid, and will condense into a cloud of small flakes of
iodine which will fall in a shower on the plate.

2. It is scarcely soluble in water.

Place a crystal or two of iodine in a test-tube, pour on them
about 5 or 6 c.c. of water, and shake the mixture.   Hardly any
of the substance will dissolve; only sufficient to colour the
liquid a pale yellow.

3. It readily dissolves in alcohol.

Pour off the liquid from the crystals used in the last experiment, and add to them a little alcohol.   A red solution will be at
once formed, rapidly deepening in colour, when the tube is

---

[1] Very fine crystals of iodine are often formed by spontaneous sublimation on the stopper of a bottle containing it.

shaken, until it becomes almost opaque. Now fill up the tube with water; most of the iodine will be reprecipitated as a black powder.

4. It also dissolves in solution of potassium iodide.

Put about half a gramme of crystallised potassium iodide into a test-tube, add 5 or 6 c.c. of water and then a few crystals of iodine. The latter will dissolve even more readily than in alcohol; but it will not be reprecipitated on filling up the tube with water.

5. It has little or no bleaching properties.

This may be tried by dipping a piece of blue litmus paper into the solution of iodine just obtained.

*6. It forms a deep blue compound with starch.

This is the most characteristic and delicate test for iodine, when it is free and uncombined: mere traces of it being recognisable.

Pour 5 or 6 drops only of the solution of iodine obtained in experiment 4 into a beaker, add 100 c.c. of water and then 8 or 10 c.c. of a *freshly made* solution of starch [1]. A deep blue colour will be produced, even in so highly dilute a solution of iodine.

Pour a little of the liquid into a test-tube, add a few drops more of the solution of iodine, to deepen the colour, then heat it over a lamp. When it has nearly reached the boiling point the blue colour will quickly disappear, showing that the compound of iodine with starch is decomposed by heat. Cool the liquid by holding the tube in a stream of water (or in a jug of cold water); the blue colour will soon reappear and become as intense as at first [2]. Hence in testing for iodine by this method care must be taken that the liquid is quite cold.

[Conversely, iodine may be used as a test for starch. If a few drops of a very dilute solution of iodine are put on a piece of writing paper, a blue colour will be produced on account of the starch present in the size used in making the paper. Similarly, a slice of raw potatoe may be shewn to contain starch.]

---

[1] For the method of making this see Appendix B.

[2] Notice that the blue colour reappears first in the *lowest* part of the liquid. It reappears, of course, where the liquid is coldest; and the fact that it appears at the bottom is a proof that cold water is denser than hot water.

27. IODIDES.

[Typical examples,—Hydrogen iodide (HI).
Potassium iodide (KI).
Phosphorus tri-iodide (PI₃).]

The formation of one or two of these compounds of iodine with other radicles has been illustrated already. Thus iodine was shown to combine readily with phosphorus and with mercury in Exercise 8 (pp. 72, 73).

Preparation of Hydrogen iodide.

Apparatus required—Large test tube; corks; cork borers; elbow tubes as used in Ex. 11; Bunsen's holder; wooden blocks; two small gas bottles; one large do.; two cylindrical gas jars, 20 x 5 cm.; glass disc; taper on wire. Potassium iodide; strong hydrogen sulphate; red phosphorus; iodine; litmus paper; bleaching powder; strong hydrogen chloride.

This substance cannot be obtained in a pure condition by a reaction analogous to that by which hydrogen chloride was obtained, viz. by decomposing potassium iodide by strong hydrogen sulphate; since part of the hydrogen iodide is itself decomposed by the hydrogen sulphate [1], and iodine set free.

To illustrate this place a few crystals of potassium iodide in a test-tube, pour on them some strong hydrogen sulphate, and heat gently. A gas will be given off which forms white fumes in the air, like hydrogen chloride, but it will be mixed with violet vapours of iodine.

Hence to prepare the gas recourse is usually had to the action of water on phosphorus tri-iodide. This is of the following nature,—the iodine combines with half the hydrogen in

---

[1] Thus :—

Hydrogen iodide.		Hydrogen sulphate.		Hydrogen sulphite.		Water.		Iodine.
2 HI	+	H_2SO_4	=	H_2SO_3	+	H_2O	+	I_2

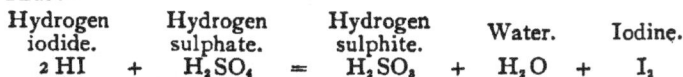

Observe that this reaction is the reverse of what takes place when cold solutions of iodine and hydrogen sulphite are mixed together.

three molecules of water, while the phosphorus together with the rest of the hydrogen and the oxygen form hydrogen phosphite (phosphorus acid) [1].

It has been already seen (p. 72) that phosphorus unites readily with iodine, but with a violence which is not easy to control. Hence it is best to use the modification of phosphorus called 'red' or 'amorphous' phosphorus, the affinities of which are much less strong than those of ordinary

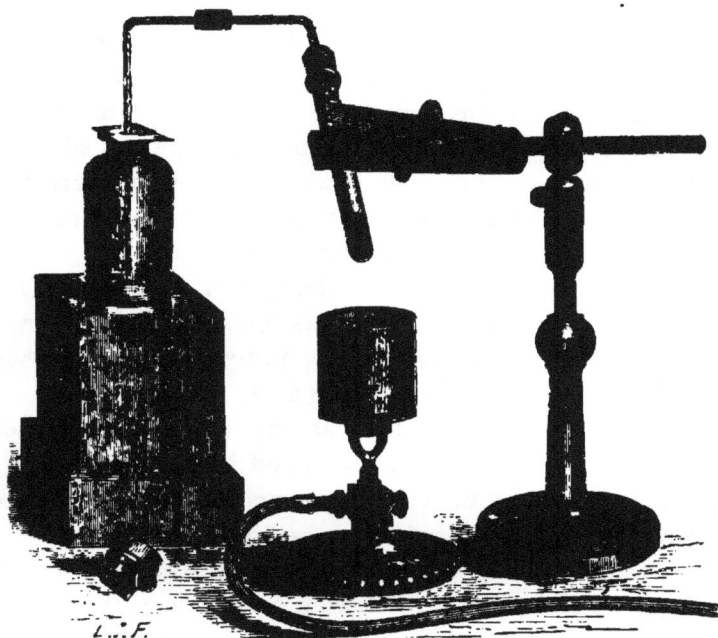

Fig. 58.

phosphorus. In any case, however, the process requires care; and strict attention must be paid to the directions given.

Take a large test-tube, about 2·5 cm. in diameter and 16 or 18 cm. in length; adapt to it a cork fitted with an elbow tube and right-angled delivery tube connected by a cork joint, such as was used in the preparation of carbon dioxide, p. 155, and support it in an inclined position in a Bunsen's holder thus, Fig. 58.

[1] $PI_3 + 3 H_2O = 3 HI + H_3PO_3.$

Weigh out half a gramme of red phosphorus in powder [1], and transfer it to the test-tube. Pour upon it 1 c.c. (not more) of water, and then (having previously placed a beaker of cold water within reach,) add 6 grms. of iodine, and mix the whole by shaking the tube, cooling it as soon as the action begins by dipping it into the beaker of water. The iodine acts on the phosphorus readily but not violently if the temperature is kept down, and the water decomposes the phosphorus tri-iodide, as above explained. We thus obtain a very strong solution of hydrogen iodide, from which the gas will be given off abundantly when it is very gently heated.

Replace the cork with delivery tube, and re-adjust the apparatus in the holder, placing under the end of the delivery tube a small gas bottle covered with a card. Now heat the mixture in the test-tube very carefully with a lamp; the gas will soon begin to pass over into the bottle, and from its very high density ($4\frac{1}{2}$ times that of air) it may most readily be collected by downward displacement, its position in the bottle being tested, as in the case of hydrogen chloride, by a lighted taper. Two small bottles and one cylindrical gas jar should be filled with the gas, the flame of the lamp being just waved under the test-tube if the evolution of gas becomes slow.

The unstable nature of the gas will be at once noticed; a slight deposit of iodine will be found in the bottles, even if kept only for a short time; and vapours of iodine will be seen whenever the lighted taper is immersed in the gas.

In most other respects hydrogen iodide strongly resembles hydrogen chloride, and the following properties of the gas may be examined in exactly the same manner as the corresponding properties of hydrogen chloride described in Ex. 21. p. 191, which should be referred to for the details of the experiments.

1. Its pungent smell and the white fumes it produces in moist air.

[1] It is usually sold in a state of powder. If however you have obtained it in lumps, it must be ground with care to a coarse powder in a mortar, covering it with water while crushing it with the pestle, in case any small particles of ordinary phosphorus may be present which might inflame by the friction and set the whole mass on fire; then pour off the water and transfer the moist powder to the test-tube.

2. Its extremely high density.
3. Its acid reaction on litmus[1].
4. Its relation to combustion.
5. Its solubility in water.
*6. Its decomposition by chlorine.

This reaction serves to distinguish it from hydrogen chloride, and also illustrates well the relative affinities of chlorine and iodine for hydrogen.

Place a large wide-mouthed gas bottle on a sheet of white paper before you. Pour into the bottle a little hydrogen iodide from one of the small bottles, then make a little chlorine by pouring a few drops of strong hydrogen chloride upon a little bleaching powder in a test-tube, and pour some of the gas (by inclining the tube) into the large bottle containing the hydrogen iodide. Violet clouds of iodine will be formed when the gases mix; the chlorine, from its stronger affinity, combining with the hydrogen and liberating iodine[2].

[After the experiment, wash out the test-tube containing the chlorine with plenty of water, that the gas may not escape unnecessarily into the room.]

Tests for iodides.

[A solution of potassium iodide containing 1 grm. of the pure colourless salt[3] dissolved in 30 c.c. of water may be used.]

1. They are, with few exceptions, decomposed by hydrogen sulphate.

This has been already shown in a previous expt., p. 206.

*2. They give a pale yellow precipitate, insoluble in ammonia, when tested with silver nitrate.

---

[1] It will be noticed that a brown deposit of iodine is formed on the litmus and that the colour of the latter is eventually destroyed, although iodine has by no means the same strong bleaching powers as chlorine and bromine.

[2] $2\,HI + Cl_2 = 2\,HCl + I_2$.

[3] Potassium iodide is very apt to decompose on keeping, especially if exposed to the light, or if it contains any potassium iodate, becoming yellow owing to liberation of iodine.

Pour a few drops of the' solution of potassium iodide into
a test-tube, dilute with water, and add one or two drops of
solution of silver nitrate. A yellow precipitate of silver iodide
will form, which should be divided into two portions. To the
one add some dilute hydrogen nitrate, which will fail to dissolve
it; to the other add excess of ammonia, which will also fail
to dissolve it; as may be proved by pouring off some of the
clear liquid and adding to this some dilute hydrogen nitrate
which will produce no precipitate.

*3. They are decomposed by many oxidising agents, with
liberation of iodine.

(a) Add one drop of solution of chlorine to a very dilute
solution of potassium iodide (about 3 drops of the solution to
10 c.c. of water). Iodine will be liberated and will colour the
solution yellow. On addition of 2 or 3 c.c. of solution of
starch, the characteristic deep blue compound will be formed.
Now add some stronger chlorine water, and notice that the
starch reaction will entirely disappear, owing to the formation
of a colourless iodine chloride which does not act upon starch.
Hence in applying this test it is important not to use an excess
of chlorine. It is, in fact, best not to use chlorine at all for
the purpose of liberating iodine, but hydrogen nitrite, as described in the next experiment.

(b) Add a few drops of dilute hydrogen sulphate to a very
dilute (see last experiment) solution of potassium iodide, and
then a drop of a recently-made solution of potassium nitrite.
This will decompose the hydrogen iodide formed by the action
of the hydrogen sulphate, and liberate iodine (see p. 141).
Shake up the yellow solution with a little carbon disulphide, and
allow it to stand for a minute. The carbon disulphide will be
found to have withdrawn the iodine from the solution, collecting into a deep purple globule at the bottom of the tube.

[It will be well to repeat the similar experiment with potassium
bromide, given at p. 203, and compare the colours of the two globules,
the bromine imparting a bright orange, the iodine a fine purple colour
to the carbon disulphide. Notice also that potassium nitrite will
not decompose hydrogen bromide, but that chlorine water must be
used for the purpose. If an excess of strong solution of chlorine is

added to the contents of both tubes, and the mixture shaken and again allowed to settle, the globule containing iodine will be found to have lost its colour, while the globule containing bromine remains unchanged.]

28. FLUORIDES.

[Typical examples,—Hydrogen fluoride (HF).
Calcium fluoride (CaF$_2$).]

We have in this case to deal with the compounds of an element which has itself never been isolated; its affinities being so strong that chemists have only succeeded in transferring it from one combination to another.

Its most important compound is hydrogen fluoride, a substance having many analogies to hydrogen chloride, and obtained, like the latter, by the action of a strong acid on other fluorides such as calcium fluoride (fluor spar).

*Formation of Hydrogen fluoride, and its action upon glass.

* Take a small cup of lead or platinum (a leaden ink-well answers perfectly)[1] and put into it as much powdered fluor spar as will lie on the end of a spatula. Obtain a piece of sheet glass about 6 cm. square, warm it gradually before the fire or over a lamp, and when it is thoroughly hot, rub over it a piece of bees'-wax (or the end of a wax-candle) and incline the plate in different directions, so that the melted wax may run over it and form an even coating. Set it up edgeways to cool, on a piece of paper, and then place it with the coated side upwards on the table, and trace on it with a pointed piece of wood (a match cut like a pencil answers very well) any letters or device which may occur to you, taking care that the lines are drawn quite through the wax so as to expose the bare

---

[1] A cup of lead may be made by turning up the edges of a circular piece of thin sheet lead about 7 or 8 cm. in diameter. It should be placed in a mortar just large enough to hold it, and moulded into shape by strong pressure with the pestle.

glass. Pour a little strong hydrogen sulphate upon the fluor spar in the leaden dish, mix the two thoroughly by stirring with a glass rod (which should be washed immediately afterwards) and cover the dish with the piece of glass, the coated side being downwards. Heat the dish very gently by placing it on some warm sand, and leave it for a short time undisturbed, taking especial care not to use so much heat as to melt the wax. The hydrogen sulphate will decompose the fluor spar, forming calcium sulphate, while gaseous hydrogen fluoride will be evolved [1], which will act upon the glass where it is unprotected by the wax. In about ten minutes the glass may be removed from the dish; the pungency and the acid reaction on litmus paper of the fumes of hydrogen fluoride should be noticed, and the contents of the dish washed away at once, in order that the glass bottles in the room may not be damaged. The wax may be removed from the glass by warming the glass and rubbing it with a cloth, and the device will be found deeply and permanently etched into the glass.

In the above reaction one of the chief products formed is silicon tetrafluoride, resulting from the union of the silicon of the glass with fluorine. This gas is very readily decomposed by water, with formation of silicon hydrate (silica), which appears as a white crust on the glass if the action is carried very far. The preparation and properties of this substance will be more fully described under SILICATES, but its formation may be illustrated on a small scale by placing a finely powdered mixture of equal parts of fluor spar and glass or fine sand in a leaden dish, adding some strong hydrogen sulphate, heating gently, and covering the dish with a bit of glass on the centre of which a drop of water has been deposited. The drop will soon become coated with a film of silica, owing to the decomposition of the silicon fluoride by moisture.

Before proceeding further it will be advisable to gain some more experience in analysis; taking, as before, single salts

[1] $CaF_2 + H_2SO_4 = CaSO_4 + 2HF.$

of an alkali-metal (such as sodium) united with one of the radicles already treated of, and examining each of them for the latter constituent. A short course, including all the radicles hitherto examined, is given in Appendix C.

29. SULPHUR.

[Formula of molecule, S_2
Weight of molecule,   64 hydrogen atoms.]

Sulphur is usually obtained from volcanic districts where it is found 'native' or uncombined; but large quantities of it are also procured from iron pyrites, a very common mineral, which consists of iron united with sulphur. When this is heated (air being excluded) it gives off one-third of the sulphur it contains, as the following experiment will show.

Powder a small fragment of iron pyrites in a porcelain mortar and place it in an ignition tube. Heat it to redness in the lamp-flame, and notice that a sublimate of sulphur is formed in yellowish-brown drops in the cool part of the tube. It may be proved to be sulphur by cutting off the sealed end of the tube (which may be effected by touching it while still hot with a drop of water) and heating the sublimate gently while the tube is held obliquely so that a current of air may rise through it. The sulphur will be oxidised to sulphur dioxide, (the gas formed by burning sulphur in oxygen, p. 102) which may be recognised by its odour and by its acid reaction on a piece of moist blue litmus-paper held at the upper end of the tube.

Allotropic forms of Sulphur.

An account of these will be found in any text-book on chemistry; the following experiments will illustrate the modes of preparing them.

1. S_a (octohedral sulphur). This is the most stable form, and the one in which sulphur nearly always crystallises from solutions of it.

Place about a gramme of 'flowers of sulphur' in a dry test-tube, and pour over them about 5 c.c. of carbon disulphide. Leave the tube, loosely corked, in a beaker of water for a few minutes, shaking it occasionally. Care must be taken that there is no lighted lamp or fire near at hand, on account of the inflammability of the carbon disulphide. There will be a residue left, not necessarily because there is not sufficient disulphide to dissolve the substance, but because the flowers of sulphur consist of two varieties of the element, of which one only is soluble in carbon disulphide. Filter the fluid quickly into another tube, through an ordinary paper filter (which must be perfectly dry). Pour about 2 c.c. of the solution into a watch-glass, and leave it on the table to evaporate. The carbon disulphide, from its great volatility, will quickly pass off, and small transparent crystals of sulphur will be formed and rapidly increase in size. They may be easily recognised, especially if a magnifier is used, to be right rhombic octohedra, Fig. 59.

Fig. 59.

2. S_β **(prismatic sulphur).** This is the form in which sulphur is obtained by crystallisation from a melted state.

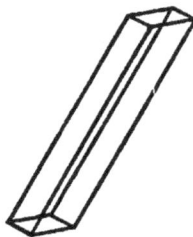

Place some pieces of roll sulphur or flowers of sulphur in an evaporating dish[1], and heat it gently on a sandbath over a lamp until the sulphur is just melted. Allow it to cool until a crust has just begun to form on its surface; then, taking up the dish, pour out the portion which is still fluid into a large dry test-tube (for use in the next experiment). The interior of the dish will be found lined with needle-shaped transparent crystals, Fig. 60, which are referred to the oblique prismatic system, a system totally distinct from that to which the crystals obtained

Fig. 60.

---

[1] The experiment is best made on a rather larger scale, a common clay crucible about 10 cm. in height being used instead of the dish.

from solution in experiment 1 belong, viz. the right prismatic, Fig. 59. In a few days the prismatic crystals lose their transparency, and are spontaneously converted into aggregations of minute octohedra.

3. S_γ (plastic sulphur). This is the most remarkable form of the three, and is obtained by the action of heat upon ordinary sulphur.

Take the large test-tube containing the sulphur poured off from the crystals in the last experiment, add a little more sulphur and support it on a piece of wire gauze in the retort stand, resting it in the smallest retort ring. Lay a cork loosely on its mouth (to prevent entrance of air) and heat it, gently at first, by an Argand burner. While it is being gently heated, fill a jug with cold water, and place in it mouth downwards a large funnel. The sulphur will first melt to a clear pale-yellow fluid, almost as mobile as water. Pour a few drops of it into some water in a dish : it will solidify into the usual yellow, brittle mass. Heat the remainder more strongly; as the temperature rises higher it will darken in colour and become thicker and thicker, until it has so far lost its fluidity that the test-tube may be inverted for a moment without spilling any of it. When the heat is further raised it becomes again fluid, but not so much so as at first. When it has reached this point, take hold of the test-tube with a cloth, or paper holder (p. 60, note), and pour its contents in a thin stream into the cold water, round the stem of the funnel. By being thus suddenly cooled, the sulphur will be preserved in the allotropic condition into which it has been converted by heat, and on lifting the funnel out of the water, the threads of sulphur which surround it will be found to be semi-transparent, soft, almost as elastic as india-rubber, and scarcely soluble in carbon disulphide. If the residue in the test-tube be watched as it cools, it will be seen to undergo the same changes as when it was being heated, but in a reverse order, becoming thick, then losing its dark colour and becoming fluid again, and finally solidifying to a yellow, crystalline, brittle mass. It is only when the temperature has been raised nearly to its boiling-point that the above modification, called 'plastic sulphur,' is obtained. The elastic threads should be dried with a cloth, and put aside in a bottle. In a few days they will be found to have lost both their transparency and their plasticity, (being, in fact, reconverted into the ordinary form,) but will still be in a great measure insoluble in carbon disulphide.

Combination of sulphur with metals.

An experiment showing that sulphur will when heated combine with copper has already been tried, p. 73. It will also combine with iron, but a higher temperature is required.

Mix about 2 grms. of iron filings with an equal weight of sulphur, and heat it strongly in a small test-tube. The sulphur will melt again and finally boil, filling the lower part of the tube with its dark yellow vapour. When the bottom of the tube becomes redhot, a bright glow will spread through the iron filings as the metal combines with the sulphur (just as it did with oxygen, p. 103,) to form iron protosulphide. Allow the tube to cool, and keep the bronze-coloured residue for use in the next exercise (p. 217).

30. SULPHIDES.

[Typical examples,—Hydrogen sulphide (H_2S).
                        Iron protosulphide (FeS).]

Preparation of Hydrogen sulphide.

This gas is so extremely poisonous and offensive in smell that all experiments with it must be made in a draught cupboard, or in the open air. The best method of getting rid of any of the gas which escapes into the room is to place a little bleaching powder on a plate and add a few drops of dilute hydrogen chloride. This will liberate chlorine, as already explained, which will diffuse into the air and decompose the hydrogen sulphide. Do not, however, add too much acid to the bleaching powder, lest the remedy should prove less endurable than the evil it is intended to cure.

Hydrogen sulphide is obtained by the action of acids on many sulphides; the hydrogen of the acid uniting with the sulphur at the moment of its liberation. Iron protosulphide (the substance you lately obtained by the union of iron with sulphur) is, on the whole the most convenient for the purpose.

Take the test-tube containing the iron protosulphide made in the last experiment (p. 216) and detach the substance from it by an iron wire (this will probably result in breaking the tube, since the substance becomes fused into the glass by the heat evolved in its formation). Powder the lump of iron sulphide and put it into a moderate-sized test-tube, to which should be fitted by a cork a short bit of glass tubing about 4 or 5 mm. in diameter (which answers better than a smaller jet). Pour 4 or 5 c.c. of dilute hydrogen chloride upon the iron sulphide, and immediately fit the cork into its place and support the test-tube upright resting in the smallest ring of the retort stand in a draught cupboard. Hydrogen sulphide will be at once given off with effervescence : the action being an ordinary double decomposition, iron sulphide and hydrogen chloride giving iron chloride and hydrogen sulphide [1].

The following properties of the gas may be examined :—

*1. Its extremely fetid smell, like that of rotten eggs.

These, in fact, give off the gas; since they contain sulphur which unites with hydrogen during the process of decay.

2. Its acid reaction on litmus paper.

Hold a bit of wetted litmus paper close to the tube from which the gas is issuing. It will be reddened, but not very strongly.

3. Its inflammability.

Apply a light to the end of the tube. The gas will catch fire and burn with a blue flame, uniting with the oxygen of the air to form water and sulphur dioxide [2]. To prove this, hold a small dry gas bottle over the flame : moisture will be de-

---

[1] $FeS + 2HCL = FeCl_2 + H_2S.$
[2] $2H_2S + 3O_2 = 2H_2O + 2SO_2.$

posited on it, and the presence of sulphur dioxide in it will be proved by its characteristic odour, and the strong acid reaction shown when a piece of blue litmus paper is put into the bottle.

*4. Its action on compounds of lead, turning them black.

Pour a drop of solution of lead acetate upon a bit of white blotting paper and hold it in the stream of gas issuing from the tube. It will immediately turn black, owing to the formation of lead sulphide.

This is the most delicate test for the presence of hydrogen sulphide, and will serve, for instance, to detect the presence of traces of it in ordinary coal gas.

5. Its solubility in water.

The method of making the solution of hydrogen sulphide for use in the laboratory is given below; and in the following experiments a solution of the gas, and not the gas itself, may be conveniently used.

6. Its action upon other radicles forming sulphides.

The chief use of it in the laboratory depends on this property, since the sulphides thus formed are often of very characteristic colours. To illustrate this, take dilute solutions of the following metal-salts (3 or 4 drops of the ordinary laboratory solution to 5 c.c. of water), and add to each several drops of solution of hydrogen sulphide.

(*a*) Copper sulphate. A *black* precipitate will be formed, of copper sulphide.

(*b*) Arsenic chloride [1]. A *yellow* precipitate will be formed.

(*c*) Zinc sulphate. A *white* precipitate will be formed.

7. Its decomposition by chlorine, and other oxidising agents.

(*a*) Add a few drops of solution of chlorine to some solution of hydrogen sulphide and heat it. A white milky precipitate of sulphur will be formed, the chlorine having united with the hydrogen and liberated the sulphur.

A solution of bleaching powder will act in a similar way, and

---

[1] This may be made by dissolving a very small quantity of arsenic tri-oxide (white arsenic) in a few drops of hydrogen chloride.

this illustrates its action as a 'disinfectant' in destroying the hydrogen sulphide which is evolved from putrefying organic substances.

(*b*) Add to another portion of the solution of hydrogen sulphide a few drops of strong hydrogen nitrate, and warm it gently. A precipitate of sulphur will be formed in this case also, since the oxygen of the acid unites with the hydrogen of the gas.

Hence we learn that in the course of an analysis hydrogen sulphide should never be added to solutions which contain much free hydrogen nitrate.

Similarly, the solution of the gas is decomposed by the oxygen of air; and hence it is best to keep it in a closely-corked bottle, inverted in a glass of water.

Tests for Sulphides.

*1. They are, in general, decomposed by acids, giving off hydrogen sulphide.

(*a*) Add a few drops of dilute hydrogen sulphate to a little solution of ammonium sulphide, and warm the mixture. Hydrogen sulphide will be given off, and may be recognised by its odour and its action on a strip of blotting paper moistened with a drop of solution of lead acetate. Moreover, if the solution of ammonium sulphide has been made some time and has become yellow, a white, milky precipitate of sulphur will be produced[1]. We learn from this that ammonium sulphide (a test very commonly employed in the laboratory) should never be added to a solution containing a free acid.

(*b*) Add 2 or 3 c.c. of strong hydrogen chloride to a small quantity of antimony sulphide. No decomposition will in this case take place until the liquid is boiled, when hydrogen sulphide will be given off, and may be recognised in the usual

---

[1] The reason is, that ammonium protosulphide slowly absorbs oxygen from the air, forming ammonia, water, and sulphur,—

$$2 (H_4 N)_2 S = 4 H_3 N + 2 H_2 O + S_2 ;$$

and the latter combines with another portion of the sulphide to form ammonium disulphide, from which it is separated when an acid is added.

way.   This is a process which is sometimes used with advantage for obtaining the gas.

*2. They are decomposed by fusion with potassium hydrate.

Powder in a porcelain mortar a small bit (about the size
of a pea) of iron protosulphide or iron pyrites ; add a bit of
potassium hydrate[1] about the same size, and grind the whole
together.   Fill the bulb of an ignition tube with the mixture,
and heat it to redness for half a minute.   The potassium
hydrate will decompose the iron sulphide, with formation of
potassium sulphide.   Before it cools dip the bulb into a little
water in an evaporating dish, when it will crack to pieces and
the potassium sulphide will dissolve in the water.   The presence
of a sulphide in the solution may be proved by the blackening
which takes place when a drop of the liquid is put upon a clean
silver coin, or a piece of blotting paper moistened with lead
acetate.

This is the best way of detecting the presence of a sulphide
in minerals such as galena (lead sulphide), blende (zinc sulphide), or the golden yellow scales which often occur in coal,
and which are iron persulphide ; and the experiment should be
tried with at least one of the above.

**3. Their solutions give a black precipitate when tested
with silver nitrate.**

Dilute 2 or 3 drops of solution of ammonium sulphide with
5 c.c. of water and add a drop of solution of silver nitrate.   A
black precipitate of silver sulphide will be formed, which will
not dissolve on addition of dilute hydrogen nitrate.

Additional Experiment.

Preparation of solution of Hydrogen sulphide.

Adapt sound corks to two bottles with moderately wide mouths,
holding about 200 c.c.   Prepare (if not already at hand) four elbow
tubes of glass tubing about 5 mm. in diameter: two of them with
branches of equal length, about 6 cm.; the other two with unequal
branches, the shorter about 6 cm., the longer about 18 cm. in length,

---

[1] Sodium carbonate will also answer the purpose, but the temperature
required for the decomposition is higher.

or of sufficient length to reach, when fitted into the cork, nearly to the bottom of the bottle. Bore two holes in each cork and fit into it one long and one short elbow tube.

Fit up a flask with funnel and elbow tube, similar to that used in Ex. 13. Place in it about 20 grms. of iron protosulphide, previously broken up with a hammer into lumps about as large as peas. Pour into the flask enough water to cover the iron sulphide, and fill the bottles about three-fourths full of distilled water ; replace the corks

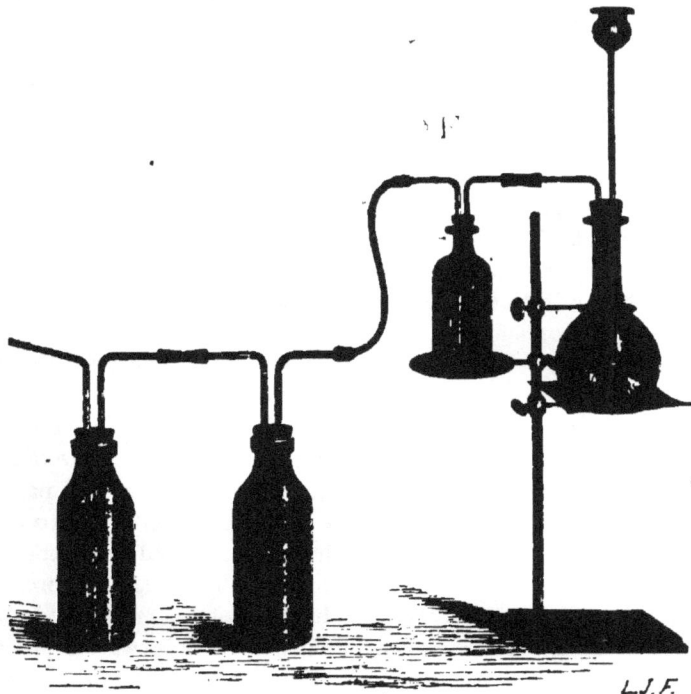

Fig. 61.

and connect the tubes, as shown in Fig. 61, with short pieces of india-rubber tubing, so that the gas may bubble through the water in each bottle successively.

[The above apparatus will serve for general use when a liquid is to be saturated with a gas. It will be necessary in some cases, and indeed not inexpedient in all, to interpose a small wash-bottle containing a little water between the generating flask and the first bottle, for the purpose of retaining any acid or other impurities which may come over with the gas. This bottle may be supported upon a glass (or wooden) disc placed on one of the rings of the retort-stand, as shown in the figure.

It will also be usually necessary to provide some means of getting rid of any excess of gas which may escape absorption by the water. This may be done best by putting the whole apparatus in a draught cupboard or by attaching to the elbow tube proceeding from the last bottle a long tube passing through the window or, better, into a chimney or stove-flue ; or, the gas may be led to the bottom of a tall jar filled with lumps of pumice or charcoal, moistened with a solution of caustic soda. One or other of these methods should always be employed when dealing with such injurious gases as hydrogen sulphide or chlorine.]

Having ascertained that the joints are tight, pour into the flask a little strong hydrogen chloride. The evolution of hydrogen sulphide commences without the application of heat (although some little time may elapse before the action begins), and, if the stream is not too rapid, the greater part of the gas is absorbed by the water in the bottles. You will observe, however, that some gas passes through the water unabsorbed. This consists mainly of hydrogen, the evolution of which is due to the presence of uncombined iron in the sample of iron sulphide.

When a steady stream of hydrogen sulphide has passed through the bottles for five or six minutes, disconnect the first bottle, and connect the other directly with the flask. Ascertain whether the water is saturated with the gas, in the following way. Take out the cork and tubes, close the bottle tightly with the thumb or palm of the hand, and shake it briskly for a few seconds in order to bring the water thoroughly in contact with the gas. Hold the bottle in a slanting position with its mouth downwards, and relax the pressure of the hand, noticing whether any bubbles of air enter the bottle, or whether on the contrary the liquid is forced outwards. In the latter case the water is fully saturated with the gas; in the former case the bottle must be again connected with the flask as at first, and more gas passed through it.

When no more gas is absorbed by the water in the first bottle, it may be withdrawn, and a little more gas passed through the water in the second bottle, which is already partially saturated by the excess of gas which has passed through the first bottle. The whole of the solution should then be poured into a larger bottle, which should be kept, well corked, in an inverted position, resting in the angle at the corner of a shelf, or in a tumbler or test-glass half full of water.

Finally, take the apparatus to pieces, at once, in the open air, and throw away the contents of the flask, washing it out thoroughly with water.

COMPOUNDS OF SULPHUR AND OXYGEN.

31. SULPHUR DIOXIDE.

[Formula of molecule, SO_2.
Weight of molecule, 64 hydrogen-atoms.]

Apparatus required—Flask, elbow-tubes, &c., used in Ex. 11. p. 155; one large and three small gas bottles; taper on wire; glass disc; deflagrating jar and cup; basin of water; small strips of sheet copper; strong hydrogen sulphate; blue litmus-paper; solution of logwood; a few flowers, such as violets or pansies.

Sulphur dioxide is the product invariably formed by synthesis when sulphur is burnt in oxygen or air, as seen already, p. 102. It is, however, usually prepared by a process of analysis; viz. by the action of mercury or copper upon hydrogen sulphate. The chemical change is analogous to that which occurs when copper acts upon hydrogen nitrate (p. 142), hydrogen being first displaced from the acid by the metal, and this 'nascent' hydrogen acting upon another molecule of the acid taking away some of its oxygen with formation of sulphur dioxide and water [1].

The gas is very soluble in water, and should therefore be collected by displacement, which, from its great density, may be readily accomplished. It is very corrosive and poisonous, and hence it should be prepared and all experiments made with it in a draught cupboard.

Arrange an apparatus similar to that used in preparing carbon dioxide [2]. Place in the flask about 10 or 12 small strips of sheet copper [3], add 60 c.c. of strong common hydrogen sul-

---

[1] $Cu + 2 H_2 SO_4 = Cu SO_4 + 2 H_2 O + SO_2$.

[2] It is advisable, though not necessary, to interpose a small wash-bottle containing a little water, to retain any spray of acid which may come over.

[3] It is slightly preferable to use about 20 grms. of mercury instead of the copper, as the action is more regular; but the expense is greater.

phate, and heat cautiously. The action scarcely commences until the boiling point of the acid is reached (about 320° C.) and the gas is rendered cloudy at first by particles of the acid which are thrown up as spray; these, however, soon subside, and their presence will not interfere with the experiments. Collect one large and three small bottles of the gas, using a lighted taper, as in the case of hydrogen chloride, to ascertain when the bottles are full.

[Reserve the large bottle for use in Exercise 34.]

The following properties of sulphur dioxide may now be examined ;—

*1. **Its suffocating odour** will have been already sufficiently noticed.

2. **Its high density, more than twice that of air.**

3. **Its relation to combustion.**

These two properties may be illustrated as in the case of hydrogen chloride, by placing a lighted taper in an empty gas jar, and pouring sulphur dioxide upon it.

*4. **Its solubility in water, forming an acid.**

Replace the stopper of a bottle of the gas by a glass plate, invert it, and withdraw the plate under water in the supplementary pan of the pneumatic trough (or a basin). On shaking the bottle so as to wet the sides, the gas will be readily, though not suddenly, absorbed.

In this case, as already noticed, p. 103, an acid, hydrogen sulphite, is formed[1], as may be proved by dipping a piece of blue litmus paper into the solution.

5. **Its combination with metallic oxides or hydrates to form sulphites.**

Pour 10 c.c. of solution of potassium hydrate into a bottle of the gas, inserting a strip of paper between the neck and the stopper, and shake up the solution with the gas. Combination will readily take place, and the odour of the gas will entirely disappear, potassium sulphite being formed[2].

*6. **Its bleaching action on organic colours.**

(*a*) Take 20 c.c. of a rather dilute solution of logwood or

---

[1] $SO_2 + H_2O = H_2SO_3$.

[2] $SO_2 + 2 KHO = K_2SO_3 + H_2O$.

cochineal, or of the dye known as 'magenta' (made by boiling a little of the substance in water for a few minutes); pour about half of it into a bottle of sulphur dioxide, reserving the other half in the tube for comparison, put a strip of paper between the neck and the stopper, and shake the bottle for half a minute: then pour out the liquid into a test tube. It will have nearly or quite lost its colour.

(*b*) Put a deflagrating jar on a plate containing a little water, and place in it some flowers, such as roses, violets and pansies. Heat a little sulphur in a deflagrating cup, until it begins to burn in air, then place it in the jar. The sulphur dioxide produced will more or less quickly take away the colours from the flowers; the reds and the blues being soon bleached while the greens resist a prolonged action of the gas. It does not, however, discharge the colour so completely as chlorine, and if the flowers are dipped in very dilute hydrogen sulphate, and left for some hours, their colour will be in most cases restored.

32. SULPHITES.

[Typical examples,   Hydrogen sulphite (H_2SO_3.)
                Sodium sulphite (Na_2SO_3).

A solution of sodium sulphite (freshly made) containing 1 grm. of the salt in 40 c.c. of water may be used.]

*1. They give off sulphur dioxide when acted on by acids.

Add a few drops of strong hydrogen sulphate to a little of the solution of sodium sulphite placed in a test-tube, and warm the mixture. Sulphur dioxide will be given off, and may be recognised by its smell. If a slip of paper moistened with solution of lead acetate is held within the tube it will not be altered in colour.

2. They give up oxygen when acted on by nascent hydrogen, forming hydrogen sulphide.

Pour 5 or 6 c.c. of dilute hydrogen sulphate upon a bit of granulated zinc in a test-tube; hydrogen will, of course, be given off. Now add a drop (not more) of solution of sodium sulphite, and test the escaping gas with moist lead acetate paper, which will be blackened. The hydrogen has decomposed the hydrogen sulphite (formed by double decomposition when the sodium sulphite was added to the acid), with formation of water and hydrogen sulphide [1].

*3. Their solutions readily absorb more oxygen, with formation of sulphates.

This property renders hydrogen sulphite one of the most useful reducing agents for laboratory use: but solutions of sulphites cannot be kept long unchanged since they absorb oxygen from the air.

(*a*) To some of the solution of sodium sulphite add a drop or two of solution of silver nitrate. A white precipitate of silver sulphite will be formed, which will on being warmed turn gray owing to reduction of the silver to the metallic state: part of it being deposited as a bright mirror on the sides of the tube.

(*b*) Acidify a little of a very dilute solution of potassium iodide with hydrogen chloride, and add one drop of chlorine water, or of solution of potassium nitrite in order to liberate iodine, which, on addition of some solution of starch, will give a deep blue solution (p. 205). Add to this a few drops of solution of sodium sulphite; the solution will immediately become colourless, showing that the iodine has entered into combination. The change really consists in the decomposition of water; its oxygen uniting with the hydrogen sulphite, while its hydrogen unites with the iodine to form hydrogen iodide [2].

33. HYPOSULPHITES.

[Typical example,—Sodium hyposulphite ($Na_2S_2O_3$).

A solution of sodium hyposulphite containing 1 grm. of the salt in 20 c.c. of water may be used.]

These are of importance, both in volumetric analysis and in photography. A reference to the formula will show that the

[1] $H_2SO_3 + 3H_2 = 3H_2O + H_2S.$
[2] $I_2 + H_2O + H_2SO_3 = 2HI + H_2SO_4.$

molecule of a hyposulphite only differs from that of a sulphite in containing one atom more of sulphur. This addition is easily effected by digesting sulphur for some time with a solution of a sulphite.

*1. They are decomposed by heat, giving off sulphur.

Heat a small quantity of crystallised sodium hyposulphite in an ignition tube. It will fuse and give off water of crystallisation (which should be absorbed by a twisted strip of blotting-paper); and finally a yellow sublimate of sulphur will form in the tube.

*2. They are decomposed by acids, giving off sulphur dioxide and depositing sulphur.

Add to a portion of the solution of sodium hyposulphite a few drops of strong hydrogen sulphate, and warm gently. Sulphur dioxide will be given off (recognisable by its smell, &c.) and a milky deposit of sulphur will be formed. (Compare the reaction of sulphites, Expt. 1, in the last exercise, in which no deposit of sulphur was formed.)

*3. Their solutions readily absorb more oxygen.

This property (which renders them of great use in volumetric analysis, especially as their solutions are much more stable than those of sulphites) may be illustrated by adding a few drops of a solution of sodium hyposulphite to a solution containing free iodine made as in Expt. 3 *b* of the last exercise. The colour will disappear, as in the case of sulphites, and for a similar reason.

*4. They give a white precipitate, quickly turning black, when tested with silver nitrate.

To another portion of the solution add some solution of silver nitrate. A precipitate is produced (soluble in excess of the hyposulphite), which is at the first moment white, but rapidly becomes yellow, brown, and finally black. These changes are due to the instability of the silver hyposulphite which is first precipitated; in presence of water it is rapidly converted into silver sulphide [1].

5. Their solutions dissolve many compounds of silver.

This is due to their tendency to form double salts of silver

---

[1] $Ag_2S_2O_3 + H_2O = Ag_2S + H_2SO_3$.

which are soluble in water. For instance,—Form a little silver chloride by mixing a few drops of solutions of silver nitrate and ammonium chloride, wash it twice by decantation, then pour upon it some solution of sodium hyposulphite. This will form silver hyposulphite, which will unite with more of the sodium hyposulphite to form a double salt, sodium-and-silver hyposulphite, which dissolves readily in water. (The use of the hyposulphites in photography will be illustrated under SILVER.)

34. SULPHATES.

[Typical examples,—Hydrogen sulphate (H_2SO_4).
Sodium sulphate (Na_2SO_4).
Calcium sulphate $(CaSO_4)$.]

1. Hydrogen sulphate.

The details of the English process for making hydrogen sulphate will be found in the text-books on general chemistry. The principle of the process may be illustrated on a small scale as follows :—

Take the bottle of sulphur dioxide reserved in Ex. 31, pour into it 2 or 3 drops of strong hydrogen nitrate, and shake it so as to wet the sides. Orange vapours will be formed, consisting of nitrogen tetroxide ; the sulphur dioxide taking away oxygen from the hydrogen nitrate to form hydrogen sulphate and the lower nitrogen oxide[1]. If the sulphur dioxide is moderately free from moisture, part of it unites with hydrogen nitrate to form a white crystalline substance, and this when shaken up with 2 or 3 c.c. of water is decomposed into hydrogen sulphate and nitrogen tetroxide, as above.

[In actual practice a large excess of sulphur dioxide is used and this reduces the nitrogen tetroxide still further, viz. to nitrogen dioxide, which latter (as has been proved, p. 144) combines readily with the oxygen of the air to form the tetroxide. And it is easy to see that, if more sulphur dioxide and water are supplied, the

---

[1] $2 HNO_3 + SO_2 = H_2SO_4 + N_2O_4$.

nitrogen tetroxide would be again reduced to the dioxide, more hydrogen sulphate being formed. Thus by continually supplying sulphur dioxide water and air to a small quantity of hydrogen nitrate, an indefinitely large quantity of hydrogen sulphate may be produced.]

The solution in the bottle should be examined by adding to it a drop of solution of barium chloride to prove that a sulphate has really been formed; if so, a white precipitate will be obtained (see Expt. 1, p. 230).

Properties of Hydrogen sulphate.

1. Its high density (1.84).
This has been already examined, p. 49.
2. Its strong acid reaction.
Put a single drop of hydrogen sulphate into 100 c.c. of water, and observe that even this extremely dilute solution reddens litmus paper.
3. Its intense affinity for water.
This is shown by the heat evolved when the acid is diluted with water, which will have been sufficiently observed already (see pp. 132, 185).

Owing to this property it decomposes many organic substances which contain the elements of water, combining with the latter and separating carbon, as the following experiments will show.

(*a*) Dissolve sufficient white sugar (about 3 ordinary lumps) in 20 c.c. of water to form a syrup, place it in a beaker standing on a plate, add to it an equal volume of strong common hydrogen sulphate, and stir the mixture. It will become hot and blacken, and will finally swell up into a porous pasty mass of charcoal. Sugar is a combination of carbon with hydrogen and oxygen, the two latter being present in the proportions in which they form water, and hydrogen sulphate combines with these latter, leaving the carbon.

(*b*) Dip the end of a glass rod into dilute hydrogen sulphate (the ordinary laboratory solution), and draw letters with it upon

a piece of writing paper; then dry the paper by holding it
before a fire, or at a little distance above a lamp. The acid,
as it becomes concentrated, will act upon the paper (which is
of nearly the same chemical constitution as sugar) in the
manner above explained, and the letters will appear in black,
owing to the separation of carbon.

4. **Its action on substances, forming sulphates.**

Several illustrations of this will have been noticed already in
the course of work; for instance, its action on zinc (p. 104),
on potassium nitrate (p. 133), on sodium chloride (p. 184). One
other example of its action may be noticed, viz. its action on
metal-oxides, which is applied to the preparation of numerous
salts.

Put a little copper oxide into a test tube, add 2 or 3 c.c. of
water, and then the same volume of hydrogen sulphate, and
heat it to boiling over a lamp. The copper oxide will dissolve,
copper sulphate and water being formed[1]. If the blue solution is concentrated by evaporation in the usual way (p. 55),
crystals of copper sulphate (blue vitriol) will be obtained.

Tests for sulphates.

[A solution of calcium sulphate (the ordinary laboratory solution),
although it is scarcely stronger than lime water, will answer for the
following tests.]

*1. **They give a white precipitate, insoluble in acids, when
tested with barium chloride.**

Add to a portion of the solution of calcium sulphate a drop
or two of solution of barium chloride. A white precipitate
of barium sulphate will be formed (even in this weak solution)
which will not dissolve on the addition of a little strong hydrogen
chloride, even on boiling the mixture.

This is the most delicate and characteristic test for sulphates.

*2. **They give a white precipitate when tested with lead
acetate.**

---

[1] $CuO + H_2SO_4 = CuSO_4 + H_2O.$

Test another portion of the solution with a drop or two of solution of lead acetate. A white precipitate of lead sulphate will be formed, nearly insoluble in cold dilute hydrogen chloride, but decomposed and dissolved by it on boiling.

*3. They are decomposed when heated with carbon, forming sulphides.

Mix together a very little calcium sulphate (plaster of Paris) with about twice as much sodium carbonate in a mortar ; place some of the finely-powdered mixture in a cavity made in a piece of charcoal (see p. 90), and heat it strongly in the reducing blowpipe-flame for about a minute. The calcium sulphate will thus be reduced to sulphide [1], and this is decomposed by the sodium carbonate, forming sodium sulphide. When it is cool, detach the ignited mass from the charcoal, place it in a test-tube, and pour on it a little dilute hydrogen chloride. The presence of a sulphide will be shown by the evolution of hydrogen sulphide, recognisable by its smell and the blackening of a piece of blotting-paper moistened with solution of lead acetate.

35. PHOSPHORUS.

[Formula of molecule, P_4.
Weight of molecule, 124 hydrogen-atoms.]

[It is hardly necessary to repeat that whenever phosphorus is to be used for an experiment, it must be handled with the greatest care, owing to the low temperature at which it catches fire. It should be cut under water and should never be touched by the fingers when dry ; a pair of crucible tongs being used to transfer it from one place to another. In cutting it, not even the smallest piece should be allowed to get under the nails or to remain upon the table or floor. All residues, after the experiments are over, should be covered with water until they can be thrown away in a safe place. A jug of water should always be at hand, in case the phosphorus should, in spite of precautions, catch fire [2].]

---

[1] $CaSO_4 + C_2 = CaS + 2CO_2$.
[2] As the burns produced by phosphorus are usually severe and slow to heal, it may be worth mentioning that lint soaked in water containing a

The properties of phosphorus which can be safely tried are the following :—

1. **It readily dissolves in carbon disulphide.**

Place about 2 c.c. of carbon disulphide in a test-tube, add a small fragment of dry phosphorus about half the size of a pea, and cork the tube. Leave it to digest for a few minutes, until wanted for use in Expt. 3, shaking it occasionally to promote the solution of the phosphorus.

2. **It melts easily (at 44°) and catches fire when heated slightly above its melting point.**

Place a very small fragment of phosphorus (not larger than a grain of wheat), carefully dried by blotting-paper, on a watch-glass, and float it on hot water in a basin. The phosphorus will melt and inflame spontaneously, giving off white fumes of phosphorus pentoxide, and leaving a small residue consisting of 'red' phosphorus, an allotropic form of the element.

3. **It inflames spontaneously in air when in a finely divided condition.**

Pour the clear solution obtained in Expt. 1 upon a piece of blotting-paper placed in a plate (best in a draught cupboard). The phosphorus will, as the carbon disulphide evaporates, be left in a state of fine division upon the paper; it will emit white fumes, and appear faintly luminous in the dark, and will shortly catch fire and burn away. The paper itself will be found to be only superficially charred, owing to the protective action of the phosphorus pentoxide formed.

4. **It reduces many metals from their salts.**

Place a small bit of phosphorus in a test-tube, pour on it about 1 c.c. of solution of silver nitrate, and leave it undisturbed for a few hours. A bright crystalline coating of metallic silver will be formed on the phosphorus, the latter having decomposed the nitrate owing to its affinity for oxygen, while the metal is separated.

'Red' or 'amorphous' phosphorus, the allotropic form ob-

little common washing soda is the best application at first. After the hydrogen phosphate is by this means neutralised, nothing is better than lint soaked in a mixture of equal volumes of glycerine and water, the whole being covered with oilskin to prevent evaporation.

tained by the long-continued action of heat on ordinary phosphorus, differs remarkably from the latter, especially in having much less strong chemical affinities. Thus,—

Put a little red phosphorus on a watch-glass floating on nearly boiling water, as. in Expt. 2. It will not melt or catch fire, or show any alteration. If, however, it is touched with a red-hot wire, it will inflame, burning with the same bright light, and emitting the same white fumes of the pentoxide as ordinary phosphorus.

It is, in fact, reconverted into the ordinary transparent form by a high temperature, as may be proved in the following way :—

Place a little red phosphorus in an ignition-tube, connect the outer end of the tube by an india-rubber connector with a glass jet (the object being to check admission of air), and heat the phosphorus very carefully over a lamp. It will sublime without previously melting, and will condense in the cooler part of the tube in amber-coloured transparent drops, which are easily recognised as being ordinary phosphorus.

COMPOUNDS OF PHOSPHORUS WITH HYDROGEN AND OXYGEN.

36. HYDROGEN PHOSPHIDE.

[Formula of molecule, H_3P.
Weight of molecule, 34 hydrogen-atoms.]

37. HYPOPHOSPHITES.

[Typical example,—Potassium hypophosphite (KH_2PO_2).]

Both hydrogen phosphide and potassium hypophosphite are formed by heating phosphorus with a solution of potassium hydrate. In this case the water is decomposed; part of its hydrogen unites with some phosphorus to form hydrogen phosphide, while the rest of the hydrogen together with the oxygen unites with potassium and phosphorus to form potassium hypophosphite [1].

The following experiment in illustration of this must be made in a draught-cupboard or in the open air, since hydrogen phosphide is not only an extremely poisonous and offensive-smelling gas, but also spontaneously inflammable.

Put some solution of potassium hydrate (5 grms. dissolved in 20 c.c. of water) into a small beaker, add a piece of phosphorus about as large as a pea, and heat the solution nearly, but not quite, to boiling on a sand bath. Hydrogen phosphide will be given off, as above explained, and each bubble as it reaches the surface of the liquid will catch fire spontaneously, combining with the oxygen of the air, and forming a white smoke-wreath of hydrogen phosphate [2].

After allowing the action to go on for about half an hour, set the beaker aside (in the draught-cupboard) to cool. When it is quite cool, decant the contents into a test-tube, and immediately fill up the beaker with water, to prevent the remaining phosphorus catching fire before it can be thrown away. Add to the solution of potassium hypophosphite enough dilute hydrogen sulphate to render it slightly acid to test paper (this will cause a white precipitate of potassium sulphate which should be filtered off), and use it for the following experiment, which illustrates one of the chief characteristics of hypophosphites, viz.

* **Their reducing action on copper salts.**

Add to a portion of the solution a few drops of solution of copper sulphate, and heat the mixture gradually to boiling. A greenish-yellow turbidity will soon appear, rapidly changing to a dull red precipitate of metallic copper.

This reaction serves to distinguish hypophosphites from phosphites and phosphates, neither of which latter reduce copper salts. The yellow substance formed at first is copper

[1] $3 KHO + 3 H_2O + P_4 = H_3P + 3 KH_2PO_2.$
[2] $H_3P + 2 O_2 = H_3PO_4.$

hydride, and is one of the few examples of the combination of a metal with hydrogen [1].

38. PHOSPHORUS PENTOXIDE.

[Formula of molecule, P_2O_5.
Weight of molecule, 142 hydrogen-atoms.]

39. PHOSPHATES.

[Typical examples,—Hydrogen phosphate (H_3PO_4).
Sodium phosphate (Na_3PO_4).
Sodium and hydrogen phosphate (Na_2HPO_4).
Sodium-ammonium-and- } ($Na(H_4N)HPO_4$.)
hydrogen phosphate }

Phosphorus pentoxide is always produced when phosphorus is burnt in a full supply of oxygen or air. It has an extremely strong affinity for water; so much so that it cannot be exposed for even a few minutes to ordinary air without combining with the moisture present and becoming liquid. The substance thus formed is hydrogen phosphate (ordinary 'phosphoric acid'), and from it other phosphates can be obtained in the usual ways. For instance, by addition of just sufficient sodium carbonate to neutralise it, part of the hydrogen is replaced by sodium, and thus sodium and hydrogen phosphate (common 'phosphate of soda') is obtained. In illustration of these facts,—

Place a small piece (about twice as large as a pea) of carefully dried phosphorus upon a plate: set fire to it with a match, and cover it with a deflagrating jar, slightly tilted by a short bit of glass rod or tube placed under its lower edge, so as to allow free access of air. The stopper should be laid across the mouth in order to check the upward draught, which would carry off a portion of the product. Dense white fumes of phosphorus pentoxide will be formed, and will be deposited partly on the sides of the jar, partly on the plate. Remove the

[1] $4\,CuSO_4 + 3\,H_3PO_3 + 6\,H_2O = 3\,H_3PO_4 + 4\,H_2SO_4 + 2\,Cu_2H_2$.

jar and scrape together, with a spatula, the snow-white powder on the plate. Pour a few drops of water upon it; it will hiss like a red-hot iron from the intensity of the combination, and a solution of hydrogen phosphate will be formed. A piece of blue litmus-paper dipped into the solution will be at once reddened. Notice also that the phosphorus pentoxide deposited on the sides of the jar will, after a few minutes exposure to the air, deliquesce into an oily liquid. From its affinity for water, which has been thus illustrated, phosphorus pentoxide is one of the most efficient substances for drying gases. Rinse the contents of the jar and the plate into a porcelain dish, and evaporate it until the liquid becomes syrupy, then add a little more water, and boil the solution again for a short time; finally filter it, heat the greater portion of the filtrate again in the dish (reserving a little, to be added in case too much sodium carbonate may be inadvertently mixed with the rest) and add solution of sodium carbonate to the hot liquid, until the addition of it no longer produces effervescence and the liquid is neutral to test-paper. You have thus obtained a solution of sodium-and-hydrogen orthophosphate, which may be examined by tests 2 and 3.

Properties of Orthophosphates.

[A solution of sodium-and-hydrogen phosphate, containing 3.5 grms. of the salt dissolved in 50 c.c. of water, may be used.]

*1. **They give a white precipitate, soluble in hydrogen acetate, when tested with calcium sulphate.**

Add to a little of the solution of sodium-and-hydrogen phosphate about half its volume of solution of calcium sulphate. A white precipitate of calcium phosphate will be formed, which will readily dissolve on addition of a little hydrogen acetate (this distinguishes it from calcium oxalate, p. 176).

*2. **They give a yellow precipitate when warmed with excess of a strongly acid solution of ammonium molybdate.**

Pour some solution of ammonium molybdate, acidified with hydrogen nitrate, into a test-tube, and add a drop or two of the solution of sodium phosphate. The solution will become yellow, and on warming it (it should not be boiled) a yellow precipitate will fall.

This is an extremely delicate and characteristic test for phosphates, if properly used. The precautions necessary in applying it are, (*a*) the solution of ammonium molybdate must be used in excess, (*b*) it must be *strongly* acid, (*c*) it should not be heated to boiling, otherwise a precipitate of hydrogen molybdate is liable to form.

*3. They give a white crystalline precipitate when tested with magnesium sulphate, in presence of ammonium salts.

To another portion of the solution of sodium phosphate add one-fourth its volume of solution of ammonium chloride, then a little ammonia, and test it with a drop of solution of magnesium sulphate. A white crystalline precipitate of magnesium and ammonium phosphate will be formed, which will readily dissolve in dilute hydrogen nitrate.

4. They give a yellow precipitate, soluble in hydrogen nitrate, when tested with silver nitrate.

To another portion add a drop of solution of silver nitrate. A canary-yellow precipitate of silver phosphate will be formed. Pour off one half of the fluid in which the precipitate is suspended into another test-tube, and add a few drops of dilute hydrogen nitrate; to the remaining half add the same quantity of ammonia; in both cases the precipitate will be redissolved.

*5. They give a yellowish white precipitate, insoluble in acetates, when tested with iron perchloride.

Add to another portion of the solution about one-third its volume of solution of sodium acetate [1] and then one drop (not more) of solution of iron perchloride. A dull white precipitate of iron phosphate will be produced, which will dissolve when excess of iron perchloride is added (hence the reason for only adding a little at first).

This reaction is important, since it enables us to decompose,

---

[1] If this is not at hand, a little potassium acetate (made by adding a slight excess of hydrogen acetate to about 2 c.c. of solution of potassium hydrate) will do quite as well.

in the course of an analysis, certain phosphates (such as those of barium, strontium, and calcium) which are soluble in an acid, but are reprecipitated unchanged when the solution is neutralised.

Additional experiments.

The action of phosphorus pentoxide on water is not very simple. A molecule of it may combine with 1, 2, or 3 molecules of water forming three distinct acids,—hydrogen metaphosphate, hydrogen pyrophosphate, and hydrogen orthophosphate[1], from each of which series of salts are derivable, having very different properties. For a full account of these a general text-book must be consulted; it will be sufficient to observe here that the salts examined above belong to the series of orthophosphates, and that from them pyrophosphates and metaphosphates can be obtained by selecting a salt from which it is possible to drive off just the requisite quantity of the elements of water by heat.

Pyrophosphates.

[Typical example,—Sodium pyrophosphate ($Na_4P_2O_7$).]

Sodium pyrophosphate is obtained by heating the ordinary sodium-and-hydrogen orthophosphate, with which experiments were made just now. When strongly heated, two molecules of this latter salt give off one molecule of water, leaving a residue of the pyrophosphate.

Place a few crystals of the salt in a platinum capsule, or in a cup of platinum foil made as directed, p. 8, and heat it gently at first, afterwards to redness. The salt will melt and give off its water of crystallisation, and at a higher temperature two molecules of it will give off one molecule of water, leaving a white earthy-looking residue consisting of sodium pyrophosphate[2]. This should be powdered, dissolved in about 40 c.c. of *cold* water, and examined by tests 3 and 4 in exactly the same way as directed in the case of the orthophosphate. The following results should be obtained:—

3. No precipitate will be obtained (if enough solution of ammonium chloride has been added): but on warming the liquid a white precipitate of magnesium pyrophosphate will fall.

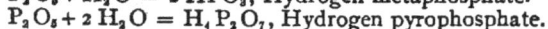

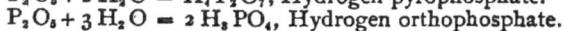

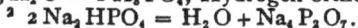

---

[1] $P_2O_5 + H_2O = 2\,HPO_3$, Hydrogen metaphosphate.
$P_2O_5 + 2\,H_2O = H_4P_2O_7$, Hydrogen pyrophosphate.
$P_2O_5 + 3\,H_2O = 2\,H_3PO_4$, Hydrogen orthophosphate.
[2] $2\,Na_2HPO_4 = H_2O + Na_4P_2O_7$.

4. A perfectly white granular precipitate of silver pyrophosphate will be obtained, soluble in hydrogen nitrate and ammonia.

Metaphosphates.

[Typical example,—Sodium metaphosphate ($NaPO_3$).]

To obtain this it is necessary to drive off a larger amount of the elements of water from an orthophosphate, and hence a salt must be selected which contains more hydrogen as a basic radicle, and less of a non-volatile radicle, such as sodium. Such a salt is sodium-ammonium-and-hydrogen phosphate, or microcosmic salt, the formula of which has been given already, p. 235.

Heat a little of this salt in a platinum cup, exactly as directed in the preparation of the pyrophosphate. The substance froths up very much, and gives off ammonia and water [1]; finally a transparent, glassy residue of sodium metaphosphate will be left. This should be powdered, dissolved in 50 c.c. of *cold* water, and examined by tests 3 and 4 as above.

3. No precipitate; if enough ammonium chloride has been added.

4. A white precipitate, soluble in hydrogen nitrate and ammonia.

It will be seen that these results scarcely distinguish metaphosphates from pyrophosphates, but the following test is highly characteristic of the metaphosphates; viz. their property of coagulating albumen (white of egg).

Add to some of the solution of sodium metaphosphate 6 or 8 drops of hydrogen acetate, and then a few drops of a clear solution of albumen. A white precipitate of coagulated albumen will form at once.

Conversion of hydrogen metaphosphate and pyrophosphate into the orthophosphate.

This change is easily effected by boiling either of them with water; the additional amount of water being quickly taken up. To illustrate this,—

Add a few drops of hydrogen nitrate to some of the solution either of sodium pyrophosphate or of sodium metaphosphate (this

[1] $Na(H_4N)HPO_4 = H_3N + H_2O + NaPO_3$.

will liberate the corresponding 'acid), and boil for ten minutes in a beaker. Hydrogen orthophosphate will be reproduced, as may be proved by neutralising the solution with sodium carbonate and examining it by tests 3 and 4.

40. BORATES.

[Typical examples,—Hydrogen borate (H_3BO_3).
Sodium diborate ($Na_2B_4O_7$).]

These are almost the only compounds of boron which are commonly met with. They are related to boron trioxide, which may be obtained from them, as the following experiment will show.

Make a concentrated solution of sodium diborate (borax) by boiling about 8 grms. of the crystals with 10 c.c. of water in a test-tube until no more is dissolved. Pour off the warm solution from the crystals, add about half its volume of strong hydrogen chloride, and cool the mixture by holding it in a stream of water. Shining scales of hydrogen borate will be precipitated, which may be separated by filtration from the solution of sodium chloride, and left to drain on a porous tile[1]. Meanwhile the other experiments may be proceeded with.

When the hydrogen borate is moderately dry, place a little of it on a piece of platinum foil formed into a cup (p. 8) and heat it over a lamp. It will fuse, give off water, and leave a clear, glassy residue of boron trioxide.

Tests for borates.

[A solution of sodium diborate, containing 1.5 grm. of the salt in 20 c.c. of water, may be used.]

*1. They give a white precipitate, soluble in ammonium chloride, when tested with calcium sulphate.

---

[1] This is often a useful method of drying substances, the solution being quickly drawn by capillary action into the pores of the tile. After use, the tile should be soaked for some time in several changes of water, and then dried in an oven.

Add to a portion of the solution of sodium diborate about half its volume of solution of calcium sulphate. A white precipitate of calcium borate will be formed, which will readily dissolve on addition of a little solution of ammonium chloride (resembling calcium tartrate, p. 174, in this respect).

*2. They impart a green colour to flame.

This cannot well be observed in the presence of volatile substances, such as sodium, which themselves impart a colour to flame. Hence in applying the test, hydrogen sulphate must be first added, to retain such substances in a less volatile condition than the hydrogen borate produced by double decomposition.

Place about 1 c.c. of the solution of sodium diborate in a porcelain dish, mix it with a few drops of strong hydrogen sulphate, then add about three times the volume of common alcohol, stir the mixture with a glass rod and set fire to it. The edges of the flame will be tinged yellowish green, owing to the presence of hydrogen borate, liberated by the action of the hydrogen sulphate, and volatilised in the alcohol vapour.

*3. Their solutions, when rendered acid, tinge turmeric paper red.

Place a little solution of sodium diborate in a watch-glass, and add enough hydrogen chloride to make the solution give an acid reaction to test-paper. Dip into the solution one-half of a slip of turmeric-paper, and dry it at a gentle heat, by holding it before the fire; the dipped portion will be coloured red. If a drop of solution of sodium carbonate be placed on the reddened portion of the turmeric-paper, the colour will be changed to bluish-black.

[For experiments illustrating the use of borax in blowpipe work, see pp. 92, 93.]

41. SILICON DIOXIDE.

[Formula of molecule, SiO_2.
Weight of molecule, 60 hydrogen-atoms.]

42. SILICATES.

[Typical examples,—Hydrogen silicate (H_2SiO_3).
Sodium silicate (Na_2SiO_3).
Aluminium silicate ($Al_2Si_2O_7$).]

A very large number of minerals consist of or contain these compounds of silicon, and they are often difficult to decompose and identify.   A few are acted on by strong hydrogen chloride, which separates hydrogen silicate as a gelatinous precipitate and combines with the basic radicle to form a chloride.   But the great majority of silicates are best decomposed by fusion with excess of sodium carbonate, which forms sodium silicate by double decomposition, while the basic radicle is converted into a carbonate or oxide.

Preparation of Sodium silicate.

To prepare this, any one of the numerous forms of silicon dioxide, such as flint[1] or white sand, may be strongly heated with sodium carbonate.   In this action, the silicon dioxide simply takes the place of carbon dioxide, which latter gas escapes with effervescence[2].

(*a*) Make a ring, as described in p. 92, at the end of a piece of platinum wire; moisten it with water, dip it into some powdered sodium carbonate, and fuse the adhering salt in the hottest part of the blowpipe flame.   It will give a bead which

---

[1] If flint or quartz is to be used, it should be previously heated to full redness in a fire and quickly dropped into water.   It will then form an opaque white mass, which can be more easily pulverised.

[2] $Na_2CO_3 + SiO_2 = Na_2SiO_3 + CO_2$.

is clear while hot, but opaque when cold. Now add to this bead some finely powdered white sand or flint (silica), and again heat it to fusion; an effervescence will take place, owing to the liberation of carbon dioxide, as above explained. You will, when enough silica has been added, obtain a bead which retains its transparency when cold.

A sodium silicate, such as you have here obtained, is the basis of ordinary glass; and if the bead is dipped into solution of cobalt nitrate and again strongly heated, it will be coloured blue, illustrating the mode in which various colours are imparted to glass by fusion with the oxides of certain metals.

(*b*) Although ordinary glass is practically insoluble in water, yet if excess of an alkali is used in preparing it, the resulting silicate dissolves without difficulty in water; forming, in fact, the 'soluble glass' which is now much used in varnishes and paints.

Grind about 1 grm. of flint (previously ignited, see note, p. 242) or of white sand to an extremely fine powder in a porcelain mortar, add to it about 4 grms. of sodium carbonate[1], and grind the whole thoroughly together. Put some of the mixture (about enough to half fill it) into a platinum capsule or a piece of platinum foil formed into a cup, and heat it to full redness over a gas blowpipe. It is an advantage to lay over the capsule a bit of platinum foil as a cover, in order to check loss of heat by radiation. When the mass ceases to swell up (owing to evolution of carbon dioxide as above explained) add some more of the mixture, and heat again until all action is over, and a fused glassy residue remains in the capsule. This should, when cool, be detached from the capsule (which may generally be done by pouring a little water into the capsule and boiling it rather quickly over a lamp), and ground with a little water in a mortar, then transferred to a beaker and boiled for some time with about 20 c.c. of water. It should almost entirely dissolve, if the fusion has been properly effected; and

---

[1] It is preferable to use a mixture of potassium and sodium carbonates (2.6 grms. of potassium carbonate with 2 grms. of sodium carbonate) if it is at hand; since this mixture fuses at a much lower temperature than sodium carbonate.

the solution of sodium silicate may be used for some of the following experiments.

Tests for Silicates.

*1. They are, in some cases, decomposed by strong acids, giving a gelatinous precipitate of hydrogen silicate.

(*a*) Add a few drops of hydrogen chloride to a portion of the solution of sodium silicate, obtained in the last experiment. A dense gelatinous precipitate will be formed, and if the solution of sodium silicate is strong the whole will appear to solidify.

(*b*) Repeat the last experiment, using a much less strong solution of sodium silicate (about 5 or 6 drops of the original solution mixed with 5 c.c. of water). In this case no precipitate will be formed. Now add a little of the clear liquid to some solution of ammonium chloride mixed with one-fourth its volume of solution of ammonia, and apply heat. A gelatinous precipitate will now be formed; since hydrogen silicate, though soluble in water under certain conditions, is quite insoluble in solution of ammonium chloride.

Hence in testing for silicates, the acidified solution should always be added to a mixture of ammonium chloride and ammonia, and heated, before deciding that no silicate is present. .

*2. They are decomposed by fusion with sodium carbonate.

This will usually be found the best way of dealing with insoluble silicates, even though they may be slowly decomposed by acids.

Grind a small bit of slate or earthenware to very fine powder in a mortar, mix it intimately with about 4 times its weight of sodium carbonate, and fuse the mixture in a platinum cup precisely as directed above, in the preparation of sodium silicate. After detaching the fused mass from the crucible, and grinding it with a little water in a mortar, transfer it to a porcelain dish, add about 10 c.c. of dilute hydrogen chloride, and evaporate the whole to perfect dryness on the sandbath. The hydrogen silicate which is separated by the action of the hydrogen chloride, gives off water when heated above 100°, with formation of silicon dioxide which is quite insoluble in water and acids. Continue the heat, stirring the residue until it loses all gelatinous appearance; then add 10 or 12 drops of strong hydrogen chloride and incorporate it with the residue by gentle rubbing with a pestle. By this treatment all the basic radicles (such as aluminium, iron, &c.) are converted into chlorides, while the silicon dioxide remains unaltered. Lastly, warm the residue with a little water, and pour the whole on a filter. The filtrate will contain aluminium, &c. as chlorides, and need not now be regarded. The white residue on the filter consists of silicon dioxide, and should, after being washed (until the filtrate is no longer acid), be tested as follows ;—

(*a*) Boil a portion of it with a little solution of potassium hydrate. It will dissolve pretty readily.

(*b*) Test another portion in a bead of microcosmic salt, as directed in the next experiment. It will float in the bead undissolved.

*3. They are decomposed when heated in a bead of microcosmic salt, silicon dioxide being formed which floats undissolved.

Make a bead of microcosmic salt (sodium-ammonium-and-hydrogen phosphate) in the same way as a borax bead, p. 92[1].

When a perfectly clear bead (of sodium metaphosphate, p. 239) is obtained, add to it a very minute quantity of powdered porcelain, slate, or brick, and heat it for half a minute as strongly as possible. The substance will be decomposed, and a white, semitransparent residue of silicon dioxide will be seen floating in the clear globule. It will be best to examine this with a magnifier while the bead is hot, since in many cases it becomes turbid when cold.

---

[1] This requires rather more care than in making a borax bead, since microcosmic salt becomes so liquid during its decomposition that it is apt to drop off the ring. It should be heated gently at first, until the white fumes of ammonia have gone off, and the ring should be held so that its plane is horizontal, not vertical, in the flame.

Additional experiment.

Preparation of Hydrogen silicofluoride.

[Formula of molecule, H_2SiF_6.]

The reaction by which this substance (also known as 'hydro-fluosilicic acid') is formed, viz. the decomposition of silicon fluoride by water, has been tried on a small scale already, p. 212. A solution of it may now be made on a larger scale, since it is occasionally used as a test for barium, and moreover the process illustrates a good method of acting upon difficultly decomposable silicates, and of obtaining pure silica.

Apparatus required—Florence flask; glass tubing, 7 or 8 mm. in diameter; corks; cork-borers; fish-tail burner; argand burner; retort-stand; wire-gauze; gas jar, 20 x 5 cm.; tube-funnel; funnel, 10 cm. in diameter; filter, or circular piece of linen, about 20 cm. in diameter; bottle, holding 200 c.c.; test-tube; fluor spar; sand; hydrogen sulphate; mercury; strontium nitrate; barium chloride; distilled water.

Take a piece of glass tubing about 60 cm. long, and of rather larger internal diameter than the ordinary gas delivery tube. Having slightly rounded the ends in the gas-flame, bend it twice to a right angle in the same direction, making the first bend about 10 cm. from one end, and the second about 30 cm. from the other end, so as to give the tube the following shape, ⊓. The end of the longer branch should be bevelled off, or notched with a file, to enable the bubbles of gas to escape more freely.

Select a Florence flask of stout glass (since it is liable to be corroded by the hydrogen fluoride, and will soon give way if the bottom is thin), fit the shorter branch of the delivery tube to it, by a perforated cork, and support it in the retort-stand with a piece of wire-gauze under it. Place under the outer extremity of the delivery tube a cylindrical gas jar containing a stratum of mercury about 12 mm. deep, into which the end of the delivery tube must dip. This stratum of mercury is quite essential, since it acts as a valve and prevents the entry of water into the tube, which would then soon be obstructed by the silica which is one of the products of the decomposition of the silicon fluoride.

Reduce 10 grms. of fluor spar to a coarse powder in the mortar, mix it intimately with twice its weight of powdered glass or fine sand (that which is called 'silver sand' is the best), and transfer the mixture to the flask. Pour on it through a funnel 60 c.c. of strong hydrogen sulphate, and shake the whole with a circular motion until it is thoroughly incorporated. Replace the cork and delivery tube, and after seeing that the end of the latter is completely covered by the mercury, fill up the jar or test-glass with distilled water, pouring it gently down the sides of the vessel so that none may enter the delivery tube through a disturbance of the surface of the mercury. Heat the flask very slowly with an argand burner or spirit lamp. The reaction between the fluor spar and the acid requires only a moderate temperature, and if too much heat be applied, the mixture froths up inconveniently, and may pass over into the water [1]. When the gas begins to come over, each bubble as it emerges from the mercury becomes coated with a film of silica, which is left behind when the bubble reaches the surface of the water, as a shrunken, collapsed bag [2].

Increase the heat slightly if the stream of silicon fluoride becomes slow, and continue it until no more gas comes over [3]: then loosen the cork and take the delivery tube quickly out of the solution by raising the retort-stand, and filter the liquid from the gelatinous silica through a paper, or more conveniently, a fine linen filter; care being taken that none of the mercury falls on the filter. [The silica thus obtained is very pure, and hence may be worth the trouble of washing, an operation which takes some time, and must be done very thoroughly. After a final rinse with warm water, the filter containing the silica may be left to dry, either spontaneously or in a hot-air cupboard.]

It is best to keep the solution of the acid, not in glass, but in a gutta-percha bottle, since the former is always acted upon to a certain extent. Before using it as a test for barium, examine it by the two following tests, to make sure that it contains no hydrogen sulphate, which may have been carried over mechanically, if the action has proceeded too rapidly.

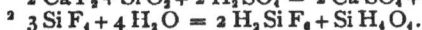

---

[1] $2\,CaF_2 + SiO_2 + 2\,H_2SO_4 = 2\,CaSO_4 + 2\,H_2O + SiF_4$.

[2] $3\,SiF_4 + 4\,H_2O = 2\,H_2SiF_6 + SiH_4O_4$.

[3] Do not, however, apply a strong heat, or hydrogen sulphate may distil over, and render the solution useless for analytical purposes. If a strong solution of hydrogen silicofluoride be wanted, it is a good plan to set up another similar apparatus, with a delivery tube leading into the same jar of water. The time necessary to obtain the required solution is thus diminished by one half.

(*a*) Dissolve a few small crystals of strontium nitrate in 10 c.c. of water, and add to the clear solution about 5 c.c. of the acid which you have prepared. If a precipitate or turbidity is produced immediately, or within a few minutes, hydrogen sulphate is present.

(*b*) If the result of the above test is satisfactory, add a few drops of the acid to a little solution of barium chloride in a test tube. An immediate crystalline precipitate should be produced; otherwise the solution of the acid is not sufficiently strong, and should have more silicon fluoride passed through it, in the manner above described.

[The student is recommended, before passing on to the study of the different metals, to refer to the account of the course of analysis given at the commencement of Part II, and to practise himself in the analysis of single salts containing some one of the foregoing non-metallic radicles associated with an alkali-metal.]

SECTION III.

PREPARATION AND EXAMINATION OF THE PROPERTIES OF THE PRINCIPAL METALLIC RADICLES AND THEIR COMPOUNDS.

Group 1.

Metals which are separated from solutions by Hydrogen chloride[1].

SILVER, MERCURY, LEAD[2].

1. SILVER.

[Symbol of atom, Ag (argentum).
Weight of atom, 108 hydrogen-atoms.]

[Liquids containing silver, such as the mother liquor from the crystals of silver nitrate and the residues from the following experiments, should on no account be thrown away, but reserved in a bottle labelled ' Silver Residues[3].'

It must be borne in mind that solutions of silver salts, if allowed to touch the skin or clothes, stain it black. The quickest way of removing these stains is to moisten them first with a solution of potassium iodide, and then with strong solution of potassium cyanide. Great care must, however, be taken in using the latter, as it is very poisonous: all traces of it should be washed away as soon as it has removed the stains.]

Preparation of silver nitrate from an alloy such as silver coin.

The English silver coinage consists of silver alloyed with about one-thirteenth its weight of copper. This alloy dissolves

---

[1] For an account of the principle on which the metals are classified into groups, see Part II. Sect. I.

[2] Mercury is only thus separated when in the monatomic condition. Lead is only partially separated, since lead chloride is not quite insoluble in water.

[3] For the method of recovering the silver from these residues, see Appendix B.

readily in hydrogen nitrate, but the silver alone is precipitated as chloride when hydrogen chloride is added to the solution.

Place a small silver coin in a large test-tube, pour on it 8 or 10 c.c. of concentrated hydrogen nitrate, and warm the tube very gently on a sand-bath. A strong action will commence, and nitrogen oxides will be evolved, the fumes of which should not be allowed to escape into the room[1]. In a short time a clear bluish green solution will be formed, containing silver nitrate and copper nitrate. Dilute the solution with four or five times its volume of water, and add about 10 c.c. of common hydrogen chloride. The silver will be completely precipitated as chloride, while the copper chloride, being soluble in water, remains in the liquid. Close the mouth of the tube and shake it for a few seconds : the suspended silver chloride will then separate in curdy masses, and after adding a few more drops of hydrogen chloride to make sure that the precipitation is complete, the liquid may be poured off. The precipitate should next be thoroughly washed on a filter with several changes of water, until the liquid which runs through the filter ceases to redden blue litmus-paper. While the last portions of water are draining off, take a strip of sheet zinc about 1 cm. broad and 12 or 14 cm. long, cover its surface with a thin film of mercury[2] by immersing it for a minute in an acid solution of mercury perchloride (2 c.c. of the ordinary laboratory solution with 20 c.c. of water and 5 c.c. of dilute hydrogen sulphate), and then wash it with clean water. Transfer the silver chloride to a large test-tube, rinsing it through a hole made in the filter : then fill the tube half full of water, add about 5 c.c. of dilute hydrogen sulphate, and put in the strip of zinc, taking care that it comes closely into contact with the mass of silver chloride at the bottom of the tube. A galvanic action will be set up, and

---

[1] The action is similar to that of hydrogen nitrate on copper in the preparation of nitrogen dioxide (see p. 142).
$$3 \, Ag_2 + 8 \, HNO_3 = 6 \, AgNO_3 + 4 \, H_2O + N_2O_2.$$
[2] This is advisable, though not absolutely necessary, since it prevents 'local action' (see under zinc) and avoids the separation of the particles of carbon, lead, &c. which always occur in ordinary zinc, and are rather difficult to separate from the reduced silver.

the hydrogen liberated in contact with the silver chloride will combine with the chlorine and thus the silver will be separated in the metallic state as a fine black powder, which, as the molecules aggregate together into larger masses, will become of a light brown, or gray. In about half an hour the reduction will be complete, and the zinc may then be taken out, and the liquid (which contains hydrogen chloride, as may be proved by testing a small portion of it with silver nitrate) poured off from the spongy mass of reduced silver. Wash the latter thoroughly by decantation, (breaking up the lumps by stirring with a glass rod,) boiling it with several successive changes of water, until the washings are no longer acid to test-paper. Lastly, after pouring off the water, add a little strong hydrogen nitrate, a few drops at a time, and warm the mixture; the metal will readily dissolve, forming silver nitrate. The solution should be filtered (if necessary) after addition of a little water to prevent action on the filter-paper, and then evaporated to the crystallising-point, and left to cool. Flat rhombic crystals of silver nitrate will be formed, which should be placed in a funnel (the neck being partially obstructed by a bit of broken glass or porcelain to prevent the crystals falling through), slightly washed by pouring a little water over them to remove adhering acid, and left to drain and dry.

Tests for Silver salts.

[A solution of silver nitrate containing 1 grm. of the salt in 30 c.c. of water may be used.]

*1. With hydrogen chloride they give a white precipitate, soluble in ammonia.

Put 2 or 3 drops (not more, in order that the delicacy of the test may be observed) of the solution of silver nitrate into a test-tube, dilute with 5 or 6 c.c. of water, and add a drop or two of dilute hydrogen chloride. A white, curdy precipitate of silver chloride will be produced, quickly subsiding if shaken, and turning gray (owing to its reduction) when exposed to daylight.

(*a*) Pour off half the liquid containing the precipitate in suspension into another tube, add a little strong hydrogen chloride and a drop or two of hydrogen nitrate (so as to form aqua regia) and boil the mixture. The precipitate will not dissolve or be altered in any way (being thus distinguished from the precipitate of mercury protochloride, p. 263).

(*b*) To the remaining portion add solution of ammonium hydrate, which will readily dissolve it. If excess of dilute hydrogen nitrate is added to the clear solution, silver chloride will be re-precipitated.

*2. With potassium chromate they give a crimson precipitate.

Add to another portion of the solution a drop or two of solution of potassium chromate. A dark crimson precipitate of silver chromate will be formed, readily soluble in dilute hydrogen nitrate.

*3. Reduced on charcoal before the blowpipe, they yield a brilliant metallic globule, but no incrustation.

Powder a small crystal (no larger than a grain of wheat) of silver nitrate, mix it intimately in the mortar with about twice as much sodium carbonate ; then place some of the mixture in a cavity cut in a piece of charcoal (see p. 90), and heat it strongly in the blowpipe flame, as directed, p. 92. Bright globules of metallic silver will be formed, which may with a little care be fused together into one mass. This will remain untarnished in the oxidising flame, and no incrustation will be formed on the cooler parts of the charcoal, since silver has but little tendency to unite with oxygen [1].

Reduction of Silver salts.

Some methods of effecting this have been already described ; e. g. by nascent hydrogen, p. 251, and by sodium carbonate in the experiment last made. The following processes are also of great practical importance.

---

[1] The slight white ash which charcoal leaves when burnt, must not be mistaken for an incrustation of a metal-oxide.

1. By organic substances.

(*a*) Draw letters on a piece of paper with a glass rod dipped in a solution of silver nitrate, and heat the paper in front of a fire, or at some distance above the flame of an argand burner, until it is nearly hot enough to become scorched. The letters will appear in an indelible black; the hydrogen and carbon having decomposed the salt, combining with its oxygen, while metallic silver is separated, and adheres closely to the fibre of the paper.

This illustrates the use of marking ink, which is simply a solution of silver nitrate thickened with gum. It also explains why silver salts stain the skin black.

(*b*) Tartrates (as seen already, p. 174) and some other organic salts, readily reduce silver from its combinations; the action being in general similar to that of the paper in the experiment just made. Mirrors for telescopes are thus covered with a firm deposit of silver which will even bear careful polishing.

Place some very dilute solution of ammonium hydrate (2 drops of the ordinary laboratory solution in 5 c.c. of water) in a test-tube, and add solution of silver nitrate, a few drops at a time, shaking the mixture after each addition. A brown precipitate of silver oxide will be formed, which will at first redissolve readily; but finally a point will be reached when the addition of a drop of silver nitrate will cause a permanent turbidity. Now add to this slightly turbid liquid a drop of solution of sodium-and-hydrogen tartrate, mix thoroughly, and warm it by placing the tube in a jug or beaker of hot water. The liquid will turn black, and a brilliant film of metallic silver will be deposited on the sides of the tube.

A watch-glass may be silvered in a similar way by first cleaning it thoroughly with a few drops of solution of potassium hydrate, and rinsing it with several portions of distilled water; then embedding it in sand, filling it with the solution made as above directed (or, better, of twice the strength) and heating the sand very slowly. A fairly good convex mirror may be thus obtained; and a concave one **may be made by** floating a watch-glass upon the surface of the solution, warmed in a porcelain dish.

2. By the action of light.

It has been already noticed that silver chloride becomes dark in colour when exposed to light. This is due to the separation of the silver and chlorine, first a subchloride and then metallic silver being produced. The reduction takes place much more readily when organic matter and a large excess of silver nitrate are present: the liberated chlorine then forms more silver chloride from the nitrate, which is reduced in its turn, and thus a black of considerable intensity is produced. This is the principle of the process for obtaining photographic prints which is still very largely employed.

> [Solutions required,—
> Ammonium chloride, 1 grm. in 50 c.c. of water.
> Silver nitrate, 4 grms. in 50 c.c. of water.
> Sodium hyposulphite, 20 grms. in 100 c.c. of water.]

Pour the solution of ammonium chloride into a flat plate or dish. Take a piece of fine drawing-paper about 10 or 12 cm. square, make a small pencil-mark in one corner, then taking it up by the two opposite corners lower it gently upon the solution in the plate, keeping the marked side downwards, so that the centre touches the liquid first, the two corners being lowered afterwards, in order that no air-bubbles may be retained between the solution and the paper. Allow the paper to float upon the solution for five or six minutes, then take it up quickly by one corner, and pin it to a shelf or the back of a chair, to drain and dry. The next stage of the preparation of the paper, since it involves the formation of a compound sensitive to light, must be performed by candle-light, and therefore in a cellar, or a room with closed shutters, or in the evening. Pour the solution of silver nitrate into a perfectly clean plate, carry it into a room lighted only by a candle, and float upon it the piece of salted paper (with the marked side downwards), taking particular care to avoid air-bubbles. The silver nitrate in the solution will act upon the ammonium chloride in the paper, and ammonium nitrate and silver chloride will be formed at the surface of the paper. After the lapse of three minutes, raise the paper by one corner and pin it up, as before, to dry in a dark cupboard or drawer. Obtain two pieces of flat window-glass, rather larger than the piece of prepared paper; lay one of them flat on the table in a room lighted by a candle only, or at any rate in the darkest corner of a room in which the blinds are drawn down. Place upon the glass a piece of black velvet or thick brown paper; lay upon the latter the piece of prepared paper, having the marked side upwards; place upon the paper a piece of black lace or a fern leaf, or a small engraving (which should be on thin paper), and on the top of all lay the other plate of glass. Bind the whole together by two letter clips or india-rubber bands, one on each side, and bring the extempore printing-frame into full daylight. You will observe the portions of the prepared paper, which are not protected by the lace or the engraving, to pass through shades of red and purple, finally becoming black and bronzed, if looked at obliquely by reflected light. When this is the case, remove the apparatus again to the dark room, and take out the prepared paper. The lace, or other opaque object, will be found to have protected the paper from the action of the light, and an image of it will be formed, white on a black ground. Of course, if the paper be now brought out to the daylight, it will blacken all over, and the image will be obliterated. If it be thought worth preserving, it must be soaked in common water to remove all excess of silver nitrate, and then placed in the solution of sodium hyposulphite. This solution will dissolve away all the unaltered silver chloride (see p. 228), leaving the reduced silver untouched. After the print has remained about ten minutes in the solution it should be removed, and washed in several changes of water; after which it may be brought out to the light without damage. The black portions will have become reddish brown, and will be reduced in intensity, since no toning process has been employed; but the experiment will serve the purpose of illustrat-

ing the action of light on silver chloride, and the principles of
the art of photography.

2. MERCURY.

[Symbol of atom, Hg (hydrargyrus).
Weight of atom, 200 hydrogen-atoms.

It must be borne in mind that mercury and most of its compounds
are very poisonous, whether taken internally or absorbed by the
skin. Moreover, the metal should not be spilled about on the table
or floor, or thrown down the sink, since it combines with lead and
soon destroys the pipes.]

Mercury is chiefly obtained from its sulphide, the mineral
cinnabar, which is chemically the same as vermilion. This
yields the metal when heated with quicklime, the calcium and
oxygen of which combine with the sulphur[1].

Powder a small bit of cinnabar or vermilion (the ordinary red
paint), and mix it intimately with an equal quantity of quicklime: then put some of the mixture into an ignition-tube and
heat it strongly. Bright globules of a liquid metal will condense
on the tube, running together into one drop when scraped off
the glass with a splinter of wood, such as a match. These
must be mercury, since it is the only metal which is liquid at
ordinary temperatures.

Properties of Mercury.

1. Its liquidity and cohesion.

Pour a few drops of mercury on a clean plate; if the metal
is pure the drops will preserve their roundness as they roll
about, since their cohesion to each other is much greater than
their adhesion to the surface on which they rest[2].

---

[1] $4 HgS + 4 CaSO_4 = 3 CaS + CaSO_4 + 4 Hg.$
[2] If the metal contains lead or other impurity, its liquidity is impaired,
and the drops will become elongated when the plate is inclined, and will
move about sluggishly, leaving portions behind them. In order to purify
it, pour it into a plate or shallow dish in sufficient quantity to form a thin

2. Its high density, 13.6 times that of water.

This may be ascertained in the manner described in pp. 48, 49; 5 c.c. of mercury being carefully measured and then weighed. Owing to the strong cohesion of mercury compared with its adhesion to the glass, the curve of its surface in the tube will be convex, thus ⌢, instead of concave, ⌣, as in the case of water; and hence, in order to obtain an approximately correct measure, the 5 c.c. mark should coincide with the line where the liquid touches the glass (and not from a tangent to the curve, as usual).

3. Its volatility (boiling point, 360°).

Place a small drop of mercury in a dry test-tube (using a pipette to transfer it), and heat it over a lamp, holding the tube nearly horizontal. The metal will boil below a red heat, and, if pure, will wholly volatilise, its vapour condensing in the cooler parts of the tube, and forming a bright metallic ring, which, when looked at through a magnifier, will be seen to consist of minute liquid globules like dew. These, if touched with a glass rod, may be made to run together into one large drop.

Preparation of Compounds of Mercury.

Mercury forms two distinct series of compounds, in one of which a given weight of it is combined with twice as much of the electro-negative radicle as is present in the other. It is found, in fact, that one atom of mercury [Hg] will saturate and form a definite crystallisable compound with either (i) one atom of the nitrate radicle [NO_3], to form a molecule of mercury

stratum at the bottom, and pour over it enough dilute hydrogen nitrate (1 vol. of the strong acid to 20 vols. of water) to cover it. Leave it in contact with the metal for three or four hours, stirring it occasionally with a glass rod to expose a fresh surface of the metal to the acid; then pour off the acid, wash the mercury with a stream of water poured from a jug, and finally, after pouring off as much water as possible, absorb the remainder with a clean cloth or blotting-paper. Fit a filter in a funnel, and make a fine hole with a pin at the point, support the funnel over a beaker or bottle, and pour into it the dry mercury. The metal which runs through will be found very nearly pure; it may be further purified by distillation, but, except for special purposes, the gain in purity will scarcely compensate for the trouble and risk.

protonitrate [$HgNO_3$], or (ii) with two atoms of the same radicle to form a molecule of mercury pernitrate [$Hg(NO_3)_2$].

Either one of these can be formed, to the exclusion of the other, by varying the conditions under which hydrogen nitrate acts on mercury.

1. Preparation of mercury protosalts.

[Typical examples,—Mercury protonitrate ($Hg_2(NO_3)_2$).
Mercury protochloride (Hg_2Cl_2)[1].]

Mercury protonitrate is prepared by taking hydrogen nitrate slightly diluted and allowing it to act on excess of the metal, the temperature being kept low.

Place a globule of mercury about as large as a pea in an evaporating dish, and pour upon it a mixture of 5 c.c. of strong hydrogen nitrate with 3 c.c. of water. Allow the action to go on for ten minutes or so (while other experiments are proceeded with) stirring the liquid frequently, but applying no heat; then pour the solution together with the remaining mercury into a test-tube, dilute it with water to 50 c.c., and reserve it (labelled ' mercury protonitrate ') for experiments, p. 262.

2. Preparation of mercury persalts.

[Typical examples,—Mercury pernitrate ($Hg(NO_3)_2$.
„   perchloride ($HgCl_2$).
„   periodide (HgI_2).
„   persulphide (HgS).]

The conditions necessary for preparing mercury pernitrate are, strong acid taken in excess, and a high temperature.

Put a small globule of mercury (about half as large as was taken in the last experiment) into an evaporating dish, pour on it 4 or 5 c.c. of strong hydrogen nitrate, and heat the whole on a sand-bath, in a draught cupboard if possible. Allow the action to go on until all the metal is dissolved, then add a drop of strong hydrogen chloride (to convert any remaining protosalt

---

[1] The above formulæ for the molecules are used instead of the simpler ones, $HgNO_3$, and $HgCl$, because the protosalts show a tendency to separate into metallic mercury and a persalt; indicating that there may be more than one atom of mercury in the molecule.

into persalt[1]) and drive off most of the excess of acid by eva-poration[2]. Pour the concentrated solution into a test-tube, dilute it with water to 50 c.c., and reserve it (labelled ‘Mercury pernitrate’) for experiments, p. 263.

3. Formation of mercury periodide.

This substance has been prepared already (p. 74), but not in a manner in which the remarkable change of colour which it undergoes could be noticed.

(*a*) Dilute 2 drops of a solution of mercury perchloride with 5 or 6 c.c. of water, and add 1 drop of solution of potassium iodide. The precipitate of mercury iodide which forms is, in this dilute solution, bright yellow at first, but changes in the course of a minute or more into pink and finally scarlet.

The reason of the change of colour appears to be this ;— Mercury periodide exists in two different allotropic conditions, which differ in crystalline form : the one occurring in oblique rhombic prisms which are yellow, the other in square octohedra which are scarlet. Of these the most stable is the latter, and hence the yellow precipitate, which consists of the prismatic modification, changes to scarlet, the conversion being quicker in strong solutions.

If a few more drops of solution of potassium iodide are added to the precipitate it will readily dissolve, a double salt, mercury-and-potassium iodide, being formed[3].

(*b*) Put a little of the scarlet mercury iodide obtained in exercise 8, p. 74, into a small dry test-tube, and heat it slowly over a lamp. It will melt and volatilise, condensing in yellow prismatic crystals in the cool part of the tube. The yellow sublimate, however, is not permanent ; on rubbing it with a

---

[1] It does this, owing to the chlorine which is liberated, and which from its strong affinities determines the change in the saturating power or ‘atomicity’ of the metal more readily than hydrogen nitrate.

[2] If the evaporation has been unintentionally carried to dryness, a little dilute hydrogen nitrate must be used to dissolve the residue; otherwise a basic salt insoluble in water will be formed.

[3] The use of this as a test for ammonium salts will be described under AMMONIUM.

By dissolving a large quantity of mercury iodide in a saturated solution of potassium iodide, a liquid may be obtained which is of so high a density that glass will float in it.

glass rod scarlet lines will appear and the whole will soon pass again into the more stable modification.

[If you possess a moderately good microscope, you will find it interesting to watch the progress of the change by subliming a little of the scarlet mercury iodide from a watch-glass to one of the ordinary microscope slides, placing it on the stage of the instrument, examining the crystals with a one-inch power, and lightly touching one of them with a needle. The change will begin in the crystal which was touched and gradually spread through the whole, while its progress may readily be watched; the yellow rhombic prisms becoming aggregations of red octohedra with a square base, belonging to a different crystallographic system[1].]

4. Formation of mercury persulphide.

This substance, which is identical with vermilion, affords another example of the change of colour due to difference of allotropic form which is often observable in the compounds of mercury.

Weigh out half a gramme of sulphur and 3 grms. of mercury, place them in a small test-tube and apply heat. As the temperature approaches the boiling point of mercury, a strong action will commence, the sulphur and mercury combining to form a black mass of mercury sulphide. Continue heating this for half a minute, observing its volatility, on account of which it can be driven about from one part of the tube to another : then, after allowing it to cool, detach some of it from the tube and grind it to powder in a mortar. Lastly, shake the powder out on a sheet of writing paper, and rub it strongly with the pestle. Observe that it changes during the grinding and rubbing from black to a dull red, leaving a streak of that colour upon the paper. The same change, which is due to the black, amorphous sulphide becoming crystalline, is effected by a very

---

[1] To obtain good crystals for the microscope, it is generally necessary to resublime the crystals first obtained. To do this, place over the slip of glass on which the crystals have been deposited, another similar slip of glass, separating the two by strips of card placed between them at each end. Hold the plates in this position between the finger and thumb, and heat both of them over a spirit lamp. When they are pretty hot, apply a higher temperature to the one on which the mercury iodide is deposited, waving the lamp-flame to and fro under it. The sublimation thus takes place slowly, and large crystals are formed.

slow process of sublimation, as in the Chinese method of making vermilion.

If, however, a little vermilion or cinnabar is heated in an ignition-tube it forms a black sublimate of the amorphous sulphide.

General properties of Mercury salts.

*1. They are volatile at a temperature below redness.

(*a*) Place a small crystal of mercury perchloride ('corrosive sublimate') in an ignition tube, and heat it gently over the lamp. The salt will fuse, and at a rather higher temperature volatilise entirely, forming a white crystalline ring in the cool part of the tube.

(*b*) Heat a little mercury protochloride ('calomel') in a similar way. It will be found to volatilise without previous fusion.

*2. Heated with sodium carbonate in an ignition tube, they yield a sublimate of liquid metallic globules.

Mix a little mercury protochloride, or perchloride, with an equal quantity of anhydrous sodium carbonate in a mortar, introduce it into an ignition tube in quantity just sufficient to half fill the bulb, and cover it with a layer of the dry sodium carbonate. Wipe the tube clean with a twisted strip of blotting-paper, and heat it very gently over the lamp. At first, if the substances are not carefully dried, some moisture will be given off, which must be absorbed by strips of blotting-paper, before a stronger heat is applied; otherwise it will be difficult to obtain a well-defined sublimate. When no more water condenses in the tube, increase the heat, until the bulb is nearly red-hot. A metallic sublimate will form in the tube which, when examined with a magnifying glass, will be seen to consist of small round globules of metal, very different in appearance from the crystalline crust of arsenic, p. 280. To remove any doubt as to the character of the deposit, introduce a thin slip of wood (such as the end of a match) into the tube, and employ it to scrape the sublimate from the sides of the tube. The small globules will run together into one large globule, which from its metallic lustre and fluidity can only be mercury [1].

*3. They are reduced by copper.

Place a drop of solution of mercury perchloride on a clean strip of copper. Allow it to remain for half a minute, and then wash it off. A dull white stain will be left on the copper, which when rubbed with a cloth will become bright and silvery. The mercury salt has been decomposed, copper nitrate being formed, while the mercury is deposited on the more electro-negative metal, the copper, with which it forms an alloy or amalgam.

If the copper is heated, but not to redness, the bright deposit will disappear.

Distinctive properties of the protosalts and persalts of Mercury.

[For the following experiments the solutions of mercury proto-nitrate and mercury pernitrate, which have been already prepared (p. 258), may be used.

It will be convenient to apply each test successively to the proto-salt and the persalt, and to record the results in parallel columns in the note-book.]

A. Mercury protosalts.

*1. With hydrogen chloride they give a white precipitate.

Pour a little of the solution of mercury protonitrate into a test tube, and add a few drops of dilute hydrogen chloride. A white precipitate of mercury protochloride will be formed.

(*a*) Pour off a portion of the liquid, in which the precipitate is suspended, into another tube; add a little more hydrogen chloride, and then a drop or two of strong hydrogen nitrate, and boil. The chlorine evolved from the aqua regia thus

---

[1] Mercury iodide is the only commonly occurring mercury compound which under the above conditions volatilises unreduced, in spite of the superposed layer of sodium carbonate. It may, of course, be easily identified by its colour and its behaviour when heated alone, p. 259.

formed (see p. 193) will convert the mercury protochloride into perchloride [1], and the latter will dissolve.

(*b*) Add to the remainder of the precipitate some ammonia. It will turn black, owing to the formation of mercury protoxide.

*2. With hydrogen sulphide they give a black precipitate.

Test another portion with solution of hydrogen sulphide. A black precipitate of mercury protosulphide will be formed. Pour off a little into another tube; render it alkaline with ammonia, add some ammonium sulphide, and warm the mixture; the precipitate will remain undissolved.

3. With potassium hydrate they give a black precipitate.

Add to a third portion some solution of potassium hydrate. A black precipitate of mercury protoxide will be formed, insoluble in excess.

*4. With tin protochloride they give a dark gray precipitate.

Add to another portion some solution of tin protochloride. A dark gray precipitate of finely divided metallic mercury will be at once formed [2].

B. Mercury persalts.

[The corresponding tests should be applied in precisely the same way as directed for the protosalts.]

1. With hydrogen chloride they give no precipitate.

*2. With hydrogen sulphide they give at first a white precipitate, changing to yellow and finally black.

If the hydrogen sulphide is added drop by drop, and the liquid shaken, a precipitate will be produced which is at first white, but on further addition of hydrogen sulphide becomes yellow, brown, and finally black. The cause of this characteristic reaction is, that a combination of mercury persulphide and nitrate is first thrown down, which is converted entirely

---

[1] $Hg_2Cl_2 + Cl_2 = 2\,HgCl_2.$

[2] This may, but not without trouble, be made to aggregate into larger globules by allowing it to subside, washing it by decantation from traces of nitrate, and then boiling it with a little solution of tin protochloride, to which a few drops of hydrogen chloride should be added.

into mercury persulphide by 'excess of the precipitant. The mercury persulphide, like the protosulphide, will be found insoluble in ammonium sulphide, and also in hydrogen nitrate, but decomposable by aqua regia.

3. With potassium hydrate they give a yellow precipitate.

It is noticeable that this precipitate which consists of mercury peroxide, has precisely the same composition as the 'red oxide' of mercury already used for experiments, p. 69.

*4. With tin protochloride they give a white precipitate, turning gray when more of the test is added.

The white precipitate formed by the addition of one drop of solution of tin protochloride is mercury protochloride. This turns gray on addition of more of the reagent, owing to its reduction to metallic mercury.

3. LEAD.

[Symbol of atom, Pb (plumbum).
Weight of atom, 207 hydrogen-atoms.]

1. Density of lead $(=11.45.)$

This, if not ascertained already, may be taken now by the method given in p. 51.

2. Action of water on lead.

Fill two test-tubes, one half full of distilled water, the other half full of common hard spring water, and immerse in each a strip of lead, cleaned by scraping it with a knife. After the lapse of an hour or less, a slight white precipitate will be formed in the test-tube containing the distilled water, and in the course of a day some quantity of this precipitate will have been collected; while in the test-tube containing the common hard water there will be little or no perceptible action.

The reason is this: The oxygen dissolved in the water attacks the lead, forming lead oxide, which dissolves in the water, but is to some extent precipitated as carbonate by its combination with the carbon dioxide present in air. If, however, the water contains sulphates and carbonates (as common hard water does, p. 66), these form a crust of insoluble lead sulphate and carbonate which protect the lead from corrosion.

The presence of lead in solution in the test-tube containing the distilled water should be proved by applying to it. test 2, p. 266.

Preparation of compounds of lead.

[Typical examples,—Lead protoxide (Pb O).
    „   red oxide ($Pb_3 O_4$).
    „   dioxide ($Pb O_2$).
    „   nitrate ($Pb(NO_3)_2$).
    „   acetate ($Pb(C_2 H_3 O_2)_2$).]

1. Lead nitrate.

Cut off a small piece of sheet lead, weighing about 2 grms., with a strong knife, and hammer it thin on an anvil, noticing its malleability, softness, and want of elasticity. Fold up the thin sheet thus obtained, place it in a porcelain dish, pour on it about 5 c.c. of strong hydrogen nitrate diluted with half its volume of water, and heat gently on the sand-bath. Orange vapours of nitrogen tetroxide will be given off, and lead nitrate will be formed and dissolve[1]. Evaporate the solution just to dryness, to drive off excess of acid; dissolve the residue by warming it with a little water, then add more water until the whole measures 50 c.c., and reserve it for future experiments, p. 266.

2. Lead red oxide ('red lead').

Mix 2 grms. of lead protoxide (litharge) with half a gramme of potassium chlorate, grinding the whole thoroughly together. Heat the mixture in an iron spoon or capsule until all effervescence due to escape of oxygen has ceased, and then allow it to cool. The lead protoxide will under these circumstances combine with more oxygen, and a bright red residue will be obtained, containing 'red lead.'

3. Lead dioxide.

Pour some dilute hydrogen nitrate on a little red lead in a test-tube, and warm the mixture. A chocolate-coloured powder

---

[1] $4 HNO_3 + Pb = Pb(NO_3)_2 + 2 H_2O + N_2O_4$.

will remain undissolved, which is lead dioxide. The acid has separated the red lead into lead protoxide, which dissolves as nitrate, and lead dioxide, which is insoluble in the acid[1].

Properties of salts of Lead.

[The solution of lead nitrate already made (p. 265) may be used.]

*1. With hydrogen chloride they give a white precipitate, unaltered by ammonia.

Add a few drops of dilute hydrogen chloride to some of the solution of lead nitrate. A white crystalline precipitate will be formed, insoluble in excess of the acid[2]. Divide the fluid containing the precipitate in suspension into three portions, and test them as follows :—

(*a*) Add to the first portion three or four times its volume of water. The precipitate will gradually redissolve. Owing to this solubility in water, it is not formed at all in weak solutions of lead salts.

(*b*) Heat the second portion to boiling. It will in this case also redissolve, but will be reprecipitated as the solution cools, in snow-like flakes, which, when examined by a magnifier, will be seen to consist of groups of slender prisms.

(*c*) To the last portion add ammonia. The precipitate will not dissolve or alter in colour.

*2. With hydrogen sulphide they give a black precipitate.

Test another portion of the solution of lead nitrate with solution of hydrogen sulphide. A black precipitate will be produced, insoluble in hydrogen chloride but decomposed on being boiled with excess of hydrogen nitrate. It will be remembered that this property of lead, of forming a black compound with sulphur, furnished a very delicate test of the presence of a sulphide (p. 218).

---

[1] $Pb_3O_4 = 2\,PbO + PbO_2$.

[2] A large excess of dilute hydrogen chloride will dissolve it, the effect being due, however, not to the action of the acid, but to the action of the water, as in (*a*).

*3. With hydrogen sulphate they give a white precipitate.

Test another portion of lead nitrate with dilute hydrogen sulphate. A white precipitate will be formed, insoluble in excess of the acid, but readily decomposed and dissolved if warmed with solution of ammonium acetate [1].

Since lead sulphate is slightly soluble in hydrogen chloride, the best method of applying this test is as follows:—

* Make a very dilute solution of lead nitrate by mixing 5 c.c. of water with 2 drops of the ordinary solution, and add 2 or 3 drops of ammonia. This will produce a white precipitate of lead hydrate. Now add to the liquid excess of dilute hydrogen sulphate; this will decompose the hydrate, forming lead sulphate which will remain undissolved.

It is always less easy to dissolve a precipitate than to prevent its formation.

*4. With potassium chromate they give a yellow precipitate, soluble in potassium hydrate.

Add to another portion of the solution of lead nitrate a drop of solution of potassium chromate. A yellow precipitate will be formed, consisting of lead chromate (the 'chrome yellow,' much used as a paint). Divide the liquid into two portions: to the first add some dilute hydrogen nitrate ; the precipitate will remain undissolved. To the second portion add some solution of potassium hydrate, which will readily redissolve the precipitate.

*5. Reduced before the blowpipe on charcoal they yield a malleable bead, and yellow incrustation.

Mix a very little lead acetate with about twice as much potassium cyanide, and heat the mixture on charcoal as directed in Expt. 2, p. 94. A metallic bead will be obtained, the malleability of which should be tested as there directed ; and the surrounding parts of the charcoal will be covered with a light yellowish incrustation of lead oxide.

---

[1] This may be made, if none is at hand, by putting about 3 c.c. of solution of ammonia into a test tube, and adding hydrogen acetate until the liquid smells of it and is acid to test paper.

Additional Experiment.

Precipitation of Lead from its compounds by Zinc.

This depends upon the superior affinity of zinc for the radicle combined with the lead; and the action, once begun, is aided by the galvanic current set up between the zinc and the portions of lead first thrown down (compare the reduction of silver by zinc, p. 251).

Make a solution of lead acetate by dissolving 10 grms. of the salt in 200 c.c. of water, with addition of a few drops of hydrogen acetate. Place the filtered solution in a wide-mouthed bottle, and suspend in it a strip of thick sheet zinc, or, better, a rod of the metal the size of a pencil[1]. A short piece of glass tubing should be laid across the mouth of the bottle, from which the zinc may be suspended by a piece of string, so that it may not touch the bottom of the bottle. Leave the whole undisturbed for twelve or fourteen hours: the rod of zinc will soon be covered with thin brilliant plates of pure metallic lead (the so-called 'Lead Tree'), while a proportional quantity of zinc is dissolved. This is a good illustration of the replacement of one metal by another in a combination; and, if we ascertained the weight of zinc dissolved and the weight of lead deposited, we should find that the former weight was to the latter as 65 : 207, numbers which are the accepted atomic weights of zinc and lead respectively.

In a day or two the whole of the lead will have been withdrawn from the solution, which may then be examined by test 3, p. 267, to prove the absence of lead.

---

[1] As the shape of the rod is immaterial, it may be easily made by melting some zinc in a ladle, and casting it in a shallow trough formed by pressing a thick pencil or a piece of glass tube into a mass of fine slightly moistened sand, or clay.

ᴄ

Group II.

Metals which are separated from solutions containing hydrogen chloride by hydrogen sulphide.

MERCURY (diatomic), LEAD, COPPER, CADMIUM, BISMUTH, ARSENIC, ANTIMONY, TIN, GOLD, PLATINUM.

1. COPPER.

[Symbol of atom, Cu (cuprum).
Weight of atom, 63.5 hydrogen-atoms.]

Compounds of Copper.

[Typical examples,—Copper oxide (CuO).
  „ nitrate ($Cu(NO_3)_2$).
  „ sulphate ($CuSO_4$).]

1. Combination of copper with oxygen, when heated.

Clean a strip of sheet copper with emery paper, and hold it in the upper part of the flame of a Bunsen's burner. As it gets hot, brilliant-coloured films of copper oxide will form on it, where it is exposed to the current of hot air rising with the flame [1]. Finally, when it has been heated to redness for a minute, plunge it into a beaker of cold water, when the oxide will detach itself in scales (since it does not contract, as the temperature falls, in the same degree as the metal), and the surface of the metal will show the characteristic red colour of copper.

___

[1] Observe that not only does the part of the copper which is actually immersed in the flame remain bright, but when the blackened part is brought into the middle of the flame the oxide is reduced and the bright red metal reappears. In fact, by holding the strip edgeways in the flame we get a good section of the oxidising and reducing parts.

The iridescence of the films of copper oxide which are formed first, is due to the interference of the rays of light reflected from the upper and under surfaces of the film, respectively: due to the same cause, in fact, as the colours of a soap bubble.

2. Preparation of copper nitrate.

Dissolve 1 grm. of fine copper wire in about 5 c.c. of strong hydrogen nitrate diluted with half its volume of water, heating the tube gently if the action becomes slow. Red vapours of nitrogen tetroxide will be seen above the liquid (the chemical action has been already explained, p. 142), and a blue solution of copper nitrate will be obtained. Evaporate this solution nearly to dryness in a dish, to drive off any excess of hydrogen nitrate, dilute it with water to 50 c.c., and use it in the following experiments.

Tests for Salts of Copper.

*1. With hydrogen sulphide they give a black precipitate.

Acidify a portion of the solution with hydrogen chloride, and add solution of hydrogen sulphide. A black precipitate will be formed, consisting of copper sulphide. Allow this to subside (it will separate more readily if the solution is warmed), pour off the clear fluid, and wash the precipitate once or twice by decantation. Divide it into two portions :

(*a*) To one of these add some ammonium sulphide, and warm it gently ; the precipitate will not dissolve.

(*b*) Boil the other portion with strong hydrogen nitrate, which will readily decompose and dissolve it, with separation of sulphur.

*2. With potassium ferrocyanide they give a reddish-brown precipitate.

This should be tried with only one drop of the solution of copper nitrate, diluted with 5 c.c. of water, in order that the delicacy of the test may be observed.

3. With potassium hydrate they give a light blue precipitate, turning black on boiling.

To another portion add a few drops of solution of potassium hydrate. A light blue precipitate of copper hydrate will be formed, which will not dissolve in excess of the precipitant, and will turn black when the solution is heated to the boiling-point, owing to its conversion into copper oxide.

*4. With ammonia they give a light blue precipitate, soluble in excess, forming a blue solution.

To another portion add one drop of solution of ammonia. A light blue precipitate will be formed, as in the last experiment, but will readily dissolve on addition of a few drops more of ammonia, and a deep blue solution will be obtained. This contains cuprammonium nitrate, a salt in which part of the hydrogen in the radicle ammonium is replaced by copper, as shown in its formula $(H_6CuN_2)_2 (NO_3)_2$.

5. **They impart a green colour to flame.**

Put a drop or two of strong hydrogen chloride into a watch-glass, dip into it the end of a piece of copper wire, and hold the latter in the flame of a Bunsen's burner. The copper chloride which has been formed will volatilise in the flame and give it a bright bluish green colour.

*6. **Heated with borax, they give a bead which is greenish-blue in the oxidising flame, red and opaque in the reducing flame.**

Make a borax bead, add to it a trace of copper sulphate, and heat it in the oxidising flame of the blow-pipe. The bead will be coloured green while hot, becoming blue as it cools. After being heated in the reducing flame, with addition of a little more copper sulphate, it will appear red by reflected light, owing to the presence of suspended particles of copper.

7. **Heated on charcoal with sodium carbonate they give grains of metallic copper.**

Mix a minute quantity of copper sulphate with about twice as much sodium carbonate, and heat it on charcoal before the blowpipe-flame. You will obtain small grains of metallic copper, but the heat of the blowpipe-flame will be scarcely sufficient to fuse these into one globule. They will be better seen if the fused mass is placed in a mortar and washed with water until the soluble salts are dissolved; the red metallic particles may then be readily distinguished with a magnifier.

Reduction of salts of Copper.

1. **By grape sugar, with formation of copper suboxide.**

Dissolve a small lump (about the size of a pea) of grape sugar in 5 c.c. of water, filter the liquid if necessary, and add to it a drop or two of the solution of copper nitrate. If solution of potassium hydrate be now added, no precipitate of copper hydrate will be formed, or if formed (owing to the grape sugar not being present in sufficient quantity) it will be redissolved on addition of more potassium hydrate. But if the mixture is heated to the boiling-point, a yellow precipitate of copper subhydrate is obtained, which quickly changes to red copper suboxide.

*2. By iron, with formation of metallic copper.

(*b*) Pour a little of the solution of copper nitrate into a test-tube, and dip into it a strip of sheet iron cleaned with emery paper, or the blade of a knife. Metallic copper will be thrown down on the surface of the iron, while the latter is slowly dissolved, taking the place of the copper in the solution. Compare the action of zinc on lead acetate, p. 268.

3. By electricity.

Copper is one of the metals most easily reduced by electricity, and an experiment illustrating this has already been made, p. 70.

Additional Experiment.

Method of obtaining Electrotypes.

When a current of electricity is sent through a solution of a salt of copper, the metal is deposited upon the negative electrode (i.e. upon that which is connected with the zinc plate of the battery), and since it is deposited molecule by molecule, the particles form a film which adheres very closely to, and is an exact copy of, the electrode itself. In order, therefore, to obtain a copy of a seal or medal, a 'mould' or reversed impression must be first obtained, which is placed as the negative electrode in a solution of copper sulphate, and the current kept up until a sufficiently thick film of copper is deposited.

To obtain the current it will not be necessary to employ a second battery. A 'single-cell' apparatus, arranged as directed below, will answer perfectly; the mould itself forming the negative plate of the battery [1].

---

[1] For a fuller explanation of the principles of electrolysis a text-book on Electricity must be consulted.

[Solutions required,—

Saturated solution of copper sulphate (350 grms. of the salt dissolved in 1 litre of water; 20 c.c. of hydrogen sulphate added).

Solution of ammonium chloride (60 grms. dissolved in 300 c.c. of water).]

1. Preparation of the mould.

For seals nothing is better than good sealing-wax. Some of this should be melted on a card held at some height above a lamp-flame, and continually stirred with the end of the stick of wax until sufficient has been melted to form a thick circular lump rather smaller than the seal to be copied. After breathing upon the seal, bring it down with gentle pressure upon the wax, increasing the pressure as the wax gets cold and hard. It is best not to remove the seal until the wax is quite hard, otherwise the mould may lose its flatness. The card should next be cut away carefully close to the edges of the wax, and a copper wire (about No. 20 wire gauge) 30 cm. long should be attached to the margin of the mould by heating it and pressing it into the wax near (but not touching) the edge of the impression. The wire must not be completely buried in the wax, and the exposed part should be scraped clean with a knife.

For a medal, plaster of paris will answer very well, and the mould may be made according to the directions given under CALCIUM, being thoroughly saturated with wax in the manner there described. The end of a piece of copper wire, about 40 cm. long, should be passed round the rim and fastened by twisting to the main portion of the wire, so as to grasp the mould securely in a loop.

Since wax is a non-conductor of electricity the surface of the mould must be rendered conductive by covering it with a thin film of graphite (the ordinary plumbago or black lead). For this purpose it should be laid face upwards on a sheet of paper (if of sealing-wax it should be held for a few seconds over the mouth of a beaker containing a little hot alcohol[1]) and brushed over with some of the best plumbago[2]: being lightly breathed upon occasionally to promote the adhesion of the substance. The brush should be worked with quick circular strokes into all the crevices of the mould, until every part of it shows a metallic lustre when looked at obliquely. An

---

[1] The vapour of this slightly softens the surface and gives the graphite a better hold.

[2] Good stove black-lead will generally serve the purpose: but it is best to obtain a little pure plumbago from a chemist or optician. A shaving brush of badger's hair answers well for applying it.

extremely thin film is all that is required, but it must be continuous. A little plumbago must be brushed over the wire where it is attached to the mould, to make a good connection throughout. Lastly, the edges of the mould outside the wire, and about 10 or 12 cm. of the wire itself, must be varnished with 'black japan' or some other thick varnish[1], to prevent the deposit of copper upon them when immersed in the solution.

[If the mould is of fusible metal (see p. 278) a portion of its edge should be moistened with solution of ammonium chloride and a clean piece of copper wire heated and pressed against it until completely imbedded. The back and edges of the mould must be thickly varnished to prevent deposit of metal upon them; and it is advisable to brush a trace of black-lead over the impression to prevent too close adhesion of the deposit.]

2. **Construction of the apparatus.**

Place a small jar of porous earthenware about 14 or 15 cm. high and 4 cm. in diameter in a wider jar of glass or china about 11 or 12 cm. in diameter and the same or rather more in height[2]. Put into the porous cell a rod of zinc about 18 or 20 cm. long, and at least as thick as a pencil. Hang the mould in the outer jar by bending the wire round the rim of the jar as shown in fig. 62, the bend being made thus: otherwise the solution may creep up between the wire and the side by capillary action, and overflow. The other end of the wire should be twisted firmly round the zinc rod, so as to support it in the porous cell.

Fig. 62.

Lastly, fill the porous cell about three-fourths full of the solution

---

[1] Sealing-wax varnish may be made by powdering 10 grms. of black sealing-wax, and digesting it at a gentle heat (best applied by a pan of hot water) with 10 c.c. of methylated spirit, stirring occasionally.

[2] For the porous cell, a common flower-pot, though unnecessarily wide, answers well, the hole at the bottom being stopped with a cork.

For the wider jar nothing is better than the lower part of a large common green glass bottle, the neck being cut off by the method given in p. 42.

A preserve jar or gallipot will answer, but the solution is apt to penetrate the glaze, unless the material is stoneware.

of ammonium chloride[1], and the outer jar to the same level with the solution of copper sulphate.

In the course of a minute a bright deposit of copper will be seen spreading over the mould. If it does not appear, there must be a break in the conducting surface, and the mould must at once be taken out, dried, and blackleaded with more care. When all is going on well, hang a small muslin bag containing crystals of copper sulphate just below the top of the solution in the outer jar, tying it to a glass rod laid across the jar: and leave the whole undisturbed for about a day, examining it occasionally to see that the deposit is forming regularly on the mould. The latter should not, however, remain out of the solution for more than half a minute, lest it should tarnish and the subsequent deposit fail to adhere to it[2].

[The defects which are most liable to occur are,—

1. *Gas given off at the surface of the mould and a sandy, dull-red, non-adherent deposit of metal.*

This is caused by the current being too strong in proportion to the strength of the copper sulphate solution. The remedy will be to pour off about two-thirds of the solution of ammonium chloride in the porous cell and add an equal volume of plain water to the remainder. The solution of copper sulphate must be kept fully saturated, fresh crystals being placed in the bag as the others dissolve.

2. *The deposit forms very slowly; its surface being rough and bristling with crystals of copper.*

This is due to the current of electricity not being strong enough; the action of the solution of ammonium chloride on the zinc having slackened. The spent solution must be poured away, and the porous cell filled up with fresh solution of the same salt. The same deficient action occurs in cold weather, and the apparatus should be kept in a warm room.

When the conditions are right, the deposit appears of a bright pink colour, even in the surface, and tough in texture.]

In about 24 hours a sufficiently thick film of metal will have been deposited, and the mould may be taken out and the copy detached from it by pulling the two apart very carefully. If there is much

---

[1] If the mould is small, it will be advisable to use a much weaker solution of ammonium chloride (the strong solution diluted with an equal volume of water), at any rate for the first half hour, or until the impression is completely covered with copper.

[2] A black deposit forms upon the zinc, and should be occasionally brushed off. It consists of copper, and is due to the copper sulphate slowly diffusing into the porous cell, and being reduced by the zinc.

difficulty in separating them, the point of a knife may be inserted at the edges, and the copper film heated for a moment over a lamp.

The thin, fragile deposit should next be strengthened by backing it up with tin. To do this, moisten the back of it with solution of ammonium chloride, lay it face downwards on a piece of wire gauze, and heat it over a lamp. When the ammonium chloride begins to volatilise in white fumes, press upon the surface a strip of metallic tin (or, better, soft solder). This will melt and spread all over the copper; enough should be applied to make a layer at least 1 mm. thick.

Lastly, the edges should be trimmed by cutting away all superfluous copper, and smoothing irregularities with a file.

If the object is a medal, an electrotype of each side may be taken, and the two placed back to back and soldered together (or simply cemented by shellac) so as to reproduce the original.

If a copy of a seal has been taken, it may be soldered to the head of a brass screw, and the latter inserted into a wooden handle such as those which are sold for bradawls.

[A rather preferable method of depositing copper is to employ a separate battery, such as a single cell of Smee's battery; the mould being connected with the zinc, and a thick strip of copper with the platinised silver of the battery. Both should then be immersed in a jar containing a solution of copper sulphate (150 grms. dissolved in 1 litre of water; 30 c.c. of hydrogen sulphate added). No addition of crystals of copper sulphate should be made: the copper being dissolved from the positive pole (the strip of copper) as fast as it is deposited on the negative pole (the mould).]

2. CADMIUM.

[Symbol of atom, Cd.
Weight of atom, 112 hydrogen-atoms.]

Tests for compounds of Cadmium.

[A solution of cadmium sulphate, containing 1 grm. of the salt in 25 c.c. of water, may be used.]

*1. With hydrogen sulphide they give a bright yellow precipitate insoluble in potassium hydrate.

Add to a portion of the solution a few drops of dilute hydrogen chloride, and then solution of hydrogen sulphide. A yellow precipitate of cadmium sulphide will be produced. Add to the solution in which the precipitate is suspended excess of solution of potassium hydrate, and warm the mixture. The precipitate will remain unaltered, differing in this respect from the two other yellow sulphides, those of arsenic and tin, which are dissolved under the same circumstances.

*2. With ammonia they give a white precipitate, soluble in excess; this solution yields a yellow precipitate on addition of hydrogen sulphide.

Add a drop of ammonia to another portion. A white precipitate of cadmium hydrate will be formed, which will readily dissolve on addition of more ammonia. Add to the clear solution some hydrogen sulphide. A yellow precipitate will be formed, of cadmium sulphide, since the latter is insoluble in ammonia and ammonium sulphide.

*3. Reduced on charcoal they give a reddish-brown incrustation.

Mix a small quantity of cadmium sulphate with an equal amount of sodium carbonate, and heat it on charcoal before the blowpipe. No metallic globule will be obtained, owing to the volatility of cadmium, but the surface of the charcoal will be covered with a characteristic reddish-brown coating of cadmium oxide.

3. BISMUTH.

[Symbol of atom, Bi.
Weight of atom, 210 hydrogen-atoms.]

Compounds of Bismuth.

[Typical examples,—Bismuth nitrate $(Bi(NO_3)_3)$.
    ,,   chloride $(Bi Cl_3)$.
    ,,   oxychloride $(Bi Cl O)$.]

Preparation of Bismuth nitrate.

Place a small fragment of bismuth in a strong porcelain mortar, and strike it with the pestle. It will not spread out, like lead, into a thin plate, but will be crushed into a crystalline powder. Dissolve about 0.5 grm. of the metal in 5 c.c. of hydrogen nitrate, as directed in the case of lead. It will be found to dissolve easily, and the solution of bismuth nitrate (after evaporation to half its bulk, in order to drive off excess of acid) may be reserved for testing, after dilution with about an equal volume (not more) of water.

Preparation of 'fusible metal.'

Although bismuth itself only melts at a temperature of 260°, yet an alloy of it with tin (melting-point 230°) and lead (melting-point 330°) fuses below the temperature of boiling water, viz. at 98°.

Weigh out 4 grms. of bismuth, 2 grms. of lead, and 2 grms. of tin. Melt the bismuth in a clean iron spoon, and add to it the lead and the tin, stirring the melted alloy with a glass rod or piece of iron wire. Pour it out on a clean iron plate or on a tile, and allow it to cool.

Meanwhile, heat some water in a beaker, and when it is boiling put into it a piece of the alloy suspended in a loop of thread. It will readily melt, and drop off the thread.

A still more fusible alloy is made by adding 1 grm. of cadmium to the above ingredients. This melts at a temperature of 80°, or a little below.

[Since the above alloy expands in solidifying, it is well adapted for taking casts of seals, medals, &c. for electrotyping. For this purpose, some of it should be melted and poured out on a clean plate. The edge of a card should be lightly drawn over its surface, to remove dross, and the moment it is observed to become pasty the seal should be brought down quickly upon it. A little practice will be required to hit the exact moment for applying the seal.]

Tests for compounds of Bismuth.

*1. They are, especially bismuth chloride, decomposed by water, forming insoluble basic salts.

(*a*) Pour a few drops of the solution of bismuth nitrate into a test-tube, and add eight or ten times its volume of water. The solution will gradually become cloudy, and deposit a precipitate of bismuth oxynitrate.

(*b*) To another small portion of the solution add a drop of dilute hydrogen chloride, and then a large quantity of water, as above. The liquid will at once turn milky, owing to the formation of bismuth oxychloride[1]. Divide the liquid into two parts,—

(*a*) To one portion add strong hydrogen chloride drop by drop. The precipitate will readily dissolve. [Reserve the solution for test 2.]

(*b*) To the other add an equal volume of solution of sodium-and-hydrogen tartrate. The precipitate will remain undissolved: a property which distinguishes it from the similar precipitate formed by antimony, p. 288.

[The best method of applying this test is, to evaporate a few drops of the solution of bismuth salt to dryness in a watch-glass, to dissolve the residue in a drop of moderately dilute hydrogen chloride, and then, placing the watch-glass on a black surface, to fill it up with water from a wash-bottle. A very slight turbidity may thus be rendered evident.]

*2. With hydrogen sulphide they give a black precipitate.

Add solution of hydrogen sulphide to the clear solution containing bismuth chloride, obtained in the last experiment. A black precipitate of bismuth sulphide will be produced, insoluble in potassium hydrate.

*3. Reduced on charcoal they give a brittle globule of metal, with a light yellow incrustation.

Evaporate the remainder of the solution of bismuth nitrate to dryness, mix the residue with about an equal quantity of potassium cyanide, and reduce it on charcoal in the usual way. Observe the yellow incrustation of bismuth trioxide formed round the cavity in the charcoal, and try the brittleness of the globule by crushing it with the pestle in the mortar.

---

[1] $BiCl_3 + H_2O = BiClO + 2HCl.$

4. ARSENIC.

[Symbol of atom, As.
Weight of atom, 75 hydrogen-atoms.

It should be remembered in dealing with arsenic that the substance and its compounds are extremely poisonous. Very small quantities should be used for testing, and care must be taken not to inhale any of the vapours.]

Compounds of Arsenic.

[Typical examples,—Arsenic trioxide ($As_2 O_3$).
  „   pentoxide ($As_2 O_5$).
  „   trichloride ($As Cl_3$).
Potassium arsenite ($K As O_2$).
  „   arsenate ($K_3 As O_4$).]

Properties of Arsenic Trioxide.

This is the common 'white arsenic' of the shops, and is the source from which the other compounds of arsenic are usually derived.

*1. It volatilises readily, condensing in brilliant octohedral crystals.

Place a very small quantity of arsenic trioxide in a small dry test-tube, and heat it slowly over a lamp with a small flame. It will volatilise completely, forming a crystalline crust in the cool part of the tube. Now warm the part of the tube beyond the deposit, and when it is moderately hot carry the lamp-flame downwards, so as to volatilise the sublimate already formed. It will condense now more slowly, and form large transparent glittering crystals, which, when examined with a magnifier, will be found to resemble Fig. 63, being more or less perfect octohedra belonging to the 'regular' or 'cubic'

Fig. 63.

system. The triangular faces of these will be easily recognised.

2. Its solubility in water.

Boil a very little of it in a test-tube with 8 or 10 c.c. of water; notice the difficulty with which it is wetted by the water, the powder showing a tendency to float as a film upon the surface of the liquid, or remain at the bottom in lumps. It will, however, dissolve to a certain extent, as may be proved by setting the liquid aside to cool, when small brilliant crystals of the trioxide will be formed on the sides of the tube. The solution may be shown to contain arsenic by test 1, p. 284.

Preparation of Arsenic trichloride.

Boil half a gramme of arsenic trioxide with 5 c.c. of strong hydrogen chloride in a test-tube. The substance will readily dissolve [1], and the solution, which contains arsenic trichloride, may be kept for examination.

Preparation of Potassium arsenite.

In the last experiment we have seen that arsenic combines with the chlorine radicle like a metal. But in many respects it resembles more closely a non-metallic element; thus, arsenic trioxide combines with potassium hydrate just as nitrogen trioxide has been already proved to do (p. 140), forming a salt, potassium arsenite, analogous to a nitrite; in which arsenic, like nitrogen, forms part of the electronegative radicle.

Boil 1 grm. of arsenic trioxide with 5 c.c. of solution of potassium hydrate. A solution of potassium arsenite will be formed [2], which may be reserved for examination.

Preparation of Arsenic pentoxide.

This is obtained by oxidising the trioxide by means of hydrogen nitrate.

Place half a gramme of arsenic trioxide in a small eva-

---

[1] $As_2O_3 + 6\,HCl = 2\,AsCl_3 + 3\,H_2O.$
[2] $As_2O_3 + 2\,KHO = 2\,KAsO_2 + H_2O.$

porating dish, pour over it about 5 c.c. of strong hydrogen
nitrate, and heat the mixture (in a draught cupboard, if possible). Orange vapours of nitrogen trioxide will be evolved,
some of the oxygen of the acid being transferred to the arsenic
trioxide [1]. When the liquid has nearly evaporated, add about
2 c.c. more acid and evaporate to complete dryness. Boil the
residue of arsenic pentoxide with about 5 c.c. of solution of
potassium hydrate, dilute the solution of potassium arsenate,
thus obtained, with 20 c.c. of water, and reserve it for testing, p. 285.

Tests for compounds of Arsenic.

A. General Tests.

***1. Heated with sodium carbonate, they are reduced, yield-
ing a sublimate of metallic arsenic.**

Mix a very small quantity of arsenic trioxide with the same
amount of potassium cyanide and about twice as much anhydrous sodium carbonate. Half fill the bulb of an ignition-tube
with the mixture, and wipe off any adhering particles from the
sides of the tube with a twisted strip of blotting-paper. Heat the
bulb, gently at first, over the lamp, and warm the whole of the
tube (held in the crucible tongs) so as to drive out any moisture
which would otherwise condense in it. When the whole is
quite dry (the last portions of moisture may be absorbed by
a strip of blotting-paper, so as to leave the whole of the tube
quite clean), heat the mixture in the bulb more strongly.
A black, shining ring of metallic arsenic will soon form in
the tube. Allow the whole to cool and then cut off the bulb
with a file, retaining the metallic sublimate in the tube; hold
the latter in a slanting position (the part containing the arsenic
being lowest), and heat the ring, beginning at the upper part,
by a lamp with a very small flame. The arsenic will volatilise,

---

[1] $As_2O_3 + 2 HNO_3 = As_2O_5 + H_2O + N_2O_3.$

will be oxidised by the current of hot air ascending in the tube, and will condense near the top of the tube as arsenic trioxide, in small crystals. The heat should not be too strong, or a portion of the arsenic will volatilise unchanged and interfere with the distinctness of the crystalline ring. If this should occur, allow the tube to cool, and re-sublime the deposit as above directed.

2. **Heated with hydrogen chloride, in contact with a strip of copper, they yield a deposit of metallic arsenic.** [Reinsch's test.]

Pour about 5 c.c. of dilute hydrogen chloride into a test-tube, and add 2 or 3 drops (not more) of the solution of arsenic trichloride lately made. Put into the liquid a piece, 4 or 5 cm. long, of fine copper wire, or a narrow strip of copper foil, cleaned with emery paper, and doubled up into a bundle. On boiling the solution, the arsenic will be thrown down on the more electropositive metal, the copper, in the form of a gray metallic film, which will, if the action be continued, separate in scales from the copper. Decant the solution, and pour some fresh water upon the coated copper, to wash away the last traces of the solution; then shake the copper out of the tube upon a piece of blotting-paper and dry it by gentle pressure, taking care not to disturb the film of arsenic; complete the drying by warming the copper over a lamp-flame for a few seconds, then place it in an ignition-tube and draw out the middle of the latter in the blowpipe flame until it is contracted like the tube represented in fig. 20, p. 33. When it is cool, heat the bulb containing the copper, and notice that a ring of crystals is deposited at the contracted portion, which may with a magnifier be easily recognised as octohedra of arsenic trioxide.

This test, which is usually called Reinsch's test, has the advantage that the presence of organic substances (e.g. milk, beer, in cases of poisoning) does not interfere with the deposition of arsenic on the copper. But other metals, such as antimony, bismuth, silver, are deposited on the copper under the same conditions as arsenic. Moreover, when the deposit of arsenic is heated, only a portion of the metal is volatilised, the rest forming a non-volatile alloy with the copper. The test, therefore, is by no means so reliable and characteristic as Marsh's test, which is described in p. 289.

B. Tests for Arsenic protosalts and Arsenites.

[The solutions of arsenic trichloride and potassium arsenite, already made, may be used.]

*1. With hydrogen sulphide, in presence of hydrogen chloride, they give a yellow precipitate, soluble in potassium hydrate.

Add a few drops of the solution of arsenic trichloride to 5 c.c. of water, and test it with solution of hydrogen sulphide. A bright yellow precipitate of arsenic trisulphide (the common 'orpiment' used as a paint) will be formed. Divide the liquid into two portions.

(*a*) To one portion add excess of solution of potassium hydrate and warm. The precipitate will readily dissolve, potassium sulpharsenite ($KAsS_2$) being formed. If a few drops of dilute hydrogen chloride are now added to the clear solution, arsenic trisulphide will be precipitated.

(*b*) Allow the precipitate in the other portion to subside, pour off the liquid, and boil the precipitate with a little strong hydrogen chloride. It will remain almost unaltered, differing in this respect from antimony trisulphide, p. 288. This is given as being one of the methods for separating arsenic from antimony.

*2. With silver nitrate, their solutions, when neutral, give a yellow precipitate.

To a portion of the solution of potassium arsenite add enough dilute hydrogen nitrate to make the solution decidedly acid, and then add a drop of solution of silver nitrate.

[If the hydrogen nitrate or potassium arsenite contained any chloride, a slight white precipitate will form on addition of the silver nitrate, which should be filtered off.]

Mix in another test-tube some solution of ammonia with an equal volume of water, and pour the diluted ammonia from a pipette very slowly and carefully down the side of the test-tube containing the arsenic solution, so that the two fluids may not mix, but the lighter solution (the ammonia) may rest on the surface of the heavier. The tube containing the arsenic solution should be held in an inclined position, in order that the ammonia may flow less rapidly down the side. A yellow film of silver arsenite will be formed at the surface of contact of the two fluids.

We have in the test-tube three fluid strata : a lower one containing excess of hydrogen nitrate, an upper one containing excess of ammonia, and an intermediate one consisting of a neutral solution of ammonium nitrate. Silver arsenite is insoluble in water, but soluble in excess both of acid and of alkali, and therefore appears only where the fluid is neutral.

On shaking the fluid the yellow film will disappear altogether, unless just enough ammonia has been added to neutralise the acid.

3. With copper sulphate, their solutions, when neutral, give a bright green precipitate.

Mix a little of the solution of potassium arsenite with about 5 c.c. of water and add a few drops of solution of copper sulphate. A grass-green precipitate of copper arsenite will be formed. This is the brilliant but very poisonous paint called 'Scheele's green.'

C. Tests for Arsenates.

[The solution of potassium arsenate, already prepared, may be used.]

1. With hydrogen sulphide, they give hardly any precipitate, until reduced to protosalts.

Acidify a portion of the solution of potassium arsenate with hydrogen chloride and divide it into two parts.

(*a*) Add to one portion some solution of hydrogen sulphide.

No precipitate will be produced, and even on heating only a slight yellow precipitate will fall.

(*b*) Add to the remainder some solution of hydrogen sulphite (or of sodium sulphite mixed with hydrogen sulphate), boil the liquid until it no longer smells of sulphur dioxide, and then test with hydrogen sulphide. A yellow precipitate of arsenic trisulphide will be formed, the hydrogen arsenate having been reduced to hydrogen arsenite by the action of the hydrogen sulphite.

*2. With silver nitrate, their solutions, when neutral, give a red precipitate.

Acidify another portion with hydrogen nitrate, add a few drops of solution of silver nitrate, and pour on the solution dilute ammonia, as directed, p. 284. A dull-red precipitate of silver arsenate will form at the point where the solution is neutral.

*3. With magnesium sulphate, in presence of ammonium salts, they give a white crystalline precipitate.

Add a few drops of ammonium chloride to another portion, and then a drop of solution of magnesium sulphate. A white crystalline precipitate will be formed, very similar in appearance, and analogous in composition, to magnesium-and-ammonium phosphate, the phosphorus being replaced by arsenic.

Additional Experiment.

Marsh's test for Arsenic.

This is described in p. 289 & seq., and may be deferred until the reactions of antimony compounds have been examined.

5. ANTIMONY.

[Symbol of atom, Sb (stibium).
Weight of atom, 122 hydrogen-atoms.]

Preparation of compounds of Antimony.

[Typical examples,—Antimony trichloride (Sb Cl$_3$).

„   trisulphide (Sb$_2$ S$_3$).

„   pentoxide (Sb$_2$ O$_5$).

Antimony and potassium tartrate (K (Sb O) C$_4$ H$_4$ O$_6$).]

1. Preparation of Antimony trichloride.

Place a small fragment (about as large as a pea) of metallic antimony in a strong mortar, and strike it with the pestle. It will split to pieces and may readily be ground into a fine crystalline powder. Put a little of this powder into a test-tube, add 5 or 6 c.c. of strong hydrogen chloride and one or two drops (not more) of hydrogen nitrate, and apply heat. The metal will dissolve pretty readily, forming antimony trichloride. When the action has ceased, allow the liquid to cool, and decant the clear solution into another test-tube for use in experiments.

2. Formation of Antimony pentoxide.

When strong hydrogen nitrate is poured upon antimony, the metal is not (like most others) dissolved as nitrate, but combines with some of the oxygen of the acid, forming antimony pentoxide.

Put a little powdered antimony into a test-tube, pour over it a few drops of strong hydrogen nitrate, and heat gently. Orange vapours of nitrogen tetroxide will be given off abundantly, and the antimony will be converted into a white powder, the pentoxide, insoluble in the acid.

Properties of compounds of Antimony.

*1. They are, in some cases, decomposed by water, forming basic salts.

Pour a few drops of the solution of antimony trichloride (prepared just now) into a test-tube and add 5 or 6 c.c. of water. A white precipitate, consisting of antimony oxychloride, will be produced. Divide the liquid into two portions ;—

(a) To one add a little strong hydrogen chloride. The precipitate will dissolve, and the solution may be reserved for use in experiment 2.

(*b*) To the other add some solution of sodium-and-hydrogen tartrate. The precipitate will in this case also be redissolved (differing in this respect from the corresponding bismuth oxychloride, p. 279).

*2. With hydrogen sulphide they give an orange precipitate, soluble in potassium hydrate and ammonium sulphide, and in strong hydrogen chloride.

Add to the acid solution of antimony trichloride, obtained in the last experiment, some solution of hydrogen sulphide. A characteristic orange precipitate of antimony trisulphide will be obtained.

Allow this to subside (which it will do more quickly if gently warmed), pour off the fluid, and shake up the precipitate with a little water: then, before it has settled again, pour off half of it into another tube;—

(*a*) To this portion add a little solution of potassium hydrate, and warm it. The precipitate will readily dissolve, potassium sulphantimonite being formed.

(*b*) Pour off the clear liquid from the remainder, and boil the precipitate with a little strong hydrogen chloride. It will dissolve, differing in this respect from arsenic trisulphide, p. 284.

*3. When placed in contact with zinc and platinum, they are reduced on the latter as a black powder.

Place a piece of platinum foil in an evaporating dish; put on it a small fragment of granulated zinc, and pour into the dish a little dilute hydrogen chloride. When the evolution of hydrogen has begun from the surface of the platinum, add a drop or two of the solution of antimony trichloride. The antimony will be thrown down on the platinum as a black coating, owing to galvanic action (p. 70). If the black deposit is heated with a drop of ammonium sulphide it becomes orange[1].

*4. Reduced on charcoal, they give a brittle metallic globule, with a white incrustation.

[1] The platinum may be cleaned from the deposit by warming it with a few drops of strong hydrogen chloride, to which one drop of hydrogen nitrate may be added.

Mix a little antimony trisulphide or antimony-and-potassium tartrate (tartar emetic) with potassium cyanide, and heat the mixture on charcoal before the blowpipe. Brilliant globules of metal will be produced, which lose their lustre at once on being removed from the reducing flame, and white fumes of antimony trioxide are produced and form an incrustation on the cool part of the charcoal. Owing to the volatility of the metal, these white fumes of oxide are emitted for some time after the globule is removed from the flame, especially if a stream of air from the blowpipe is directed upon it. When cold, the globule should be detached from the charcoal, placed in a mortar, and the pestle pressed forcibly down upon it. It will, as already observed, be crushed to powder instead of spreading out into a plate.

Additional Experiment.

Marsh's Test for Arsenic and Antimony.

Apparatus required.—Small flask; thistle funnel; two elbow tubes, one with long branch; retort stand; Bunsen's holder; drying tube, filled with calcium chloride or quicklime; hard glass tubing, 4 or 5 mm. external diameter; Bunsen's burner; Herapath's blow-pipe; small porcelain dish, or broken pieces of one; three-square file; test tubes in basket; small funnel; filters.

A. Detection of Arsenic.

If a solution containing arsenic is brought into contact with nascent hydrogen, a volatile compound of arsenic and hydrogen is formed, corresponding in composition to the compound of nitrogen and hydrogen, or ammonia, and to one of the compounds of phosphorus and hydrogen. This reaction furnishes us with a most delicate test for arsenic, since the hydrogen arsenide is very readily decomposed, and thus the presence of arsenic in it may be demonstrated. The best method of applying the test, which is known as Marsh's test, is as follows :—

Place a few pieces of granulated zinc in a small flask, fitted

with an acid funnel and right-angled elbow tube, and connect
the latter with a drying tube filled with fragments of calcium
chloride or quicklime, and supported in a horizontal position by
a Bunsen's holder. Take a piece of difficultly fusible glass
tubing about 4 mm. in diameter and 40 cm. in length; soften
it in the middle by means of the blowpipe-flame, and draw the
two ends asunder, until the softened portion is contracted to
a diameter of 2 mm. When it is cool, make a scratch with a
file in the middle of the contracted portion, and break it at that
point. You have now two tubes each about 18 cm. in length

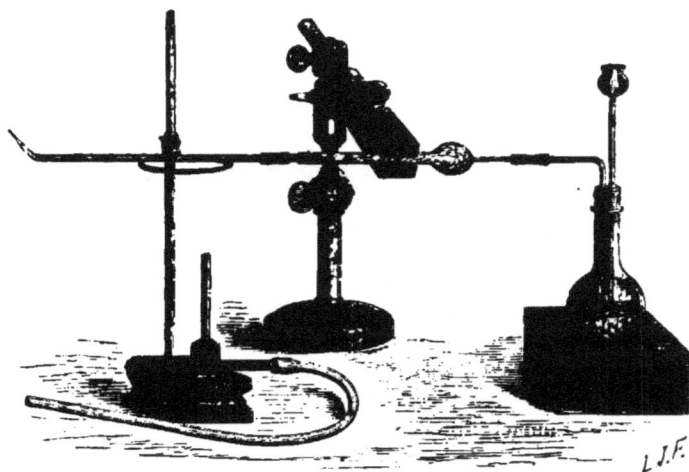

Fig. 64.

and terminated by a jet[1]. Reserve one for an experiment with
antimony, and connect the other with the drying tube, supporting it horizontally on the largest ring of the retort-stand at
such a height that a lamp may be placed underneath it. The
whole apparatus will then be arranged thus, Fig. 64.

Pour a little water on the zinc in the flask, and add a few

---

[1] The jet shown in the figure is turned up. This, which is a slightly
preferable form, may be obtained by moving the hands, in drawing out the
glass, laterally (the one hand towards, the other from the body), so as to
give the drawn-out tube the following form   . It should
then be cut in two at a.

drops of strong pure hydrogen sulphate, so as to generate a moderately rapid current of hydrogen. Allow the gas to pass through the apparatus for about two minutes, in order to sweep out all traces of air. While this is going on, you may get ready a right-angled elbow tube with one long branch, a test-tube half filled with a dilute solution of silver nitrate (5 drops of the test-solution to 10 c.c. of water), and one or two pieces of clean porcelain, such as the lid of a crucible, or fragments of an evaporating dish. Now test the purity of the gas by collecting some in a test-tube held over the jet (see p. 111). When you are quite sure that the gas is unmixed with air, you may light it at the jet and depress into the flame, for a few seconds, one of the pieces of porcelain. If any stain appears on the white surface (as may happen if the zinc or hydrogen sulphate is contaminated with arsenic) allow the stream of gas to pass a little longer, and do not proceed with the experiment until the porcelain depressed on the flame remains unstained. [If after several trials the stain still appears, the zinc is impure, and another sample must be used.] When this is the case, pour into the funnel about two drops (not more) of a solution of arsenic chloride, and wash it down into the flask with a little water. The first effect of this will be to increase the evolution of gas, and the flame of the jet will shortly become tinged with gray (not unlike the colour imparted to flame by the presence of potassium vapour). Hydrogen arsenide is now being formed, and you may proceed to decompose it in the three following ways :—

(a) **By the heat produced by its own combustion.**

Place in the flame a clean cold surface of porcelain, such as the glazed interior of an evaporating dish, depressing it until it almost touches the jet from which the gas is issuing. A black spot, consisting of metallic arsenic or of a lower hydride, will be formed on the white surface. Two or three of these should be made on different parts of the porcelain, and reserved for future examination [1].

---

[1] They are formed for the same reason that soot is deposited on a plate held in a candle-flame. The gaseous compounds of carbon and hydrogen

(*b*) **By heat applied to the tube through which the gas is passing.**

Place a lamp underneath the tube near the end at which the gas enters, and heat that part of the tube to redness. A black, shining ring of arsenic will be deposited just beyond the heated portion of the tube, while the flame at the jet will lose its gray colour, showing that the arsenic has been arrested in its course to the jet.

(*c*) **By a solution of silver nitrate.**

Detach the drying tube (and the tube containing the arsenic deposit), and fit in its place the elbow tube, the longer branch of which should reach to the bottom of the test-tube containing the solution of silver nitrate : the latter may be supported in an upright position by the smallest ring of the retort-stand. Pour, if necessary, a little more acid and solution of arsenic chloride into the flask. The solution of silver nitrate will soon become turbid, and a black deposit of metallic silver will be formed, while silver arsenite will remain in solution in the hydrogen nitrate which is formed [1].

The evolution of gas should now be stopped, by taking the cork out of the flask and pouring the solution away from the zinc, which latter, after being thoroughly washed, may be retained in the flask for a similar experiment with antimony.

You have now three results of the decomposition of the hydrogen arsenide, which are to be further examined.

(*a*) **The spots on the porcelain.**

Dip a glass rod in strong hydrogen nitrate, and moisten one of the spots with it. The metal will dissolve in the acid, especially if the porcelain be gently warmed, and on evaporation a white residue of hydrogen arsenate will be left. Add another drop of acid, and again evaporate to dryness, to ensure

---

are decomposed by the heat of combustion; the hydrogen is completely oxidised to water, the carbon is only partially oxidised to carbon dioxide, the excess of carbon being suspended in the flame and deposited on any cold surface brought in contact with it. Similarly in the present experiment the hydrogen of the hydrogen arsenide is oxidised into water, while an arsenic soot is deposited on the porcelain.

[1] $H_3As + 9 AgNO_3 + 3 H_2O = Ag_3AsO_3 + 9 HNO_3 + 3 Ag_2$.

the complete oxidation of the arsenic. Do not raise the temperature beyond the point necessary to drive off the excess of hydrogen nitrate, lest the hydrogen arsenate should be itself decomposed. Moisten the residue with a drop of solution of silver nitrate and again evaporate to dryness. A red deposit of silver arsenate will be left, readily seen on the white porcelain.

(*b*) **The deposit formed in the tube.**

Connect the tube with a drying tube (filled with calcium chloride, not quicklime), and arrange them as before, when the deposit was to be formed (placing the whole in a draught cupboard, if possible). In place of the hydrogen generating flask attach to the drying tube a test-tube fitted with a thistle funnel and elbow tube, and containing one or two lumps of iron sulphide and a little water. Pour down the funnel a few drops of strong hydrogen sulphate, just sufficient to liberate a slow stream of hydrogen sulphide gas. When the air has been expelled from the apparatus, light the gas at the jet (merely to prevent its escape into the room), and heat the metallic deposit by a spirit lamp with a very small flame, beginning at its outer border, or that farthest from the end at which the gas enters. The arsenic as it volatilises will decompose the hydrogen sulphide, combining with its sulphur and forming a yellow ring of arsenic trisulphide in the cool part of the tube, which is very volatile and may be driven about from one part of the tube to another by the heat of the lamp.

(*c*) **The solution containing silver arsenite.**

Filter this from the reduced silver, and pour gently upon its surface some very dilute solution of ammonia from a pipette (see experiment 2, p. 284). A yellow stratum of silver arsenite will be formed at the point where the solution is neutral.

B. Detection of Antimony.

Antimony, like arsenic, forms a volatile compound with hydrogen, which may be decomposed by heat and silver nitrate, but with results which enable us to distinguish it without difficulty from hydrogen arsenide.

Arrange an apparatus similar to that already described in the case of arsenic. After making sure that the hydrogen evolved is pure and leaves no stain upon the porcelain, pour into the funnel two drops of solution of antimony trichloride, and wash it down with a little dilute hydrogen sulphate. Then proceed with the experiments precisely as already directed for arsenic, and compare the results obtained in the two cases.

(a) **Decomposition of the gas by the heat of combustion.**

The dark spots formed on the porcelain are more like soot, and not so brown and lustrous as those of arsenic. They leave a white residue when treated with hydrogen nitrate, but yield no red deposit with silver nitrate [1].

(b) **Formation of a black deposit of antimony in the heated tube, and subsequent conversion of this into sulphide, by passing hydrogen sulphide over it.**

A dark orange-red film of antimony trisulphide will be formed, close to the deposit in the tube, which is only volatilised with difficulty, by the utmost heat of the lamp.

(c) **Decomposition of the gas when passed into solution of silver nitrate.**

The whole of the antimony is precipitated as silver antimonide [2]. No yellow layer or precipitate, therefore, is seen when the filtrate is neutralised by ammonia. In order to prove the presence of antimony in the precipitate, wash it on the filter, then place the portion of the filter which contains the precipitate in a beaker and boil it with a little solution of sodium-and-hydrogen tartrate together with one or two drops of hydrogen chloride, when the antimony will dissolve while the silver will remain as insoluble chloride. Filter the solution and test it with hydrogen sulphide, which will give the characteristic orange precipitate of antimony trisulphide.

It is obvious that in cases where we have both antimony and arsenic in the same solution this last test gives us a means of separating the two metals, the antimony being found in the

---

[1] A black deposit of reduced silver is sometimes formed, if a sufficient excess of hydrogen nitrate has not been added.

[2] $H_3Sb + 3 AgNO_3 = 3 HNO_3 + Ag_3Sb$.

precipitate produced when the gases are passed into the silver nitrate solution, while the arsenic remains in solution.

The two metals may also be discriminated by the difference in volatility of their sulphides. When the mixed metallic deposit in the tube is heated in a current of hydrogen sulphide, the yellow ring of arsenic trisulphide is always deposited considerably in advance of the orange sublimate of antimony trisulphide.

6. TIN.

[Symbol of atom, Sn (stannum).
Weight of atom, 118 hydrogen-atoms.]

Crystalline character of tin.

Take a strip of tin (the best tin is usually sold in strips or thin rods), bend it quickly backwards and forwards several times, holding it to the ear while doing so. A crackling sound will be heard, and the metal will become sensibly warm at the point of flexure. Cast tin although to a certain extent malleable and ductile is crystalline in structure, and the crackling sound and heat is due to the motion of the crystals over each other, and their mutual friction.

The crystalline structure may also be shown by washing the surface of a strip of tin plate (which is sheet iron coated with tin) with a little alcohol to free it from grease, and then, after rinsing it with water, applying with a piece of sponge or tow a mixture of 5 c.c. dilute hydrogen chloride with 2 c.c. of dilute hydrogen nitrate (the ordinary laboratory solutions). The bright surface will soon show a variety of irregular patches, somewhat like 'watered' silk, and should then be well washed with clean water, to stop further action of the acid. The reason is, that the film of tin upon the iron consists of numerous small crystals buried in a mass of less crystalline material. The acid attacks the latter more readily than the crystals, and lays bare the faces of the latter, showing their symmetrical arrangement, like masses of soldiers all facing the same way.

Preparation of compounds of Tin.

[Typical examples,—Tin protochloride (Sn Cl$_2$).
,,   perchloride (Sn Cl$_4$).
,,   dioxide (Sn O$_2$).]

Tin, like mercury, forms two well-defined series of salts, in one of which it is diatomic, in the other tetratomic. The conditions necessary for the formation of each series are, presence of excess of the metal for the protosalts, and excess of the acid radicle for the persalts.

1. Tin protochloride.

Place about 2 grms. of *pure* tinfoil (see note[1]) in a large test-tube, add 10 c.c. of strong hydrogen chloride, and boil for a quarter of an hour. Hydrogen will be evolved, the metal taking its place and forming tin protochloride[2]. Pour off the solution while a little of the metal still remains undissolved, dilute it with water to 30 c.c., and reserve it in a corked tube or bottle for experiments, placing in it a small bit of metallic tin to prevent its passing into the state of persalt.

2. Tin perchloride.

Put 1 grm. of tinfoil into a test-tube, pour on it 5 c.c. of strong hydrogen chloride, and two drops (not more[3]) of strong hydrogen nitrate, and heat gently. The metal will readily dissolve, and when the solution is complete the liquid should be poured into a porcelain dish and evaporated on the sand-bath, about 1 c.c. of strong hydrogen nitrate being added to convert any remaining protosalt into persalt (which it does, of course. by causing the liberation of chlorine, p. 193). To make sure that the conversion into perchloride is complete, take out a drop

---

[1] Much of what is sold as tinfoil is an alloy of tin and lead. The pure tinfoil may be known by its comparatively rough surface and greater thickness. The sheets of the alloy are much thinner and more pliable, and have a bright polished surface.

     If tinfoil is not at hand, a little tin should be granulated as directed in the case of zinc (p. 28).

[2] $2 \, HCl + Sn = Sn \, Cl_2 + H_2.$

[3] If much hydrogen nitrate is added at once, the tin may be precipitated as perhydrate.

or two of the liquid with a pipette, dilute it with water and test it with a drop of mercury perchloride. If any precipitate is produced, some tin protochloride still remains, and a few more drops of hydrogen nitrate must be added. Finally, evaporate the solution to half its original bulk (stopping the evaporation when white fumes of the perchloride appear), dilute it with water to 30 c.c., and reserve the solution for experiments.

3. Tin dioxide.

This is formed by the action of strong hydrogen nitrate upon tin, which is quite analogous to its action upon antimony, p. 287.

Place a bit of granulated tin, or of tinfoil, in a test-tube, and pour on it some strong hydrogen nitrate. Notice the violent action, the evolution of orange vapours of nitrogen tetroxide, and the formation of a white powder, tin dioxide, which is insoluble in excess of the acid.

Properties of compounds of Tin.

A. Protosalts (Stannous Salts).

[The solution of tin protochloride already made, may be used.]

*1. With hydrogen sulphide they give a dark brown precipitate, soluble in potassium hydrate and ammonium sulphide.

Add excess of solution of hydrogen sulphide to a portion of the solution of tin protochloride. A dark brown, nearly black precipitate of tin protosulphide will be formed. Divide the liquid into two portions,—

(*a*) To one add some solution of potassium hydrate, and warm it. The precipitate will redissolve, a double sulphide of tin and potassium being formed. On addition of dilute hydrogen chloride to this, the brown protosulphide will be reprecipitated.

(*b*) To the other portion add a slight excess of ammonia, then some solution of ammonium sulphide (which should be yellow, owing to the presence of higher sulphides), and warm the mixture. The precipitate will redissolve, as in the last experiment, but in this case a higher sulphide of tin is formed, viz. tin disulphide, which forms a double salt, ammonium sulphostannate. On addition of excess of dilute hydrogen chloride to the clear solution, a light yellow precipitate of tin disulphide will be formed.

*2. With mercury perchloride they give a white precipitate, quickly turning gray.

To another portion add a drop of solution of mercury perchloride. A white precipitate of mercury protochloride will be produced, the colour of which changes, however, immediately to gray (the tin salt being in excess) owing to its reduction to metallic mercury [1].

*3. Reduced in contact with zinc and platinum they yield a gray spongy deposit of metal.

Put a piece of platinum foil in a porcelain dish, add a little dilute hydrogen chloride, and drop in a bit of granulated zinc. Hydrogen will of course be evolved. Now add one or two drops of solution of tin protochloride; metallic tin will be quickly reduced as a porous, moss-like mass (similar to the 'lead-tree,' p. 268) chiefly round the zinc, and readily washed away from both the zinc and the platinum by a stream of water. Compare the similar experiment with antimony, p. 288, in which the deposit obtained was formed on the platinum alone, black, and closely adherent.

*4. Reduced on charcoal, they yield a malleable metallic globule, with a white incrustation.

The mode of performing this experiment has been already given, p. 94.

B. Persalts (Stannic Salts).

[The solution of tin perchloride (p. 297) may be used.]

*1. With hydrogen sulphide they give a yellow precipitate, soluble in potassium hydrate and ammonium sulphide.

---

[1] (i) $Sn\,Cl_2 + 2\,Hg\,Cl_2 = Sn\,Cl_4 + Hg_2\,Cl_2.$
(ii) $Sn\,Cl_2 + Hg_2\,Cl_2 = Sn\,Cl_4 + 2\,Hg.$

Add solution of hydrogen sulphide to a portion of the solution of tin perchloride, and apply heat. A light yellow precipitate of tin disulphide will be formed, increased in quantity and rendered flocculent, as the temperature rises. Divide the liquid into two portions and try its solubility in potassium hydrate and ammonium sulphide as directed in the case of the protosalts, p. 296.

*3. With mercury perchloride they give no precipitate.
*4. Reduced in contact with zinc and platinum. ⎫
*5. Reduced on charcoal before the blow-pipe. ⎬

The results with these two tests are the same as in the case of the protosalts.

7. GOLD.

[Symbol of atom, Au (aurum).
Weight of atom, 197 hydrogen-atoms.]

[Solutions containing gold should not, of course, be thrown away, but put into a bottle labelled 'Gold Residues.' For the method of recovering gold from them, see Appendix B.]

Preparation of Gold trichloride.

[Formula of molecule, Au Cl₃.]

Place a leaf of gold in a test-tube, by placing a slightly damp glass rod in contact with one edge of the leaf, rolling the latter round the rod, then transferring it to the tube and unrolling it within. Add about 2 c.c. of strong hydrogen chloride, and observe that the metal will not dissolve in the acid, even on boiling. Treat another leaf of gold with hydrogen nitrate in a similar way; it will also be found to be unacted upon. Now mix the contents of the two tubes; the metal will at once dissolve in the 'aqua regia' thus formed (see p. 193), and a yellow solution of gold perchloride will be obtained. Pour this into a porcelain dish, add two or three more leaves of gold, and a little more hydrogen chloride, and evaporate nearly, but not quite, to dryness on a sand-bath[1].

Dilute the solution with about 15 c.c. of water, and use it for the following experiments.

Tests for compounds of Gold.

*1. With hydrogen sulphide they give a black precipitate, soluble in potassium hydrate and ammonium sulphide.

Add excess of hydrogen sulphide to a portion of the solution of gold trichloride, and warm the mixture. A black precipitate (or if the solution is dilute, a deep black liquid) will be obtained, which will slowly dissolve on addition of excess of potassium hydrate, forming a nearly colourless solution of potassium sulphaurate.

*2. With iron protosulphate they yield a brown precipitate of metallic gold.

Add to another portion some solution of iron protosulphate. The liquid will rapidly darken, and metallic gold will be precipitated as a brownish powder (the liquid in which the precipitate is suspended having a blue or green colour by transmitted light), or if the solution is strong, as a spongy mass. If the quantity of precipitate is sufficient, it may be washed by decantation and dried in a watch-glass; if a portion of it be then burnished with the end of a glass rod or of a test-tube, it will acquire the metallic yellow lustre of gold.

*3. With tin chloride they yield a purple precipitate.

Test another portion of the solution of gold trichloride with solution of tin protochloride to which one drop of iron perchloride has been added (in order to form a little perchloride in the liquid). A brownish purple precipitate will be formed, called from its discoverer, the 'purple of Cassius,' containing gold, tin, and oxygen, but not, apparently, of definite composi-

---

[1] If you wish to drive off the whole of the excess of acid, the evaporation should be finished on the water-bath, since, if the temperature is raised much above 100°, an insoluble protochloride is formed. For the following experiments a slight excess of acid is of no consequence.

tion. When the solutions are very dilute, the colour of the liquid changes to brownish red, but no precipitate falls which can be separated by filtration.

Additional Experiment.

Transparency of Gold leaf.

Gold is the only metal which has with certainty been reduced to sheets so thin that light will pass through them; their thickness being, in fact, no greater than $\frac{1}{10000}$ mm.

In order to examine this, it is best to spread out the leaf flat upon a glass plate, which may be done in the following way.

Clean the glass plate, which should be about 15 cm. square, with a tuft of cotton-wool dipped in alcohol, dry it with a clean cloth, and lay it flat on the table. Slightly damp the edge of a paper-knife or a glass rod, and lay it on one edge of the leaf of gold, to which it will adhere; you can then transfer the leaf to the plate and lay it approximately flat on the glass. Detach the paper-knife from the gold by slightly rubbing it against the glass, and then, still keeping the glass horizontal, direct a very gentle stream of water from a wash-bottle under the leaf of gold. The water will spread quickly over the glass, and the leaf will float on it, losing all inequalities and wrinkles. You have now only to raise very carefully and slowly one corner of the glass, holding a glass rod at the opposite corner to guide the water in flowing off, and to absorb the remaining water with blotting-paper, finally holding the plate over a lamp until it is perfectly dry. The leaf of gold is left as a perfectly smooth film on the glass. Hold it up to the light, and notice the deep green colour of the transmitted light. Examine it with a magnifier; many holes will be seen, and much inequality of thickness, but the greater portion of the leaf is a continuous film, even under high magnifying powers.

8. PLATINUM.

[Symbol of atom, Pt.
Weight of atom, 197 hydrogen-atoms.]

[All waste solutions &c. containing platinum should be put into a bottle labelled 'Platinum Residues.' To recover the metal from them they may be treated as directed in Appendix B.]

Preparation of compounds of Platinum.

[Typical examples,—Platinum perchloride (Pt Cl₄).
Platinum-and-ammonium chloride (Pt (H₄ N) Cl₆).]

1. Platinum perchloride.

For the experiments on platinum compounds the ordinary laboratory solution of platinum perchloride may be used; but if you have any waste scraps of platinum foil or wire they may be converted into the tetrachloride in the following way.

Measure 6 c.c. of strong hydrogen chloride into a large test-tube, add 2 c.c. of strong hydrogen nitrate, drop in about 0.2 grm. of thin platinum foil or wire, and heat the tube on the sand-bath. Platinum, like gold, is not soluble either in hydrogen chloride or in hydrogen nitrate, but is readily acted upon by the chlorine evolved during the mutual decomposition of these acids, and dissolves as perchloride. Evaporate the solution in a dish, in a draught cupboard or the open air, observing the same precautions as in the case of gold, since an insoluble platinum protochloride is formed if the temperature is high. Dilute the solution with about 10 c.c. of water for use in the following experiments.

2. Platinum-and-ammonium chloride.

Put about 3 c.c. of the solution of platinum perchloride into a test-tube, add an equal volume of strong solution of ammonium chloride, and shake the mixture. A yellow, crystalline precipitate of platinum-and-ammonium chloride will be formed[1],

---

[1] The ordinary platinum foil &c. often contains iridium. If this is the case, the solution of the perchloride will be deep red, and the precipitate formed by ammonium chloride will be dull, brick-red, instead of yellow. The presence of iridium will not, however, interfere with the experiments.

the separation of which will be hastened by the addition of 5 or 6 c.c. of common alcohol (in which the precipitate is less soluble than in water). Leave the precipitate to subside for a few minutes; then pour the whole on a small filter, and wash it with water to which about one-fourth its volume of alcohol has been added, rinsing the precipitate as much as possible down to the point of the filter. Lastly, dry it on the filter at a gentle heat, while other experiments are proceeded with.

3. Spongy platinum.

This is the condition in which the metal is obtained by gently igniting platinum-and-ammonium chloride.

Take the dry (or partially dry) filter containing the precipitate of platinum-and-ammonium chloride, cut off the part of it which contains the precipitate, fold this up as closely as possible, and coil the end of a piece of platinum wire two or three times round it, so as to form a compact bundle. Heat this gradually to low redness over the flame of a Bunsen's burner, until all the paper has been burnt away and nothing remains but a gray mass of metallic platinum. During the ignition ammonia and chlorine are given off, and the skeleton of metal which remains is highly porous and has the property of absorbing large quantities of gases. It should be kept for use in the following experiments.

[Platinum may be obtained in a state of division still finer than platinum sponge—in particles so fine, in fact, that they almost cease to reflect light, and form a black impalpable powder. The readiest method of preparing 'platinum black' is to add excess of sodium carbonate to platinum perchloride, and then to boil the liquid with a little grape sugar. Carbon dioxide is evolved with strong effervescence, and a black powder is formed (gradually, if the solution is dilute) which should be washed by decantation successively with dilute alcohol, hydrogen chloride, potassium hydrate, and finally water, and then dried at a gentle heat. This powder is still more effective than spongy platinum in producing chemical combination.]

Action of platinum in causing chemical combination.

This is best shown when the metal is in a finely divided state, as in the form of spongy platinum and platinum black.

But even platinum in the ordinary compact condition shows the same property to a certain extent, as has been already illustrated in the oxidation of ammonia, p. 130. The action is thought to be due to the power possessed by platinum of condensing gases upon its surface, and thus bringing the molecules closely into contact.

1. **Combination of oxygen and hydrogen.**

(*a*) Fit a glass jet by means of a cork to a moderate sized test-tube; put into the latter a bit of granulated zinc and pour on it some dilute hydrogen sulphate; then fit the cork again into its place. Allow the stream of hydrogen to escape for at least a minute, and test its purity by collecting some in a test-tube, as directed in p. 111. When pure hydrogen is issuing from the jet, hold a little in front of it the spongy platinum already prepared, still held in the coil of wire. It will become red-hot, owing to the combination at its surface of the hydrogen with the oxygen of the air, and the gas will ignite at the jet [1].

(*b*) Hold the spongy platinum in a stream of coal gas and air (not ignited) issuing from the tube of a Bunsen's burner. The metal will become red-hot, but the gas will not ignite, since coal gas requires a higher temperature than hydrogen to kindle it.

(*c*) Place a piece of fine wire gauze on the chimney of an Argand burner, and put upright upon it a small cylinder of platinum foil, made by rolling it loosely round a pencil or small test-tube. Turn on the gas partially, and light it above the wire gauze. When the platinum has become red-hot, extinguish the flame by pinching the india-rubber connecting tube, and immediately (before the platinum has had time to cool entirely) allow the gas to flow again. The platinum will begin to glow afresh, and maintain a steady red heat. If a dry gas bottle is held close over it, moisture will be deposited, and the presence of carbon dioxide may be proved in the usual way.

2. **Combination of ether with oxygen.**

Pour a few drops of ether into a beaker or wine-glass, and cover the glass partially with a card, to the centre of which is

---

[1] If the spongy platinum has been exposed to the air for some time, it may be necessary to ignite it gently afresh, and let it cool before use.

attached the wire carrying the spongy platinum (or a coil of platinum wire) previously heated to redness in a lamp, so that it may hang down within the glass nearly to the bottom. Pungent vapours of aldehyd are at once produced, and this compound is further oxidised to hydrogen acetate. The presence of the latter may be shown by holding in the beaker a strip of blue litmus paper which will be strongly reddened [1].

Tests for compounds of Platinum.

[The solution of platinum perchloride may be used.]

*1. With hydrogen sulphide they give a black precipitate, soluble in potassium hydrate and ammonium sulphide.

Dilute a few drops of the solution with 5 or 6 c.c. of water, add solution of hydrogen sulphide, and warm it. A black precipitate of platinum persulphide will be formed, which will dissolve, but only with difficulty, in excess of solution of potassium hydrate.

*2. With potassium salts they give a yellow crystalline precipitate.

Add a few drops of dilute hydrogen chloride to a little of the solution of platinum perchloride (which should not be diluted with water), and then a drop or two (not sufficient to neutralise the acid) of solution of potassium hydrate, and stir the mixture with a glass rod if the precipitate does not form at once. A yellow crystalline precipitate will be more or less quickly formed, especially along the lines where the sides of the tube were touched by the glass rod. This consists of platinum-and-potassium chloride, analogous to the platinum-and-ammonium chloride already prepared.

---

[1] The successive stages of the oxidation may be expressed as follows :—

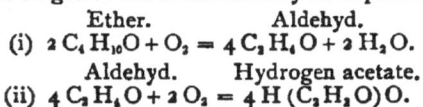

$$\text{Ether.} \qquad \text{Aldehyd.}$$
$$\text{(i)} \quad 2\,C_4H_{10}O + O_2 = 4\,C_2H_4O + 2\,H_2O.$$
$$\text{Aldehyd.} \qquad \text{Hydrogen acetate.}$$
$$\text{(ii)} \quad 4\,C_2H_4O + 2\,O_2 = 4\,H(C_2H_3O)O.$$

Group III.

Metals which are separated from solutions by Ammonium sulphide.

IRON, COBALT, NICKEL, MANGANESE, CHROMIUM, ALUMINIUM, ZINC.

1. IRON.

[Symbol of atom, Fe (ferrum).
Weight of atom, 56 hydrogen-atoms.]

Compounds of Iron.

[Typical examples,—Iron magnetic oxide ($Fe_3 O_4$).
      „   peroxide ($Fe_2 O_3$).
      „   protosulphate ($Fe SO_4$).
      „   persulphate ($Fe_2(S O_4)_3$).]

[The combination of iron with sulphur has been illustrated already, p. 216.]

1. Formation of iron magnetic oxide.

Clean a small strip of sheet iron with a file or with emery-paper, until it shows a bright metallic surface; then hold it with a pair of pliers or crucible tongs in the flame of a Bunsen's burner or spirit lamp. The surface will soon lose its lustre, becoming in succession light yellow, orange, blue, and finally dark gray, owing to the formation of an extremely thin film of oxide. If the strip of iron be maintained at a red heat for a few minutes (in a blowpipe flame or in an ordinary fire), the film of oxide will increase in thickness, and will be detached in the form of black scales when the iron is quenched in cold water. These scales consist of the same iron oxide as that which was formed when the watch-spring was burnt in oxygen gas (p. 103). When a magnet is brought near them they will be attracted by it; and this oxide, from this property and from its identity in composition with the native loadstone, is called the magnetic oxide.

2. Formation of iron peroxide (ferric oxide).

Fill the bulb of an ignition-tube with powdered iron proto-sulphate, and heat it strongly in the flame of a Bunsen's burner, holding it nearly horizontally in the crucible tongs. The salt will first melt in its water of crystallisation, then it will turn white, give off aqueous vapours, and lastly, as the temperature rises nearly to redness, it will decompose, giving off vapours of sulphur dioxide and trioxide, the latter of which will combine with the water to form an extremely strong kind of hydrogen sulphate which will redden litmus paper held at the mouth of the tube. A red residue will remain in the bulb, which consists of iron peroxide, or sesquioxide (the ordinary 'rouge' used for polishing, &c.)[1].

[This illustrates the derivation of the term 'oil of vitriol' applied to hydrogen sulphate, since the old name for iron protosulphate is 'green vitriol.' The acid thus obtained is called 'Nordhausen acid' from the place where it is made.]

3. Preparation of iron proto- and per-sulphates.

Iron forms at least two well-defined series of salts, in one of which (the protosalts) 1 atom of the metal replaces 2 atoms of a monatomic radicle, such as hydrogen; in the other (the per-salts) 1 atom replaces 3 atoms (or 2 atoms replace 6 atoms) of a similar radicle[2].

The conditions necessary to form them are, as usual, excess of the metal for the protosalts, and excess of the non-metallic radicle for the persalts. But the oxygen of the air, or that dis-solved in the water, so readily determines the change of atom-icity that it must be excluded as far as possible in preparing the protosalts.

A. Iron protosulphate (ferrous sulphate).

Take about 2 grms. of fine iron wire (or of iron filings, if

---

[1] $4\,\mathrm{Fe\,SO_4 + H_2\,O = 2\,Fe_2\,O_3 + 2\,SO_2 + H_2\,S_2\,O_7}$.

[2] This will be seen by comparing the formula of each of the two iron sulphates with that of a quantity of hydrogen sulphate containing the same number of atoms of the sulphate radicle.

Iron protosulphate, $\mathrm{Fe\,SO_4}$. | Iron persulphate, $\mathrm{Fe_2(SO_4)_3}$.
Hydrogen sulphate, $\mathrm{H_2\,SO_4}$. | Hydrogen sulphate, $\mathrm{H_6(SO_4)_3}$.

wire is not at hand), clean it from rust, if necessary, by drawing it between folds of emery paper, form it into a small close coil and place it in a moderate-sized test-tube, to which must be fitted a cork in the side of which a small nick has been cut to allow gas to escape. Pour on the iron about 10 or 12 c.c. of dilute hydrogen sulphate, and add a few drops of the strong acid and warm gently if the action appears slow. Hydrogen gas will be given off with effervescence[1], the iron taking its place and forming iron protosulphate (compare the action of zinc on hydrogen sulphate, p. 106). Allow the action to go on for 10 or 12 minutes (keeping up a brisk action by warming the liquid and adding a little more strong acid when required) while other experiments, such as those in p. 313, are proceeded with. When nearly (but not quite) all the iron has dissolved, pour half the solution into a small porcelain dish, and reserve the rest (taking especial care that some metallic iron remains in it undissolved, and that the tube is kept corked, otherwise some persalt will be formed) for experiments on the proto-salts.

B. Iron persulphate (ferric sulphate).

To prepare this from the protosulphate we have in fact to cause two molecules of the salt $[2\,FeSO_4]$ to take up one atom of the sulphate radicle $[SO_4]$ and form a molecule of the per-sulphate $[Fe_2(SO_4)_3]$. Excess of acid must therefore be present, and also some oxidising agent such as hydrogen nitrate; the oxygen from which may combine with the hydrogen of a molecule of the acid, and leave the sulphate radicle free to unite with the iron protosulphate.

Take the portion of the solution which was poured off just now into the dish, add a few drops of dilute hydrogen sulphate, and heat it gently on a sandbath. Add strong hydrogen nitrate, a drop at a time, as long as the addition of a drop causes a transient brown colour in the liquid[2]. Leave the dish exposed

---

[1] Observe the strong peculiar smell of the gas (pure hydrogen having no perceptible smell), which is due to the presence of traces of compounds of hydrogen and carbon, the latter element being always present in ordinary iron.

[2] The reason of this will be evident from the experiment already made

to a gentle heat for some minutes (in order to drive off any excess of hydrogen nitrate) while the following experiments are tried.

Properties of iron protosalts (ferrous salts).

[These are so very readily converted into persalts by mere exposure for a few moments to air or to water containing air, that it is rather difficult to observe their true reactions. In the following experiments a drop or two of the solution of iron protosulphate should be added to the tube already containing the test solution, and the cork must be replaced at once.]

1. **With hydrogen sulphide they give no precipitate or change of colour.**

Pour into a test-tube 2 or 3 c.c. of solution of hydrogen sulphide, and add about two drops of the solution of iron protosulphate. This will produce no change, showing that iron differs from the group of metals last considered in not forming a sulphide in presence of acid.

*2. **With ammonium sulphide they give a black precipitate, soluble in hydrogen chloride.**

To the clear liquid obtained in the last experiment add solution of ammonium hydrate (which will combine with the hydrogen sulphide forming ammonium sulphide). A black precipitate of iron protosulphide will be formed, which will readily dissolve on addition of a few drops of dilute hydrogen · chloride.

*3. **With potassium hydrate they give a white precipitate, quickly turning black and then reddish brown.**

Pour into a test-tube a few drops of solution of potassium hydrate, add a little water, and then a drop or two of the solution of iron protosulphate. A grayish white flocculent precipitate of iron protohydrate will be formed, which, if the tube be shaken for half a minute, will rapidly become darker, and be

with nitrogen dioxide (p. 146). The hydrogen nitrate, by the loss of oxygen, is reduced to nitrogen dioxide, which unites with the iron salt, forming a brown compound easily decomposed by heat. Compare the test for nitrates (p. 137).

finally converted into a reddish brown perhydrate, by absorbing oxygen from the air.

*4. With potassium ferrocyanide they give a white precipitate quickly turning blue.

Pour into another test-tube a little solution of potassium ferrocyanide, add water, and then a drop of the solution of iron protosulphate. A precipitate will be formed which is at the first moment nearly white, but rapidly changes to deeper and deeper shades of blue, when the test-tube is shaken, or when the solution is poured backwards and forwards several times from one test-tube to another.

*5. With potassium ferricyanide they give a deep blue precipitate at once.

Repeat the last experiment, using potassium ferricyanide instead of ferrocyanide. A deep blue precipitate will be at once formed, and will undergo no further change in the air. This forms the paint called ' Turnbull's blue.'

*6. With potassium sulphocyanate they give no change of colour or precipitate.

To some solution of potassium sulphocyanate diluted with water add a drop or two of the solution of iron protosulphate. No change will take place at first in the solution, since iron proto-sulphocyanate is colourless ; but on shaking the test-tube the liquid will become light red, and eventually of the colour of dark sherry. The appearance of this red colour is an extremely delicate test for iron persalts, as will be seen, p. 312.

*7. Heated in a borax bead they colour the bead yellow in the oxidising flame, dull green in the reducing flame.

Form a borax bead in the usual way (p. 92) and add to it a minute quantity of powdered iron protosulphate. Heat the bead again, holding it near the tip of the blue flame until the iron salt is dissolved, then bring it into the oxidising flame and hold it there steadily for a few seconds. On withdrawing it from the flame, you will find that (if the right proportion of iron has been taken) it is orange-coloured while hot, becoming light yellow as it cools. In the next place, heat it in the reducing flame for half a minute ; its colour will now be found to have changed to a dull green (like that of bottle-glass), which becomes paler on cooling.

Properties of iron persalts (ferric salts).

Pour into a test-tube the solution of iron persulphate prepared already; notice that it has a yellow colour, while the original solution was nearly colourless; add 30 c.c. of water, and examine portions of the solution with the tests given above, which may be applied in the usual way to portions of the solution, since iron persalts are not altered in the air.

*1. With hydrogen sulphide they give a white, milky precipitate of sulphur.

This is owing to their reduction to protosalts; the hydrogen of the hydrogen sulphide combining with some of the nonmetallic radicle (in this case the sulphate radicle), while its sulphur is separated [1].

*2. With ammonium sulphide they give a black precipitate.

Add ammonium sulphide to another portion of the solution. A black precipitate of iron protosulphide will be formed, soluble (with the exception of a little sulphur which may separate) in dilute hydrogen chloride.

*3. With potassium hydrate they give a reddish-brown precipitate.

This consists of iron perhydrate, and is unaltered when shaken up with air.

*4. With potassium ferrocyanide they give a deep blue precipitate.

This consists of 'Prussian blue,' the formation of which has been already mentioned, p. 181.

5. With potassium ferricyanide they give no precipitate, the liquid turning brown.

(If the solution of potassium ferricyanide has been made for some time, it is liable to contain a trace of ferrocyanide, and then the solution becomes green instead of brown.)

---

[1] $2 Fe_2(SO_4)_3 + 2 H_2S = 4 FeSO_4 + 2 H_2SO_4 + S_2.$

*6. With potassium sulphocyanate they give a dark red liquid.

This should be tried, with one (or at most two) drops of the solution of the iron persalt in order to prove the delicacy of the test.

*7. Heated in a borax bead they give the same results as the protosalts.

Conversion of salts of the one series into those of the other.

A. Protosalts into persalts.

The principle on which this may be done has already been explained, p. 308 : and one mode of effecting it, viz. by hydrogen nitrate, has been employed. Besides oxygen, however, other substances which have an affinity for hydrogen may be employed to withdraw it, such as chlorine. In this case the change is, by analogy, still termed an 'oxidation.'

Pour off the remainder of the solution of iron protosulphate into another test-tube (keeping back the undissolved iron), add a few drops of solution of chlorine, and warm the mixture [1]. The solution will now give the characteristic reactions of iron persalts, and portions of it should be examined by tests 4 and 6.

B. Persalts into protosalts.

In this case we have to withdraw a portion of the non-metallic radicle from the molecule of persalt. This may be done by various reducing agents (for instance hydrogen sulphide, as in Expt. 1, p. 311), but hydrogen in the nascent state is one of the best for the purpose.

Add a few drops of dilute hydrogen sulphate to the remainder of the solution of iron persulphate, place a fragment of granulated zinc in the solution, and close the tube with a cork having a nick cut in its side. Hydrogen will be evolved by the action of the zinc on the acid, and the solution will

---

[1] $2\,FeSO_4 + H_2SO_4 + Cl_2 = Fe_2(SO_4)_3 + 2\,HCl.$

gradually lose its colour [1], and will give the reactions of iron protosalts, viz. a deep blue precipitate with potassium ferricyanide, and a colourless solution with potassium sulphocyanate [2].

Additional Experiments.

1. Formation of ink.

Dissolve about half a gramme of iron protosulphate in 10 c.c. of water, and add a few drops of solution of iron perchloride (to ensure the presence of a persalt), and then a little infusion of nut galls or solution of hydrogen gallate (gallic acid). A bluish black precipitate of iron gallate will be formed, which will remain in suspension for a long time, owing to its finely divided state. Ordinary writing ink is thus made, a little gum being added to prevent the subsidence of the precipitate.

2. Passive state of iron.

This is the name given to that condition of iron in which it is unacted on by strong hydrogen nitrate, probably owing to the presence of a thin film of oxide upon its surface (compare the protection of lead by a film of carbonate from the action of water, p. 264).

Pour about 2 c.c. of *strong* hydrogen nitrate (see note [3]) into a test-tube and put into it a strip of sheet iron or a piece of iron wire, thoroughly cleaned from rust with emery paper. There will be a momentary action only, a few bubbles of gas being given off, and the iron may then remain for any length of time quite unacted on, and with its surface apparently bright. But if a few drops of water are added, a violent action begins at once, and the iron dissolves, with evolution of quantities of nitrogen oxides.

---

[1] $Fe_2(SO_4)_3 + H_2 = 2 FeSO_4 + H_2SO_4$.

[2] The presence of the zinc salt will not interfere with these reactions, if the solution contains hydrogen sulphate.

[3] The acid must not be of lower density than 1.35. The strong common acid, such as is used for Grove's battery, will do. If this is not at hand, add to the ordinary laboratory acid about half its volume of strong hydrogen sulphate, which will concentrate it by withdrawing water. The mixture should be cooled before the iron is put in.

Properties of Steel.

1. Annealing of steel.

Take a piece of thick watch-spring, or clock-spring, about
1 cm. broad and 10 or 12 cm. long, heat the whole of it to
redness in the flame of a Bunsen's burner and allow it to cool
slowly, withdrawing it by degrees from the flame, in the same
manner as glass, p. 30. It will be found, when cool, to have
lost its elasticity almost entirely, being pliable enough to be
bent into any shape, and soft enough to be pretty easily
scratched with a file.

2. Hardening of steel.

Straighten the piece of steel which was just now annealed,
heat it again to a uniform redness in the lamp-flame, and
plunge it quickly into a jug of cold water. This sudden cooling will be found to have had the effect of making it extremely
hard and brittle, so that no file will scratch it, and a bit may be
broken off the end almost as easily as if it was glass. It is
even harder than glass, for its edge will make a scratch on a
piece of glass.

3. Tempering of steel.

It has been seen that steel, if heated to redness and cooled
very slowly, becomes soft. If, however, it is only heated *slightly*
it loses no more than a portion of its hardness; and the
amount lost depends entirely upon the temperature to which it
has been raised. This is the principle of the process of
' tempering' steel; the temperature being usually judged rather
roughly by observing the colour of the film of iron oxide
formed in consequence of the heat. A yellow colour implies a
very thin film, and a blue colour denotes a thicker one; and
thus, by watching for the appearance of a particular colour and
stopping the rise in temperature directly it appears, any desired
degree of hardness may be obtained. To illustrate this,—
Take the strip of steel which has just been hardened, lay it flat
upon a piece of board and rub it gently with a piece of moistened whetstone (or with a bit of wood dipped in fine emery
made into a paste with water) until it becomes quite bright.

Now lay it upon a strip of sheet iron and heat it again very carefully, holding it at some distance above the flame and moving it to and fro, so as to heat it uniformly. The surface of the steel will soon begin to show the same succession of tints as the iron in expt. 1, p. 306, passing from yellow to orange and blue. When it has just reached this latter tint, plunge it into cold water, to stop any further action of the heat. You will now find that it has regained in a great measure the properties of the original watch-spring; that it is stiff and elastic, and that its surface can be filed away, though with difficulty, and much to the damage of the file. If the application of heat had been stopped when the surface became straw-yellow, the hardness would have been less reduced, and the temper would have been such as is required for razors and tools for cutting steel.

4. Relation of iron and steel to magnetism.

Iron is the most strongly magnetic of all bodies; i. e. it is attracted strongly when a magnet is held near it. This is due, in fact, to its becoming a magnet itself, with poles opposite in kind to the magnet which causes or 'induces' them.

Support a strip of sheet iron, about 6 or 7 cm. long and 1 cm. broad, vertically in a Bunsen's holder, and bring close to it one pole (e. g. the N. or marked pole) of a small strong horse-shoe magnet. The lower end of the strip of iron will now shew magnetic properties, and will attract a short bit of wire or a small iron nail held near it. This in its turn will become a magnet, and will attract another bit of wire, and thus several may be hung from the strip of iron. But as soon as the inducing magnet is withdrawn, all the magnetic properties of the iron cease to show themselves and the bits of wire drop off.

Steel, on the contrary, is not only magnetic in the above sense, but is (especially when hardened) capable of being magnetised permanently by several methods, of which the following is the simplest.

Lay the strip of steel, which was tempered just now (or a similar piece of watch-spring), on the table, and hold it down by pressing the forefinger on its centre. Bring down the N. pole of the horse-shoe magnet vertically on the middle of the strip, and move it to the extremity, rubbing it rather strongly upon the surface of the steel. Then bring it back through the air to the centre of the strip as before, and again pass it along to the extremity. Repeat this about six times, and then rub the other half of the strip with the *other* (i. e. the S.) pole in the same way and for the same number of times. Remember to move the pole along the strip always in the same direction, viz. from the centre to the end, and to bring back the magnet to the centre in a wide curve through the air, otherwise the magnetism already imparted will be weakened.

The bit of steel will now be found to be a permanent magnet, that end being the N. pole which was rubbed with the S. pole of the horse-shoe.

(*a*) Touch some small nails with it : they will be strongly attracted.

(*b*) Hang it in a small stirrup of paper, suspended by a fine bit of silk from a Bunsen's holder : it will range itself N. and S. like the compass needle, and its N.-seeking pole will be repelled by the N.-seeking pole of the horse-shoe magnet (the other pole of the horse-shoe being kept out of the way as much as possible); whereas if a strip of sheet iron is hung in the same way both ends will be attracted indifferently by the magnet.

(*c*) Lay it flat on the table, place a sheet of paper upon it, and scatter some iron filings upon the paper; then tap the paper so as to allow the filings to arrange themselves in obedience to the magnet underneath. They will place themselves in regular curves marking the direction of the lines of magnetic force.

2. COBALT.

[Symbol of atom, Co.
Weight of atom, 59 hydrogen-atoms.]

Properties of compounds of Cobalt.

[Typical examples,—Cobalt nitrate $(Co(NO_3)_2)$.
 „   chloride $(Co Cl_2)$.

A solution of cobalt nitrate (1.5 grm. of the salt dissolved in 30 c.c. of water) may be used.]

1. **With hydrogen sulphide, in acid solutions, they give no precipitate.**

Add to a portion of the solution of cobalt nitrate two or three drops of dilute hydrogen chloride and then some solution of hydrogen sulphide. No precipitate will be formed, showing that cobalt cannot, like the metals in Group II, be separated from acid solutions by hydrogen sulphide.

*2. **With ammonium sulphide, in neutral solutions, they give a black precipitate, scarcely soluble in dilute hydrogen chloride.**

Test another portion with a drop of solution of ammonium sulphide. A black precipitate of cobalt sulphide will be formed. Add to this 1 or 2 c.c. of dilute hydrogen chloride; the precipitate will scarcely dissolve, even when heat is applied. If, however, two or three drops of strong hydrogen nitrate are added (so as to form aqua regia) the cobalt sulphide will be decomposed and dissolved (except, possibly, a slight white residue of sulphur). Observe that, though the addition of a very little hydrogen chloride was sufficient to prevent the formation of the sulphide (in Expt. 1), yet a much larger quantity failed to dissolve it when once formed [1].

[This is the basis of a method of separating cobalt (and nickel) from the other metals of this group which form sulphides readily soluble in cold dilute hydrogen chloride.]

*3. **With potassium hydrate they give a light bluish precipitate, turning brown on boiling.**

Test another portion with solution of potassium hydrate. A light blue precipitate, consisting of a basic cobalt salt, will be formed; which does not dissolve in excess of the precipitant, but becomes reddish brown when the solution is boiled for a minute or two, owing to its conversion into cobalt hydrate. This change is very slow unless an excess of potash is present.

*4. **Heated in a borax bead they give a deep blue transparent bead, both in the oxidising and reducing flames.**

This may be tried in the usual way, as directed in p. 93.

---

[1] Compare what was noticed in the case of lead sulphate (p. 267).

Additional Experiments.

1. Change of colour in cobalt chloride when heated.

Cobalt chloride separates from its solutions in red crystals containing six molecules of water united with each molecule of the salt. These when slightly heated give up their water of crystallisation, leaving the anhydrous salt, which is blue. But this latter substance quickly absorbs water again, becoming once more the pink hydrated salt.

Prepare a little cobalt chloride by evaporating 5 c.c. of the solution of cobalt nitrate with about 2 c.c. of strong hydrogen chloride nearly (but not quite) to dryness[1]. Dilute the concentrated solution with 5 c.c. of water and trace letters with it on a sheet of writing paper with a camel's-hair brush or a glass rod : then allow them to dry in the air. The letters will now be only faintly visible from their pink colour (if pink paper is used they will be scarcely visible at all). Now warm the paper very gently by holding it before a fire or at some distance over a lamp. As it gradually gets warm, the letters will appear in deep blue, owing to the loss of water of crystallisation as above explained. Discontinue the heating as soon as the above result is obtained (otherwise the salt may be decomposed entirely), and breathe upon the paper or hold it in a current of steam from a kettle, or elbow-tube attached to a flask of boiling water. Water will be again absorbed by the cobalt chloride, and the original faint pink salt will be obtained, the letters almost disappearing.

This illustrates what was formerly called a 'sympathetic ink,' in which secret messages could be written which only appear when warmed[2].

*2. Formation of cobalt-and-potassium nitrite.

Add to about 2 or 3 c.c. of the solution of cobalt nitrate

---

[1] $Co(NO_3)_2 + 8\,H\,Cl = Co\,Cl_2 + 4\,H_2O + 2\,NO\,Cl + 2\,Cl_2$.

[2] It may be interesting to notice that if three or four drops of a very concentrated solution of cobalt chloride are diluted with 4 or 5 c c. of alcohol, a liquid is obtained which is red when cold, but turns deep blue when the tube containing it is put into hot water.

five or six drops of hydrogen acetate and then a little solid potassium nitrite (about as much as will lie on the end of a spatula). Warm the liquid gently (but do not boil it), and set it aside in a warm place for a day. A bright yellow precipitate of cobalt-and-potassium nitrite (the paint known as 'cobalt yellow') will gradually form, and the whole of the cobalt will eventually though slowly be separated from solution.

This illustrates a method (probably the best) of separating cobalt from nickel, since nickel forms no similar insoluble double salt under the above conditions.

3. NICKEL.

[Symbol of atom, Ni.
Weight of atom, 59 hydrogen-atoms.]

Properties of compounds of Nickel.

[Typical example,—Nickel sulphate (Ni SO₄).

A solution of the salt containing 1 grm. dissolved in 30 c.c. of water may be used.]

(These bear a very close analogy to the compounds of cobalt, and the experiments with them may be made in the same way as already directed in the case of the latter, the results being put down in parallel columns in the note-book.)

1. With hydrogen sulphide, in acid solutions, they give no precipitate.
*2. With ammonium sulphide, in neutral solutions, they give a black precipitate scarcely soluble in dilute hydrogen chloride[1].
*3. With potassium hydrate they give a light green precipitate, unaltered on boiling.

---

[1] This precipitate of nickel sulphide is very slightly soluble in ammonium sulphide, so that on filtering liquids containing nickel sulphide in presence of excess of ammonium sulphide the filtrate has a characteristic dark brown colour. The traces of nickel thus dissolved may be separated by evaporating the solution until the excess of ammonium sulphide has been decomposed and then filtering again.

*4. Heated in a borax bead they colour the bead brownish red in the oxidising flame, gray and turbid in the reducing flame.

This must be carefully distinguished from the results obtained with manganese compounds. The bead obtained with the latter is violet, and loses its colour but does *not* become turbid in the reducing flame.

[Methods of separating cobalt and nickel will be given in vol. ii.]

4. MANGANESE.

[Symbol of atom, Mn.
    Weight of atom, 55 hydrogen-atoms.]

Compounds of Manganese.

[Typical examples,—Manganese dioxide ($Mn\,O_2$).
            „   protochloride ($Mn\,Cl_2$).
        Potassium manganate ($K_2\,Mn\,O_4$).
            „   permanganate ($K\,Mn\,O_4$).]

Nearly all the compounds of manganese are obtained from the dioxide, which can be made both (*a*) to give up oxygen (and thus form salts related to lower oxides), and (*b*) to combine with more oxygen (and thus form compounds corresponding to higher oxides).

1. **Decomposition of manganese dioxide by heat.**

Place a little dry manganese dioxide in a small test-tube and heat it in a Bunsen's burner as strongly as possible. It will, at a red heat, give off a portion (one-third) of the oxygen it contains, leaving a dull reddish residue consisting of a lower oxide of manganese [1]. The presence of oxygen in the tube may be proved by a glowing cedar match in the usual way.

This reaction is of interest, as showing the method by which oxygen was formerly obtained on a large scale, before potassium chlorate could be cheaply obtained.

[1] $3\,Mn\,O_2 = Mn_3\,O_4 + O_2$.

2. Preparation of manganese protochloride.

Place 0.5 grm. of manganese dioxide in a porcelain dish, add 6 or 7 c.c. of strong hydrogen chloride, and heat the mixture gently on a sand-bath, taking great care that none of the chlorine evolved escapes into the room. The reaction has been already explained under CHLORINE, p. 186. The oxygen of the manganese dioxide is not, in this case, evolved as gas, but combines with the hydrogen of the hydrogen chloride, forming water. Part of the chlorine unites with manganese to form manganese protochloride, while the rest is evolved as gas.

When all the black oxide has disappeared, evaporate the liquid to complete dryness, and heat the light pink residue rather strongly (but not quite to redness), in order to decompose any iron chloride which has been formed, owing to the presence of iron as an impurity in the manganese dioxide. Warm the residue, when cool, with 20 c.c. of water, filter the solution and reserve it for experiments.

3. Preparation of potassium manganate.

We have in this case an example of the way in which manganese dioxide combines with more oxygen to form an electronegative radicle $[MnO_4]$, somewhat analogous to the sulphate radicle $[SO_4]$. This takes place when it is heated with an oxidising substance, such as a chlorate, in presence of a base, such as potassium hydrate, which can combine with the radicle formed.

Make a mixture of 2 grms. of manganese dioxide with 1.5 grm. of potassium chlorate, and place it in an iron capsule, or spoon. Dissolve 3 grms. of potassium hydrate in 2 c.c. of water, and add the solution to the mixture in the spoon. Stir the whole together, and evaporate it to dryness over the lamp, stirring it, as it froths up, with an iron wire ; then heat the dish nearly to redness, and keep it at that temperature for two or three minutes. The dark green, semi-fused mass which is formed consists of potassium manganate [1], and was formerly called

---

[1] $3\,MnO_2 + 6\,KHO + KClO_3 = 3\,K_2MnO_4 + KCl + 3\,H_2O.$

Y

'mineral chameleon,' from the changes of colour which its solution undergoes, as will be presently observed.

4. Formation of potassium permanganate.

This is obtained by the decomposition of the manganate, which shows a great tendency to split up into a compound containing more oxygen, viz. a permanganate, and a compound containing less oxygen, such as manganese dioxide or protoxide.

Grind some of the potassium manganate, obtained in the last experiment, to fine powder, place a little of it (not more than will lie on the end of a pen-knife) in a test-tube and add about 10 c.c. of cold water. The salt will readily dissolve, forming a dark green liquid.

(*a*) Pour about 1 c.c. of this solution into another test-tube, and add enough water to nearly fill the tube. The colour of the liquid will more or less quickly (according to the amount of free alkali present) change, first to a dusky neutral tint, and then to a fine purple, owing to formation of potassium permanganate [1].

(*b*) Pour 4 or 5 c.c. of the solution into another test-tube, and heat it to boiling. The change of colour will be much quicker in this case, and a flocculent brown precipitate of manganese dioxide, or rather, the corresponding hydrate, will be formed.

(*c*) Add to the remainder of the solution a few drops of dilute hydrogen sulphate. The formation of the purple potassium permanganate will be immediate in this case; no precipitate occurs, since manganese protosulphate is produced and remains in solution [2]. Dilute the liquid with 30 c.c. of water, observing how slightly the intensity of the rich purple colour is lessened by dilution, and keep it for use in the next experiments.

5. Properties of the permanganates.

These salts are chiefly remarkable for the readiness with which they give up oxygen, passing (when an acid is present)

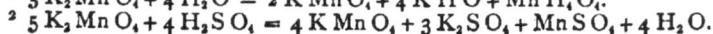

---

[1] $3 K_2 Mn O_4 + 4 H_2 O = 2 K Mn O_4 + 4 K H O + Mn H_4 O_4.$

[2] $5 K_2 Mn O_4 + 4 H_2 S O_4 = 4 K Mn O_4 + 3 K_2 S O_4 + Mn S O_4 + 4 H_2 O.$

into the condition of colourless manganese protosalts. Hence they are much used as oxidising agents in the laboratory, and also on the large scale as disinfectants ('Condy's fluid' being a solution of sodium permanganate). To illustrate this,—

(*a*) Add to a portion of the solution of potassium permanganate (containing free hydrogen sulphate) a few drops of solution of hydrogen sulphite [1]. The purple colour will immediately disappear, the permanganate giving up oxygen to the hydrogen sulphite, to form a sulphate [2].

(*b*) Add to another portion about one-third its volume of dilute hydrogen sulphate, and 6 or 8 drops of solution of ammonium oxalate, and warm it. The liquid will gradually become brown and finally colourless; the oxalate being oxidised to carbon dioxide [3] (compare the reaction of oxalates with manganese dioxide, p. 176).

(*c*) Add to another portion a few drops of a solution of iron protosulphate. The purple colour will disappear; the iron protosalt being oxidised to a persulphate. This is the basis of a process for determining the value of iron ores.

(*d*) Put about 200 c.c. of rain-water (or water from a stagnant ditch [4]) into a flask, add 5 c.c. of dilute hydrogen sulphate and about 3 drops of the solution of permanganate (not more than sufficient to colour the liquid decidedly pink); then leave it to stand for an hour or so. If the water contains organic matter (or nitrites formed by its oxidation), this will be oxidised (burnt, as it were) by the permanganate, and the pink colour will disappear. This illustrates the principle of a process for testing the wholesomeness of water to be used for drinking.

---

[1] If this is not at hand, a freshly made solution of sodium sulphite, acidified with hydrogen sulphate, may be used.

[2] $2 KMnO_4 + 5 H_2SO_4 = K_2SO_4 + 2 MnSO_4 + 2 H_2SO_4 + 3 H_2O.$

[3] $2 KMnO_4 + 3 H_2SO_4 + 5 H_2C_2O_4 = K_2SO_4 + 2 MnSO_4 + 8 H_2O + 10 CO_2.$ This reaction is used in volumetric analysis for determining the strength of a solution of potassium permanganate; a known weight of hydrogen oxalate being dissolved in water, hydrogen sulphate added, and the solution of permanganate dropped in carefully as long as the purple colour is destroyed.

[4] If no such water is at hand, bruise a leaf in a mortar, and boil it with a little water for half a minute; add 2 or 3 c.c. of the liquid to 200 c.c. of ordinary water.

Tests for Manganese protosalts.

[The solution of manganese protochloride, already prepared,
may be used.]

***1. With ammonium sulphide they give a dull pink coloured
precipitate.**

Test a portion of the solution with a drop or two of solution
of ammonium sulphide.   This will produce a pink [1] precipitate
of manganese sulphide, readily soluble on addition of a few
drops of dilute hydrogen chloride.

***2. With potassium hydrate they give a grayish precipitate,
soon turning brown.**

Test another portion with a drop of solution of potassium
hydrate.   A nearly white precipitate of manganese protohydrate
will be formed, which will soon become brown on agitation,
passing into the perhydrate owing to absorption of oxygen
from the air.

***3. Heated in a borax bead they colour the bead violet in
the oxidising flame, but it loses all colour in the reducing
flame.**

Make a borax bead, add to it a minute quantity of manganese dioxide, and heat it in the oxidising flame of the blowpipe.   The bead will be coloured violet, but when heated for
a short time in the reducing flame, it will become colourless.
The reason of this has been already explained, p. 93.

***4. Heated in a bead of sodium carbonate they colour it
bluish green.**

Make a bead of sodium carbonate (p. 242), add to it a
trace of manganese dioxide, and heat it in the oxidising flame.
The opaque bead will, when cold, be of a bluish green colour.
This is due to the formation of sodium manganate (analogous
to potassium manganate, p. 321).

---

[1] If any iron salt is present, the precipitate will be dark in colour.   The
solution of manganese protochloride must be evaporated to dryness and
ignited again more strongly, to decompose the last traces of iron salts.

5. CHROMIUM.

[Symbol of atom, Cr.
Weight of atom, 52.5 hydrogen-atoms.]

Compounds of Chromium.

[Typical examples,—Chromium sesquioxide (Cr_2O_3).

         „   sesquichloride (Cr_2Cl_6).

         „   sulphate and potassium ($CrK(SO_4)_2$).

         „   trioxide (CrO_3).

     Potassium chromate (K_2CrO_4).

         „   dichromate ($K_2Cr_2O_7$).]

Chromium, like manganese, forms several distinct series of compounds, in some of which (e. g. chromium sulphate) it is present as an electropositive radicle or metal, while in others (e. g. potassium chromate) it forms part of an electronegative radicle [CrO_4], analogous to the sulphate radicle [SO_4].

Tests for Chromium salts.

[A solution of chromium and potassium sulphate ('chrome alum') containing 0.5 grm. of the salt dissolved in 25 c.c. of water may be used.]

*1. With ammonium sulphide they give a dull greenish precipitate.

Test a portion of the solution with a few drops of ammonium sulphide. A light bluish green gelatinous precipitate will be formed, which consists of chromium hydrate and not chromium sulphide, as might be expected; since the latter is decomposed in presence of water, with evolution of hydrogen sulphide [1].

Warm the liquid, to promote the separation of the precipitate, adding another drop or two of ammonium sulphide to

---

[1] $Cr_2S_3 + 6H_2O = 2CrH_3O_3 + 3H_2S.$

complete the precipitation : then pour the whole on a filter, wash the precipitate thoroughly, and put the filter containing it to dry in a porcelain dish on the sand-bath. Observe that it shrinks in bulk considerably in drying, especially if the temperature is pretty high, giving off water and becoming chromium sesquioxide, of an olive-green colour, sometimes used as a paint. Reserve it, when dry, for future experiments, p. 328.

*2. With potassium hydrate they give a dull greenish precipitate, soluble in excess.

To another portion add, drop by drop, solution of potassium hydrate, shaking the mixture after each addition. The chromium hydrate, which is at first precipitated, will readily dissolve in excess of potassium hydrate to a clear green fluid. [Be careful to add no more potassium hydrate than is required to dissolve the precipitate.] Boil this fluid for two or three minutes ; it will become turbid, and eventually the whole of the chromium hydrate will be reprecipitated, in a form in which it is no longer soluble in potassium hydrate.

*3. Heated in a borax bead they colour the bead bright green both in the oxidising and reducing flames.

This may be tried in the usual way, a very minute quantity (not so large as a pin's head) of chrome alum being used.

Tests for Chromates.

[A solution of potassium chromate, containing 1 grm. of the salt dissolved in 30 c.c. of water may be used.]

*1. With hydrogen sulphide, in acid solutions, they give a white precipitate of sulphur, the colour of the liquid becoming green.

Add a few drops of dilute hydrogen chloride to a portion of the solution of potassium chromate. Observe that the yellow colour of the solution changes to red, owing to the formation of potassium dichromate [1].

---

[1] A portion of the potassium is withdrawn by the hydrogen chloride.
Thus, $2 K_2 Cr O_4 + 2 H Cl = K_2 Cr_2 O_7 + 2 K Cl + H_2 O.$

Now add excess of solution of hydrogen sulphide and warm the mixture. A white, milky precipitate of sulphur will be formed, and the red colour of the solution will change to green. This is due to the reduction of the chromate to chromium sulphate, some of its oxygen having united with the hydrogen of the hydrogen sulphide (compare the somewhat analogous reaction of iron perchloride, p. 311).

*2. With lead acetate they give a bright yellow precipitate.

This, it will be remembered, has been already applied as a test for lead salts, p. 267.

*3. With silver nitrate they give a purple-red precipitate.

This has been already applied as a test for silver, p. 252.

*4. Heated in a borax bead they colour the bead bright green in both flames.

Reduction of chromates to chromium salts, and vice versâ.

These changes, which afford excellent instances of the processes of oxidation and reduction, are effected by means quite analogous to those described in the case of the compounds of manganese, pp. 321, 323; and the explanations there given will apply equally to the present case.

A. Conversion of chromates into chromium salts.

This is easily effected by boiling them, in presence of an acid, with alcohol; some of the hydrogen of the latter being withdrawn by the oxygen of the chromate, forming water, while a substance called 'aldehyde' is given off [1].

Add to a little solution of potassium chromate about 1 c.c. of strong hydrogen chloride and 2 c.c. of alcohol, and warm the mixture. Vapours of aldehyde will be evolved, easily recognised by their pungent oppressive smell, and the red colour of the solution will change to green, owing to the formation of chromium sesquichloride as explained above. After boiling

---

[1] Potassium chromate. Alcohol. Hydrogen chloride. Potassium chloride. Chromium chloride. Aldehyde. Water.

$$2 K_2 Cr O_4 + 4 C_2 H_6 O + 10 H Cl = 4 K Cl + Cr_2 Cl_6 + 4 C_2 H_4 O + 8 H_2 O.$$

it for half a minute, pour off half of the solution into another tube.

(*a*) Add to one portion an equal volume of water (to prevent the precipitation of lead chloride), and test it with solution of lead acetate. No yellow precipitate will be formed; this shows that no chromate is now present.

(*b*) To the remainder add solution of potassium hydrate drop by drop, until it is in excess. The formation of a greenish precipitate soluble in excess shows that a chromium salt has been formed.

[Other methods of effecting the same change are,

1. The action of hydrogen sulphite, which is analogous to its action upon permanganates, p. 323.

2. The action of strong hydrogen chloride at a high temperature upon a solid chromate. In this case chlorine is given off (as with manganese dioxide, p. 321); and this process is sometimes used for preparing chlorine.]

B. Conversion of chromium salts into chromates.

This is effected by heating them with an oxidising agent in presence of some basic radicle which will combine with the chromate when formed.

Take the precipitate of chromium sesquioxide obtained in Expt. 1, p. 325, mix it intimately with about the same quantity of potassium nitrate and of sodium carbonate (roughly measured), and fuse the mixture in an ignition-tube. It will turn red, becoming yellow as it gets cooler; the chromium sesquioxide having combined with oxygen and potassium from the potassium nitrate to form potassium chromate[1].

Dip the bulb while still hot into a little water placed in a porcelain dish, when it will crack and the chromates will readily dissolve in the water, forming a yellow solution. Pour this solution into a test-tube, warm it and add hydrogen acetate (to decompose the excess of sodium carbonate) as long as there is any effervescence. Then divide it into two parts.

---

[1] $Cr_2 O_3 + 4 K N O_3 = 2 K_2 Cr O_4 + N_2 O_4 + N_2 O_3.$

The above equation does not, of course, express fully what takes place. Some sodium chromate is also formed, and a variable mixture of various nitrogen oxides is given off.

(*a*) Test one portion with solution of potassium hydrate. No precipitate will be produced; hence no chromium salt is present.

(*b*) Test the other portion with solution of lead acetate. A yellow precipitate will prove that a chromate has been formed.

Additional Experiments.

1. Preparation of chromium trioxide.

It has been seen already (Expt. 1, p. 326) that when an acid is added to a chromate part of the base is withdrawn, and a dichromate is formed. If, however, a great excess of strong hydrogen sulphate is added, the whole of the base is withdrawn and chromium trioxide is separated. This latter, though soluble in strong as well as in very dilute hydrogen sulphate, is insoluble in slightly dilute acid (containing about 16 per cent. of water), and may thus be almost entirely precipitated.

Dissolve 2 grms. of potassium dichromate in 15 c. c. of water heated in a small beaker. Allow it to cool until crystals just begin to áppear, and then place it on a plate (lest it should crack), add 20 c. c. of strong common hydrogen sulphate, and stir with a glass rod. The liquid will, of course, get very hot, and the beaker should be covered with a glass plate or watch glass, and set aside to cool. Crimson crystals of chromium trioxide will be gradually deposited; and when they have entirely subsided, the liquid may be poured off, and the mass of crystals may be scraped out with a glass rod upon a dry porous tile or clean brick. Cover them with an evaporating dish, and leave them until the liquid has been absorbed as far as possible by the brick. Ten minutes will generally be sufficient for this purpose, and as the chromium trioxide is very deliquescent it should be protected as far as possible from the air, and not left longer than is necessary on the brick. When it is dry the following experiments may be tried with portions of it; the remainder, if the crystals are good, may be kept as a specimen in a small stoppered bottle, or in a stout test-tube, hermetically sealed by drawing out the upper part of it in the blowpipe flame.

2. Properties of chromium trioxide.

(*a*) **It gives off oxygen when heated.**

Place a little of the chromium trioxide in a small test-tube, supported nearly horizontally, and heat it. The substance will melt, and on being further heated will decompose with incandescence, leaving a green residue of chromium sesquioxide, while oxygen will be evolved, and may be tested for by a glowing match.

(*b*) **It oxidises alcohol.**

Pour a few drops of strong alcohol into a bottle, shake it up so as to diffuse the vapour; then throw into the bottle some of the chromium trioxide. The latter will be reduced to sesquioxide, the action being occasionally so violent as to set the alcohol on fire.

3. Formation of a higher chromium oxide.

Although chromium trioxide so readily gives up oxygen, yet it is possible to get it to combine with more oxygen. For this purpose a very powerful oxidising agent, hydrogen dioxide, must be used; and the first step will be to prepare a solution of this by acting upon barium dioxide with hydrogen chloride (an action which will be more fully explained under BARIUM). Place a little barium dioxide [1] (as much as will lie on the end of a spatula) in a mortar, powder it, if necessary, pour upon it about 7 or 8 c. c. of dilute hydrogen chloride, and grind them together until the peroxide has entirely dissolved. A slight effervescence will probably be noticed, owing to the escape of oxygen, and a solution containing barium chloride and hydrogen dioxide will be obtained.

(*a*) Place about half of this solution in a test-tube, and add three or four drops of a solution of potassium chromate acidified with hydrogen chloride. The colour of the solution will change to a deep blue, but in a second or two this will disappear, while oxygen is given off with effervescence, and a pale green solution will be obtained. These changes of colour are due to the

---

[1] For the method of making this substance see p. 342.

formation of a higher chromium oxide, which is very unstable, and decomposes, when hydrogen chloride is present, into oxygen and chromium chloride.

(*b*) This blue chromium oxide (of which the constitution is uncertain) is much more stable when dissolved in ether. To illustrate this, add to another portion of the solution of hydrogen dioxide sufficient ether to form a stratum about 1 cm. in depth; pour in one or two drops of the solution of potassium chromate, close the mouth of the test-tube with the thumb, and shake the mixture. If it is now allowed to remain undisturbed for a minute, the ether will rise to the surface, forming a magnificent blue stratum, while the liquid below is nearly colourless.

Action of light upon Chromates.

When a soluble chromate is exposed to light in presence of organic matter, it is reduced to an insoluble chromium sesquioxide which adheres very closely to the substance, and, if the latter is gelatine, renders it quite insoluble also. This fact has lately found very extensive applications in the Autotype, Heliotype, and many other printing processes.

Dissolve 4 grms. of potassium dichromate in 50 c.c. of water, and filter the solution into a plate. Float upon it a piece of drawing paper 10 or 12 cm. square (as directed, p. 254), covering the whole with a larger plate or inverted tray to protect it from the action of light. [It is best, in fact, to make the experiment in a darkened room or at night by candle-light, although the dichromate is not so sensitive to light as silver chloride.] When it has soaked for five minutes, remove the paper and pin it up to dry in a dark cupboard.

When it is quite dry spread it out flat on a board, and lay upon it a piece of black lace, or a pattern cut out in thick brown or black paper: put over the whole a plate of glass to keep it flat, and expose it to full daylight. The unprotected portions of the paper will soon change in colour from bright yellow to dull brown, owing to the reduction of the dichromate, as above explained. In about half-an-hour's time the print may be taken into a dimly-lighted room, and washed with several changes of warm water. All the unchanged yellow salt will dissolve out, leaving a permanent brown picture on a white ground: the chromium sesquioxide attached to the paper being quite insoluble.

In practice the dichromate is mixed with a warm solution of gelatine or glue, and lampblack is stirred up in the mixture, which is then spread upon paper. After exposure to light, the print is washed as above described, when all the unaltered dichromate, together with the gelatine and lampblack, is removed, leaving the picture in permanent black.

[If it is desired to strengthen the print above obtained, it may, after washing, be soaked for 5 minutes in a solution of iron proto-sulphate (2 grms. in 50 c.c. water). Iron oxide is thus deposited where the chromium oxide exists; and if the print, after another thorough washing, is placed in a solution of potassium ferrocyanide (5 c.c. of the laboratory solution with 20 c.c. of water, and one drop of dilute hydrogen chloride), a deep blue picture is obtained.]

6. ALUMINIUM.

[Symbol of atom, Al.
Weight of atom, 27.5 hydrogen-atoms.]

Tests for compounds of Aluminium.

[Typical examples,—
Aluminium sesquioxide (Al_2O_3).
Aluminium-and-potassium sulphate ($Al K(S O_4)_2, 12 H_2 O$).
Aluminium-and-ammonium sulphate ($Al(H_4 N)(S O_4)_2, 12 H_2O$).

A solution of either of the two latter (which are both sold under the name of 'alum'), containing 3 grms. of the salt in 30 c.c. of water, may be used (the presence of potassium or ammonium will not interfere with the tests).]

*1. With ammonium sulphide they give a white precipitate.

Test a portion of the solution of alum with ammonium sulphide, warming the mixture. The grayish-white gelatinous precipitate thus produced consists of aluminium sesqui-hydrate, since the sulphide (like that of chromium) is not stable in pre-sence of water.

*2. With potassium hydrate they give a white precipitate soluble in excess.

To another portion of the solution add solution of potassium hydrate, drop by drop. A white gelatinous precipitate of aluminium hydrate will be formed at first, but on addition of more potassium hydrate it will readily dissolve. Divide the solution into two parts:

(*a*) To one portion add one or two drops of solution of hydrogen sulphide. No precipitate will be produced, since aluminium sulphydrate is not formed in presence of water.

(*b*) To the other add about half its volume of solution of ammonium chloride, and warm the mixture; aluminium hydrate will be precipitated.

[These two tests distinguish it from the somewhat similar precipitate formed by adding potassium hydrate to zinc salts, p. 336.]

*3. Heated on charcoal they give a white infusible residue, turning blue when ignited with cobalt nitrate.

Place a small crystal of alum in a cavity cut in a piece of charcoal, and heat it in the hottest part of the blowpipe-flame. The salt will at first swell up like borax, white fumes of ammonium sulphate (if ammonium alum is used) and hydrogen sulphate will then be given off, and finally a white infusible incandescent residue of aluminium oxide will be left on the charcoal. Allow this to cool, then moisten it with a drop of solution of cobalt nitrate (transferred from the bottle on a glass rod), and ignite it again for several seconds. The mass, when cool, will be found to have acquired a bright blue colour[1].

Additional Experiments.

1. Water of crystallisation of alum.

Heat a small fragment of alum in an ignition-tube. It will melt, and at a temperature a little above 100° will give off a large quantity (nearly half its weight) of water, which was

---

[1] This reaction, however, is not absolutely characteristic of aluminium, since some silicates, calcium phosphate, and one or two other substances behave in a similar way.

chemically combined with the salt in its crystallised form. A white porous mass will be left, which is the common 'burnt alum.'

2. Formation of 'lakes.'

These are combinations of aluminium hydrate with various colouring matters, for which it has a great affinity.

Add some solution of cochineal to about 5 c.c. of the solution of alum; then add just enough ammonia to precipitate all the aluminium as hydrate, boil, and filter the liquid. The filtrate will be quite colourless, the colouring matter of the cochineal being retained on the filter in combination with the aluminium hydrate, forming the carmine lake used in painting.

3. Use of aluminium salts as 'mordants' in dyeing.

This depends upon the property possessed by aluminium hydrate of attaching itself firmly to the fibre of cloth and calico, as well as of combining with colouring matters, as already shown. Thus it may be employed to bind the two together, as it were; the result being a 'fast' colour, i. e. one which cannot be removed from the material by washing.

In order to precipitate the aluminium hydrate on the cloth, the latter is soaked in some readily decomposable salt of aluminium, such as the sulphate or acetate, and left to dry in a warm place. By this means aluminium hydrate is deposited in close contact with the fibre.

Put about 15 c.c. of a solution of alum into a test-tube, and add solution of sodium carbonate, drop by drop, shaking thoroughly after each addition. The precipitate of aluminium hydrate which is at first formed will redissolve, and the addition of sodium carbonate should be continued until a *slight* permanent cloudiness shows itself; when about 1 c.c. of the solution of alum should be added to redissolve this. You have now a solution containing an aluminium sulphate which is readily decomposed on boiling, aluminium hydrate being precipitated. [Prove this by pouring off 2 or 3 c.c. of the liquid into another tube and heating it; a white precipitate will be formed.]

Dip one half of a strip of white calico into the solution, and dry it before a fire, or at some distance above a lamp. During the process of drying, the aluminium sulphate is decomposed, aluminium hydrate being deposited in the fibres of the calico. Warm some moderately strong solution of cochineal or log-wood in an evaporating dish, immerse the strip of calico, and heat the liquid to boiling for about ten minutes ; then take out the calico, and wash it thoroughly in several changes of water. The part of it which was soaked in the 'mordant' solution will be found to be permanently dyed, while all the colouring matter will be washed out of the other portion.

[If a little powdered gum arabic is mixed in a mortar with 2 or 3 c.c. of the solution of aluminium sulphate, and patterns or letters are drawn upon a piece of calico with a brush or glass rod dipped in it, they will be permanently printed, when treated as above de-scribed.]

7. ZINC.

[Symbol of atom, Zn.
Weight of atom, 65 hydrogen-atoms.]

Properties of the metal.

1. **Its fusibility** (at 330°) **and its oxidability** when heated in air have been already noticed in the granulation of the metal (p. 27).
2. **Action of heat in altering its tenacity.**

Take a strip of sheet zinc about 20 cm. in length, hold it by its extremities and bend it double ; notice that it is stiff, and requires some little force to bend it. Now heat it in the middle gently over a lamp, and observe that its pliability is greatly increased. If, however, it is still further heated it be-comes quite brittle as the temperature approaches its melting-point.

Compounds of Zinc.

[Typical examples,—Zinc oxide (Zn O).

    ,,   sulphate (Zn SO,).

    ,,   chloride (Zn Cl,).]

*1. Formation of zinc oxide.

Heat a small fragment of zinc on charcoal in the hottest part of the blow-pipe flame. It will, when the heat rises to full redness, begin to burn with production of zinc oxide, light flakes of which will be swept away by the current of air, but the greater part will remain on the charcoal, as an incrustation which is yellow while hot, white when cold. Moisten this incrustation with a drop of solution of cobalt nitrate on the end of a glass rod, and heat it again before the blowpipe. The mass will now acquire a fine green colour, a compound of zinc oxide and cobalt oxide being formed, which is used as a paint.

2. Zinc chloride.

3. Zinc sulphate.

[These have both been prepared already, pp. 71, 106.]

Tests for Salts of Zinc.

[A solution of zinc sulphate, containing 1 grm. of the salt in 30 c.c. of water, may be used.]

*1. With ammonium sulphide they give a white precipitate.

Test a portion of the solution of zinc sulphate with a drop of solution of ammonium sulphide. A white precipitate of zinc sulphide will be formed, which will readily dissolve on addition of a drop or two of dilute hydrogen chloride.

*2. With potassium hydrate they give a white precipitate, soluble in excess.

Add to another portion a drop of solution of potassium hydrate. A white precipitate of zinc hydrate will be formed, which will readily redissolve on addition of a few more drops of the solution of potassium hydrate. Divide the clear liquid into two parts:

    (a) To one portion add a drop or two of solution of hydrogen sulphide. A white precipitate of zinc sulphide will be formed.

    (β) To the other portion add about half its volume of solution of ammonium chloride, and warm the mixture: no precipitate will be formed[1]. Compare the reactions of aluminium, p. 333.

Additional Experiments.

1. Amalgamation of zinc.

By this is meant the process of covering the metal with a film of mercury, which forms an alloy or 'amalgam' with it. Ordinary zinc is readily acted on by dilute acids, as has been already noticed (p. 71), but this is chiefly due to the presence of particles of carbon, lead, &c., which promote the chemical action by the galvanic current set up between each of them and the zinc. This 'local action,' as it is called, is entirely prevented by covering the zinc with mercury; the latter buries the impurities, so that the acid does not reach them at all.

Put about 20 c.c. of water into a test tube, add about 1 c.c. of common hydrogen sulphate, and dip into the acid a strip of zinc about 20 cm. long and 1 cm. broad. Bubbles of hydrogen will at once be given off at all parts of the surface of the zinc. Now add about 1 c.c. of solution of mercury perchloride, and mix it thoroughly with the acid by stirring with the strip of zinc. A bright film of mercury will be deposited on the zinc (compare the deposition of mercury on copper, p. 262), and form an amalgam with it; and the evolution of gas will cease entirely. The strip of zinc in this condition is said to be 'amalgamated,' and may be reserved for use in the next experiment[2].

---

[1] If the potassium hydrate contains aluminium hydrate, or silicon hydrate, a precipitate will, of course, be formed.

[2] Another and more usual mode of amalgamating zinc plates for batteries is, first to clean its surface with dilute acid, as above, and then to lay it in a flat plate and pour over it a drop or two of mercury, rubbing the latter all over the surface with a cork. The excess of mercury should then be drained off (and kept in a bottle for this purpose alone), and the zinc plate should be thoroughly rinsed with water.

2. Use of zinc in galvanic batteries.

This depends upon the readiness with which it is acted on by many substances; the conditions in the usual galvanic batteries being, two metals immersed in a liquid which acts more strongly upon one of them than upon the other.

Put about 50 c.c. of water into a large test tube, and mix with it 5 or 6 c.c. of common hydrogen sulphate. Take the strip of zinc which you amalgamated just now, and a strip of copper of the same size; put between them a flat bit of cork about 1 cm. thick, so as to keep them parallel but slightly separated, and bind them in this position by string or a couple of small india rubber bands. Twist tightly round the upper end of each a bit of fine copper wire 18 or 20 cm. long, taking care that these wires do not touch each other in any part. Immerse the strips in the dilute acid in the test tube, and observe that no action is perceptible on either metal. Now press together the ends of the wires attached to the copper and zinc respectively; bubbles of hydrogen gas will immediately appear on the surface of the copper, while an electric current flows along the wire[1].

The action may be explained as follows:—The molecules of hydrogen sulphate between the zinc and the copper become 'polarised,' i.e. arranged in a definite direction (like a row of magnets), their electro-negative parts pointing towards the zinc; thus,

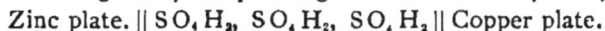

Zinc plate. $\| SO_4H_2, SO_4H_2, SO_4H_2 \|$ Copper plate.

When the metals are connected by a good conductor, such as the copper wire, the SO_4 radicle which is next to the zinc unites with it, while the H_2 unites with the SO_4 of the next molecule, and so on; the H_2 of the last molecule being liberated on the surface of the copper. Thus a molecule of zinc sulphate $(ZnSO_4)$ is formed, and a molecule of hydrogen is liberated. Then other molecules are polarised and decomposed in like manner, and thus the action continues.

It should be observed that, while the metallic connection between the two metals is maintained, the zinc alone is acted on, and not the copper, which remains bright. This illustrates the use of zinc in protecting iron from rust, as in the ordinary galvanised iron (which

---

[1] The current from so small a battery is, of course, weak; but if a part of the wire is straightened and placed in a north-and-south direction, and a small compass is held as close to it as possible, a distinct deflection of the needle will be noticed; the tendency of an electric current being to set a magnet at right angles to its own direction.

is iron coated with zinc). When a piece of this is exposed to air and moisture, the iron (though rapidly attacked when alone) remains unaltered so long as there is any zinc in contact with it.

Preparation of a set of borax beads.

Since so many of the metals hitherto studied give characteristic colours when added to a bead of borax, it will be interesting and useful to keep a set of characteristic borax beads, as a help to the memory, and as a standard of comparison. Such a set may be conveniently made as follows. Seal one end of a glass tube about 6 cm. long and 2 mm. internal diameter. Place near the blowpipe a deep porcelain dish, about 10 cm. in diameter (a porcelain mortar, if glazed inside, or large tea-cup will serve the purpose), perfectly clean and dry. After making the bead in the usual way, and while it is still fluid, tap the platinum wire somewhat obliquely on the edge of the porcelain dish. The bead will detach itself and roll round the side of the dish, finally coming to rest at the bottom. It will be found to have preserved its spherical shape; and if the colour is satisfactory, it should be placed at once in the glass tube, since its surface would soon effloresce in the air. Other beads may be formed in the same way, and dropped one by one into the tube, from which they should not much differ in diameter. When the set is complete, the tube should be heated about 5 or 6 mm. above the last bead, drawn out, and sealed, thus preserving the beads from further change.

Two beads should be preserved in each case; the one showing the colour imparted by the substance in the oxidising, the other in the reducing flame.

The following list includes nearly all the substances which impart characteristic colours to borax glass :—Chromium, manganese, nickel, cobalt, iron, copper.

Group IV.

Metals which are separated from solutions by ammonium carbonate.

BARIUM, STRONTIUM, CALCIUM.

1. BARIUM.

[Symbol of atom, Ba.
Weight of atom, 137 hydrogen-atoms.]

Tests for compounds of Barium.

[Typical examples,—Barium chloride ($BaCl_2$).
       ,,   oxide (BaO).
       ,,   hydrate (BaH_2O_2).
       ,,   dioxide (BaO_2).

A solution of barium chloride, containing 1 grm. of the salt in 25 c.c. of water, may be used.]

1. With ammonium sulphide they give no precipitate.

To a portion of the solution add a drop of solution of ammonium sulphide. No precipitate will be formed, since barium sulphide is soluble in water, a property which distinguishes it from the sulphides of the metals included in the previous groups.

*2. With ammonium carbonate they give a white precipitate.

To another portion add a drop of solution of ammonium carbonate, and warm the liquid. A white precipitate of barium carbonate will be formed, which will readily dissolve, with evolution of carbon dioxide, on addition of a few drops of dilute hydrogen chloride.

*3. With calcium sulphate they give a white precipitate immediately.

To another portion add about one-third its volume of solution of calcium sulphate. A white precipitate of barium sulphate will be immediately formed.

The same reaction has already been employed to detect the presence of a sulphate (p. 230).

*4. With hydrogen silicofluoride they give a white crystalline precipitate.

To another portion add a few drops of solution of hydrogen silicofluoride. A crystalline precipitate of barium silicofluoride will be produced.

*5. Heated on platinum wire in the Bunsen's burner flame they tinge the flame green.

Make a small loop (smaller than that intended for a borax bead, p. 92) at the end of a piece of platinum wire, dip it into strong hydrogen chloride (a few drops should be placed in a watch glass), and hold it just within the edge of the upper part of the flame of a Bunsen's burner (or, better, of a gas blowpipe) until it does not itself impart any colour to the flame. Then moisten it again with hydrogen chloride, dip it into a little barium chloride, and hold it again in the flame. The barium chloride will, as it volatilises, impart to the flame a bright green colour.

In applying this test, the substance under examination should always be moistened with strong hydrogen chloride before being brought into the flame, since the chlorides of the metals in this group are more volatile than their other salts.

Preparation of compounds of Barium.

1. Preparation of barium hydrate.

Place a small lump (as large as a pea) of barium oxide in a test-tube, and pour on it a few drops of water. If the barium oxide is pure, and has not been exposed to the air, it will swell up, evolving much heat as it combines with the elements of the water to form barium hydrate. (Compare the action of water on calcium oxide, p. 58.) Pour about 8 or 10 c.c. of water upon the mass, heat it to boiling, and filter it while hot into a clean test-tube. Crystals of barium hydrate will be deposited as the liquid cools, since the salt is much more soluble in hot than in cold water (differing from calcium hydrate in this respect). Observe also the alkaline reaction of the solution on reddened litmus and turmeric-paper.

This solution is often called 'baryta water,' and may be used in the laboratory for the same purposes as lime-water; for instance as a test for carbon dioxide.

Place a little of the solution in a test-tube; dip the long branch of an elbow-tube into it and blow gently through it for a few seconds. A white cloudiness of barium carbonate will be formed.

2. Preparation of barium dioxide.

When barium oxide is heated, and oxygen gas passed over it, each molecule of the oxide unites with one atom of oxygen to form barium dioxide. For ordinary purposes it is most convenient to evolve the oxygen in contact with the barium oxide, in the following way :—Place about a gramme of barium oxide in a mortar, grind it to powder, then add rather more than an equal quantity of potassium chlorate, and mix the two substances intimately with the pestle. Place the mixture in an iron spoon or a small porcelain crucible, and heat it rather strongly over a Bunsen's burner. It will at first melt and effervesce, and finally a low incandescence will spread through the mass, the barium oxide burning in presence of the chlorate (precisely in the same way as carbon or any other combustible body) with formation of barium dioxide[1]. Allow the semi-fused mass to cool, then detach it from the spoon, grind it to fine powder, and keep it in a corked tube for use in experiments.

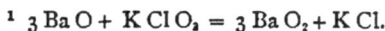

[1] $3 BaO + KClO_3 = 3 BaO_2 + KCl$.

Additional Experiments.

1. Preparation of hydrogen dioxide.

[Formula of molecule, H_2O_2.]

This can be obtained by acting upon barium dioxide with an acid, the hydrogen of which unites with all the oxygen of the substance, while the other radicle unites with the barium.

Place some powdered barium dioxide in a mortar, add a little water, and grind the whole till it forms a thin cream. Pour into the mortar about 6 or 8 c.c. of dilute hydrogen chloride, and mix it quickly with the semi-fluid mass, which ought to dissolve readily and completely; if not, a little more hydrogen chloride must be added at once. A slight effervescence will take place, owing partly to the presence of carbonate as an impurity in the barium oxide, partly to the decomposition of the dioxide with evolution of oxygen gas. If, however, the hydrogen chloride is kept in excess, the amount of decomposition is small, and a solution is obtained containing barium chloride and hydrogen dioxide[1].

2. Properties of Hydrogen dioxide.

[The solution just made may be used, since the presence of barium chloride will not interfere with the experiments.]

This substance shows a great tendency to give up one of the atoms of oxygen which its molecule contains, thus becoming ordinary water[2]. Hence it is a very powerful oxidising agent, and an experiment illustrating this property has been given already under CHROMIUM (p. 330). But it has scarcely less strongly marked reducing powers; the atom of oxygen which it gives up seeming to elicit a fellow-atom from other substances. Thus the two atoms from different sources combine to form a single molecule of oxygen gas, while both the hydrogen dioxide and the substance which it acts upon are simultaneously reduced.

A. Its oxidising powers.

Add a few drops of the solution of hydrogen dioxide to a dilute solution of potassium iodide (3 or 4 drops of the laboratory solution diluted with 5 c.c. of water). Iodine will be liberated by degrees,

---

[1] $BaO_2 + 2HCl = BaCl_2 + H_2O_2$.
Potassium chloride is, of course, also present.

[2] $H_2O_2 = H_2O + O$.

colouring the liquid yellow, and giving the characteristic blue compound if a little solution of starch is added.

In this case hydrogen iodide (formed by the action of the excess of hydrogen chloride upon the potassium iodide) is decomposed, its hydrogen uniting with oxygen, while iodine is set free[1].

B. Its reducing powers.

(*a*) Put a little manganese dioxide into a test-tube, and pour on it some of the solution of hydrogen dioxide. A strong effervescence will take place, owing to the formation of oxygen as above explained: half of it coming from the hydrogen dioxide, and half from the manganese dioxide. The latter is reduced to protoxide, which dissolves in the hydrogen chloride present, forming manganese protochloride[2].

(*b*) Add a little hydrogen dioxide to some rather dilute solution of potassium permanganate. The deep purple liquid will become colourless, manganese protochloride being formed, while oxygen is evolved with effervescence[3]. It is possible that this reaction may be strictly analogous to that of hydrogen dioxide on chromates, a higher and less stable manganese oxide being formed momentarily, which at once loses the greater part of its oxygen.

2. STRONTIUM.

[Symbol of atom, Sr.
Weight of atom, 87.5 hydrogen-atoms.]

Tests for compounds of Strontium.

[Typical examples,—Strontium oxide (Sr O).
         „   nitrate (Sr(NO$_3$)$_2$).
         „   chloride (Sr Cl$_2$).

A solution of strontium nitrate, containing 1 grm. of the salt dissolved in 25 c.c. of water may be used, and the experiments should be tried in the same way as those with barium.]

1. With ammonium sulphide they give no precipitate.
*****2. With ammonium carbonate they give a white precipitate.**

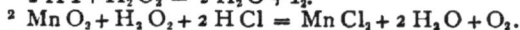

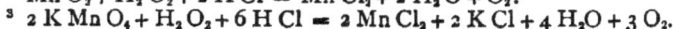

---

[1] $2 H I + H_2 O_2 = 2 H_2 O + I_2.$
[2] $Mn O_2 + H_2 O_2 + 2 H Cl = Mn Cl_2 + 2 H_2 O + O_2.$
[3] $2 K Mn O_4 + H_2 O_2 + 6 H Cl = 2 Mn Cl_2 + 2 K Cl + 4 H_2 O + 3 O_2.$

*3. With calcium sulphate they give no immediate precipitate, but a white precipitate forms in four or five minutes.

This is owing to strontium sulphate being decidedly more soluble in water than barium sulphate ; hence, when such a weak solution is used as that of calcium sulphate necessarily is, the precipitate only forms gradually. If dilute hydrogen sulphate (the ordinary laboratory solution) is added to the solution of strontium nitrate, a precipitate forms at once.

*4. With hydrogen silicofluoride they give no precipitate.

*5. Heated on platinum wire in the Bunsen's burner flame they tinge the flame crimson.

If the flame is looked at through a piece of deep blue glass, it still appears of a crimson colour, while the ordinary Bunsen's burner flame is almost invisible.

Preparation and properties of Strontium oxide.

This (and also barium oxide) is best obtained by strongly heating the nitrate, when it breaks up (like lead nitrate, p. 139) into strontium oxide, nitrogen tetroxide, and oxygen[1].

Place about 2 grms. of crystallised strontium nitrate in an iron spoon, or on a piece of a broken evaporating dish, and heat it to full redness before the gas blowpipe. The salt will fuse and effervesce, owing to the evolution of nitrogen oxides and oxygen, and, finally, a dark gray porous mass of strontium oxide will be left. When this is cool, pour a few drops of water upon it, and notice that it ' slakes' like the corresponding barium and calcium salts, and that its solution in water has an alkaline reaction on test-paper.

[1] $2 Sr(NO_3)_2 = 2 SrO + 2 N_2O_4 + O_2.$

3. CALCIUM.

[Symbol of atom, Ca.
Weight of atom, 40 hydrogen-atoms.]

Tests for compounds of Calcium.

[Typical examples,—Calcium oxide (Ca O).
                 „   sulphate (Ca SO$_4$).
                 „   carbonate (Ca CO$_3$).
                 „   chloride (Ca Cl$_2$).

A solution of calcium chloride, containing 1 grm. of the salt dissolved in 50 c.c. of water may be used. The experiments should be made in the same way as in the case of the barium and strontium salts.]

1. **With ammonium sulphide they give no precipitate.**
*2. **With ammonium carbonate they give a white precipitate.**
*3. **With calcium sulphate they give no precipitate, even on standing for ten minutes.**

This result would be naturally expected, since no more calcium sulphate can be formed than exists already in the solution, and there is obviously more than enough water present to keep that amount dissolved.

[Calcium sulphate is, however, not very soluble in water, and nearly insoluble in alcohol. To prove this,—

Add two or three drops of dilute hydrogen sulphate to some of the solution of calcium chloride, and divide the liquid into two portions.

(*a*) Set one portion aside for a time, shaking it occasionally. A white crystalline precipitate will gradually form.

*(*b*) Add to the other portion some alcohol. A white precipitate will appear at once.]

4. **With hydrogen silicofluoride they give no precipitate.**
*5. **Heated on platinum wire in the Bunsen's burner flame they tinge the flame orange red.**

If the flame is looked at through a piece of deep blue glass, it appears greenish gray.

*6. With ammonium oxalate, in neutral solutions they give a white precipitate.

This property, it will be remembered, was employed as a test for an oxalate, p. 176.

Preparation and properties of Calcium oxide.

This, which is the ordinary quicklime, is obtained by heating the carbonate to redness (as is done on a large scale in lime-kilns), when it breaks up into calcium oxide and carbon dioxide [1].

Place a piece of reddened litmus-paper in a porcelain dish; lay on the paper a small fragment of marble or chalk, about as large as a pea, and pour a few drops of water from a washing bottle over the marble. The colour of the litmus-paper will not be altered, since calcium carbonate is quite insoluble in water. Place the fragment of marble on a piece of platinum foil, fold a corner of the foil over it to prevent loss of heat by radiation, and heat it to full redness before the blowpipe for a minute or two. Place it, when cool, upon the litmus-paper, and add a drop or two of water. The paper will now turn blue, since the marble has been decomposed and calcium oxide formed, as above explained, and the latter unites with the elements of water to form calcium hydrate, which has been already shown (p. 59) to be soluble in water and to give an alkaline reaction.

Additional Experiments:

1. Hardness of water.

Water is said to be 'hard' when it refuses to form a lather or permanent froth with soap. This hardness is mainly due to the presence of calcium salts in the water, especially calcium-and-hydrogen carbonate [2] and calcium sulphate. These decompose the

---

[1] $Ca Co_3 = Ca O + C O_2$.

[2] The formation of this has been already illustrated, p. 158.

soap (which is chiefly sodium oleate and margarate) forming calcium oleate, &c. which are insoluble in water.

Dissolve 1 grm. of common yellow soap in 20 c.c. of distilled water, warming it gently.

(*a*) Put 100 c.c. of distilled water into a small gas bottle, add about 2 c.c. of the soap solution, and shake it up thoroughly. A thick, permanent layer of froth will be produced.

(*b*) Put 100 c.c. of common hard water[1] into a small gas bottle, add about 2 c.c. of the soap solution, and shake it up. If the water is moderately hard, it will become milky (owing to the precipitation of calcium oleate, &c.), and no permanent layer of froth will be formed; the bubbles bursting almost as soon as the shaking is discontinued. Add 2 c.c. more of the soap solution and shake it up again. If the water is very hard no lather will even now be formed, and more soap solution should be added in successive small portions, the mixture being vigorously shaken after each addition. A point will finally be reached at which enough soap has been added to decompose all the calcium salts, and then any further addition will produce a lather lasting unbroken for five minutes or more.

If we try different samples of hard water in the above way, and observe how much soap has to be added to each in order to get a permanent lather, we gain a knowledge of the comparative hardness of each sample. This is the basis of Clark's process for determining the hardness of waters.

2. Preparation of plaster casts.

The mineral gypsum (native calcium sulphate) contains two molecules of water of crystallisation attached to each molecule. This water is driven off by heating the powdered mineral to a temperature of 250°, and the anhydrous calcium sulphate (the common 'plaster of Paris') thus obtained combines with a considerable quantity of water, the whole solidifying by degrees into a hard porous mass. Moreover it expands in solidifying, and hence very sharp casts of medals can be obtained.

Take a medal or large coin, and having slightly greased its surface by rubbing it with a cloth moistened with oil, roll a strip of paper (about 1 cm. broad) round its edge, so as to form a shallow circular trough; then fasten down the end of the strip by sealing-wax or by a fold or two of string. Place 20 c.c. of water in a por-

---

[1] If no hard water is at hand, add 10 c.c. of solution of calcium sulphate to 90 c.c. of rain water or distilled water, and use it for the above experiment.

celain dish, and shake into it gradually sufficient fresh plaster of Paris to form a thick cream, stirring the mixture continually with a glass rod. As soon as the whole is thoroughly mixed, pour over the coin sufficient of it to fill the trough, and stir it with a feather or splinter of wood, in order to detach any bubbles of air which may remain adhering to the surface of the coin. [Wash out the dish containing the residue of plaster at once, or it will be difficult to remove the hardened material.] Leave the whole at rest for about half an hour, in which time you will find that the paste has solidified, or 'set,' as it is termed, the calcium sulphate having (as above explained) combined chemically with the water. Unroll the paper rim, and carefully detach the coin from the plaster by gently pulling them apart. You will thus obtain a copy of the coin in plaster, but reversed, the raised parts of the coin forming depressions in the plaster. The cast should be left for several hours in a drying cupboard, or in front of a fire, in order to dry it thoroughly.

When dry, it may be used as a mould from which to obtain a facsimile of the original medal. For this purpose it should first be thoroughly saturated with wax by placing it, face upwards, in a porcelain dish containing some melted wax or paraffin (such as a portion of a candle) and heated on a sand-bath. The wax will rise by capillary action through the porous plaster, and when the upper surface appears thoroughly wet, it may be taken out and placed on a plate to cool. A paper rim should then be fastened round it, and a cast made in plaster as above directed. Or it may be covered with plumbago, and copper deposited on it by the electrotype process described already, p. 272.

Group V.

Metal which is separated from solutions by sodium phosphate in presence of ammonium salts.

MAGNESIUM.

[Symbol of atom, Mg.
Weight of atom, 24 hydrogen-atoms.]

Properties of the metal.

1. It readily acts on acids, displacing their hydrogen.

Place a bit of magnesium ribbon about 15 cm. in length in a test-tube, add 5 c.c. of water, and then one or two drops of dilute hydrogen sulphate   Hydrogen gas will be evolved with effervescence, and the metal will readily dissolve [1].   [No more acid must be added than is absolutely required to dissolve it.] When the action has ceased, pour off the liquid into a small porcelain dish, and evaporate it down until it begins to crystallise.   Long prismatic crystals of magnesium sulphate (Epsom salts) will be deposited as the solution cools : these, after the liquid has been poured off from them, should be rinsed once with a little water, to remove adhering acid, and then redissolved in about 20 c.c. of water, and the solution reserved for experiments.

Observe the ready solubility of magnesium sulphate, as contrasted with the slight solubility of barium-, strontium-, and calcium-sulphates.   In fact magnesium, although its sulphide is soluble, has a greater analogy to zinc than to these latter metals.

2. It burns in air, forming an oxide.

Place a piece of reddened litmus-paper in a porcelain dish, and moisten it with water.   Cut off a piece of magnesium ribbon or wire about 10 cm. in length, hold it in the crucible tongs over the porcelain dish, and apply the lamp-flame to one extremity.   The metal will take fire and burn with an intense white light, forming white flakes of magnesium oxide, which should be allowed to drop upon the test-paper, and spread over it with a glass rod.   The blue colour of the paper will be restored, proving the alkaline character of magnesium oxide, and its solubility in water, which is, however, much less than even that of calcium oxide.

Tests for compounds of Magnesium.

[Typical examples,—Magnesium sulphate, (Mg SO$_4$).

Magnesium-and-ammonium phosphate
(Mg (H$_4$N) (PO$_4$).

The solution of magnesium sulphate, already made, may be used.]

A characteristic of magnesium salts is their tendency to form double salts with those of ammonium ; such double salts being,

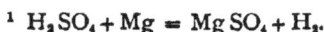

[1] $H_2SO_4 + Mg = MgSO_4 + H_2$.

as a rule, more soluble than the simple magnesium salts. Hence we can often prevent the precipitation of the latter, and redissolve them when formed, by the addition of ammonium chloride.

1. With ammonium carbonate they give no immediate precipitate.

Add 8 or 10 drops of solution of ammonium carbonate to some of the solution of magnesium sulphate. No precipitate will be formed until after the lapse of some time, when a slight crystalline precipitate of magnesium-and-ammonium carbonate will appear. The formation of this is still slower when ammonium chloride is present, and hence the metals of Group IV can be completely separated by ammonium carbonate from magnesium, if ammonium chloride has been previously added.

2. With sodium carbonate they give a white precipitate.

To another portion add some solution of sodium carbonate, and warm the mixture. A precipitate of magnesium carbonate will be formed which will readily dissolve on addition of a few drops of solution of ammonium chloride.

*3. With sodium phosphate, in presence of ammonium salts, they give a white crystalline precipitate.

Add to another portion of the solution five or six drops of solution of ammonium chloride, then a drop or two of solution of ammonia, and lastly a drop of solution of sodium phosphate. A white granular precipitate will be formed, consisting of magnesium and ammonium phosphate.

[To illustrate the delicacy of this reaction, you may place one drop of the solution of magnesium sulphate in a clean test-tube, add about 10 c.c. of water, and then successively a few drops of solution of ammonium chloride, a drop of ammonia, and a drop of solution of sodium phosphate. If no precipitate is produced at first, dip a clean glass rod into the solution, and gently rub the sides of the test-tube with it. Crystals will soon begin to form along the lines at which the rod touched the sides of the tube, especially if the liquid is agitated for a few seconds.]

*4. Heated on charcoal before the blowpipe they leave a white residue, which, when moistened with cobalt nitrate and re-ignited, turns pink.

Place a small crystal of magnesium sulphate on charcoal

and heat it strongly before the blowpipe. Moisten the white infusible residue (which consists of magnesium oxide) with solution of cobalt nitrate, and heat it again. The mass, when cool, will be found to have acquired a faint pink colour.

[Compare the results obtained with zinc oxide, p. 336 (green colour), and with aluminium oxide, p. 333 (blue colour).]

Group VI.

Metals which are not separated from solutions by any of the reagents hitherto employed.

POTASSIUM, SODIUM, AMMONIUM, HYDROGEN.

1. POTASSIUM.

[Symbol of atom, K (kalium).
Weight of atom, 39 hydrogen-atoms.]

Occurrence of potassium salts in plants.

Growing plants extract the potassium compounds which all fertile soils contain, and when they are burnt a large quantity of potassium carbonate is found in their ashes (hence, in fact, comes the name 'pot-ash').

Break up a couple of matches into small pieces, place them in a heap on wire gauze supported on a ring of the retort-stand, and set fire to them by applying the flame of a Bunsen's burner. Allow the wood to burn until only a white ash is left, then shake this into a porcelain dish and warm it with a few drops of water. In burning, all the organic matter of the wood goes off, chiefly as carbon dioxide and water, while the potassium salts present in the wood are decomposed with formation of carbonates (compare the decomposition of tartrates, p. 173, and oxalates, p. 176). The presence of an alkaline carbonate in the solution may be shown by dipping a piece of reddened litmus into it, which will be turned blue.

Properties of Potassium.

[It must be borne in mind that potassium is a rather dangerous substance. No moisture must on any account be allowed to touch it. It must always be kept immersed in the mineral naphtha sold for the purpose. Great care should be taken that neither potassium itself nor potassium hydrate gets under the finger nails.]

1. It becomes quickly oxidised when exposed to air.

Take a small piece of potassium, about as large as a pea, out of the bottle, free it from adhering naphtha by pressing it between folds of blotting-paper, and cut it in two with a knife. Notice the softness of the metal, the brilliant lustre of the freshly cut surfaces, and the rapidity with which the latter are tarnished in the air, owing to the formation of a film of potassium oxide.

2. It is lighter than water, and hence floats upon it.

3. It decomposes water, liberating hydrogen.

[Place near at hand a large funnel, to hold over the potassium while burning.]

Pour some distilled water into a plate or large porcelain dish, and wet the sides of the dish above the surface of the water by rubbing them with the moistened finger[1]. Take up one of the pieces of potassium in the crucible tongs and drop it in the centre of the dish of water. It will immediately take fire, burning with a violet flame as it floats from side to side of the dish. The water is decomposed by the metal, a portion of its hydrogen being replaced by potassium, with formation of potassium hydrate[2]. The heat produced is sufficient to kindle the hydrogen evolved, and to volatilise a small portion of potassium, which gives the flame its violet tinge. Immediately on the cessation of the flame, you will observe the globule of potassium hydrate floating red-hot upon the water. Cover it at once with the inverted funnel, and observe that

---

[1] This is done in order to prevent the globule of potassium adhering to the sides, instead of floating freely on the water. It is a good plan to rinse out the dish, just before using it, with a little solution of caustic potash, and then with clean water.

[2] $2 H_2O + K_2 = 2 KHO + H_2$.

in a second or two it disappears with a slight explosion. The globule, while red-hot, was not in actual contact with the water, but was supported on a cushion of steam. As it cooled, a point arrived at which the layer of steam was condensed, and the globule touched the water, causing a fresh burst of steam sufficient to scatter portions of the globule to a considerable distance. It is well, therefore, to cover the substance with a funnel, as above directed, lest the eyes should suffer. After burning the other piece of potassium in a similar way, you may examine the solution in the dish. It will have acquired a sharp caustic taste, and the colour of a piece of reddened litmus-paper dipped into it will be changed to blue. If the liquid is evaporated to dryness a white residue of potassium hydrate will be left.

Compounds of Potassium.

[Typical examples,—Potassium hydrate (KHO).
                ,,   carbonate (K$_2$CO$_3$).
                ,,   nitrate (KNO$_3$).]

Properties of Potassium hydrate ('caustic potash').

[The practical mode of preparing this substance will be explained under SODIUM, p. 359.]

1. It deliquesces in moist air, forming a carbonate.

Powder a small bit of potassium hydrate, spread the powder on a watch glass, and leave it exposed to the air for a day. It will speedily become moist, and finally deliquesce entirely into a thick liquid, as it absorbs water and carbon dioxide from the air, with formation of potassium carbonate.

The presence of a carbonate may be proved by pouring the liquid into a test-tube and adding 2 or 3 c.c. of dilute hydrogen chloride : an effervescence will occur, and the escaping gas may be tested with lime water in the usual way[1].

---

[1] Potassium hydrate itself does not effervesce when an acid is added. Its strong affinity for carbon dioxide has been illustrated already, p. 159.

2. It is very soluble in water, with which it combines chemically, evolving much heat.

Place a piece (about the size of an almond) of potassium hydrate in a test-tube, add 6 or 8 c.c. of water, and shake it until the whole has dissolved, which it will do very readily. . Observe the great heat produced, which shows that potassium hydrate combines chemically with the water (in the same way as sulphuric acid, p. 229). Use the solution in the following experiments, keeping the tube corked, to prevent absorption of carbon dioxide.

3. It has a strong alkaline reaction.

Potash is, in fact, the typical 'alkali' (a word meaning 'ash').

This may be proved by putting a drop of the solution on a strip of reddened litmus. The reason why the blue colour is restored is, that the colouring matter of litmus is an acid which in ordinary blue litmus is in combination with ammonium or sodium. When an acid is added to this, it liberates the colouring matter which is red. When potassium hydrate or other basic substance is added, it combines with the colouring matter, forming a blue salt.

4. It combines with oils and fats, forming soap.

Pour a little water into a test-tube and add a few drops of olive oil which will, of course, float on the water and not mix with or dissolve in it at all, even when shaken. Now add five or six drops of solution of potassium hydrate, and shake up the mixture. A thick cream-like solution of soap will be produced; the potassium hydrate having acted on the oleates and stearates present in the oil, to form potassium oleate and stearate, of which 'soft soap' is mainly composed.

5. It acts upon most organic substances, decomposing and dissolving them.

Put a small bit of flannel (or silk) into the remainder of the solution of potassium hydrate, and apply heat. It will be broken up and dissolved, but will be re-precipitated on addition of an excess of dilute hydrogen chloride.

This explains the action of caustic potash and of soap upon the skin, which it softens and dissolves.

We learn also that liquids containing much free potash should not be filtered through paper, although the action on cotton is not so strong as that on wool.

Tests for compounds of Potassium.

[A solution of potassium nitrate, containing 1 grm. of the salt dissolved in 30 c.c. of water may be used.]

Nearly all the salts of potassium are soluble in water, and hence no precipitates are formed with any of the group reagents hitherto used.

*1. With platinum perchloride they give a yellow crystalline precipitate.

Add to a portion of the solution of potassium nitrate a drop of dilute hydrogen chloride and then two drops of solution of platinum perchloride.

Divide the liquid into two portions ;—

(*a*) Set aside one portion for a few minutes, shaking it occasionally and stirring it with a glass rod. A yellow precipitate, consisting of minute octohedra of potassium-and-platinum chloride, will be gradually formed.

(*b*) Pour the remainder into a watch glass and evaporate it to dryness on the sandbath, at a gentle heat. Mix in a test-tube equal volumes (about 5 c.c.) of water and common alcohol and pour some of the mixture upon the residue, when the latter has cooled a little. Break up the dry crystalline crust with a glass rod, and stir the whole together : allow it to stand for a few moments, then decant the solution, pour on the residue some more of the weak alcohol, and stir together as before. A light yellow residue, consisting of the potassium-and-platinum salt, will remain, since the salt is insoluble in alcohol. This is the best mode of applying the test when small quantities of potassium have to be detected.

*2. With sodium-and-hydrogen tartrate they give a white crystalline precipitate.

Add to another portion a few drops of solution of sodium and hydrogen tartrate, rub the sides of the test-tube with a glass rod (p. 174), and shake the mixture. A crystalline precipitate of potassium and hydrogen tartrate will be formed.

*3. Heated on platinum wire in the Bunsen's burner flame they tinge the flame bluish lavender.

Moisten a perfectly clean platinum wire with hydrogen chloride, dip it into a little powdered potassium nitrate, and hold it in the flame. The latter will be coloured lavender, and if looked at through a piece of deep blue glass will appear crimson[1].

2. SODIUM.

[Symbol of atom, Na (natrium).
Weight of atom, 23 hydrogen-atoms.]

Properties of Sodium.

[The metal sodium requires quite as much caution in dealing with it as potassium; and the remarks on p. 353 apply with equal force to sodium and sodium hydrate.]

1. It readily combines with oxygen when exposed to air.
2. It is lighter than water.
3. It decomposes water, liberating hydrogen and forming sodium hydrate.

These properties of sodium should be examined in precisely the same way as the corresponding properties of potassium (p. 353).

It will be found that the metal has many points of resemblance to potassium, but that it does not tarnish quite so readily in the air, and that when placed upon water it causes the evolution of hydrogen, but that the action is not so violent as to inflame the gas.

If, however, a piece of blotting-paper is floated on the water and a bit of sodium is placed in the centre of it so as to confine it in one place and thus prevent the distribution of the heat,

---

[1] Glass coloured with cobalt transmits both blue and red rays, and the light emitted by ignited potassium vapour contains many of the latter.

the gas evolved will catch fire, 'and burn with an intense yellow flame, the colour being due to the volatilisation of a portion of the sodium.

4. It unites with mercury, forming an amalgam.

Place in a dry test-tube a globule of mercury about twice as large as a pea. Cut off a piece of sodium about half the size, and carefully free it from naphtha by blotting-paper, then cut it into a number of small pieces (no larger than a pin's head), and, having previously heated the mercury slightly, add to it the bits of sodium, one by one. The two metals will combine with a slight explosion, forming a semi-fluid alloy or amalgam. Reserve it, corking the tube tightly, for use in a future experiment, p. 362.

Compounds of Sodium.

[Typical examples,—Sodium hydrate (Na HO).
,, chloride (Na Cl).

Preparation of Sodium hydrate ('Caustic soda').

This is made by boiling a rather dilute solution of sodium carbonate with an excess of lime: sodium carbonate and calcium hydrate giving sodium hydrate and calcium carbonate, which latter is precipitated [1].

Powder about 3 or 4 grms. of quicklime, and mix it in the mortar with 10 c.c. of water. Place 10 c.c. of solution of sodium carbonate (the ordinary laboratory solution) in a beaker, add to it the whole of the lime and boil it for ten minutes, covering the beaker with a watch glass or glass plate, to prevent access of carbon dioxide from the air. Allow the precipitate of calcium carbonate to subside, then pour off a little of the clear solution into a test-tube, and add excess of dilute hydrogen chloride. No effervescence should occur, thus proving that all the carbonate has been decomposed. [If there is any effervescence a little more lime must be ground up with

[1] $Na_2 C O_3 + Ca H_2 O_2 = Ca Co_3 + 2 Na H O.$

water and added to the remainder of the liquid in the beaker, and the boiling continued for a few minutes longer.]

Potassium hydrate is prepared in exactly the same way, a solution of potassium carbonate (1 part by weight of the salt dissolved in not less than 12 parts of water [1]) being substituted for the sodium carbonate.

The properties of sodium hydrate closely resemble those of potassium hydrate, and need not be fully examined. Its chief use is for the manufacture of ordinary soap, which may be illustrated as follows ;—

Put about 8 or 10 c.c. of the solution of sodium hydrate into an evaporating dish, add about 2 c.c. of olive oil, and evaporate the whole (at a gentle heat) nearly to dryness, with frequent stirring. The viscid residue will dissolve readily when warmed with water, and if some strong solution of salt is added to the solution, white curdy flakes of ' hard soap ' (sodium oleate, stearate, etc.) will be precipitated, while glycerin will remain in solution.

Tests for compounds of Sodium.

[A solution of sodium chloride, containing 1 grm. of the salt dissolved in 25 c.c. of water, may be used.]

1. With platinum perchloride they give no precipitate.

Put a few drops of the solution of sodium chloride into a watch glass, add a couple of drops of platinum perchloride and evaporate to dryness. A residue, but no yellow particles, will be obtained, which will completely dissolve in dilute alcohol. (Compare the result obtained in the case of potassium, p. 356.)

2. With sodium-and-hydrogen tartrate they give no precipitate.

In fact, all the compounds of sodium (except sodium metantimonate) are soluble in water, and their presence is usually inferred from the following characteristic reaction.

---

[1] If the solution is stronger than this, the whole of the carbonate cannot be decomposed by the calcium hydrate.

*3. Heated on platinum wire in the Bunsen's burner flame, they tinge the flame intensely yellow.

This has been already observed, p. 92, but should be tried again. When a piece of deep blue glass is held in front of the flame the intense yellow light is entirely cut off, and hardly anything but the faint bluish flame of the burner is visible through the glass.

Additional Experiments.

[It will be best to do these at night, or in a darkened room.]

The light emitted by ignited sodium vapour consists almost entirely of yellow rays of one definite refrangibility; and it may be interesting to examine the character of this simple 'homogeneous' light a little more closely.

1. If the rays fall on any surface which cannot reflect that particular kind of light, that surface appears perfectly black whatever its colour may be in ordinary white light.

Melt a little sodium carbonate in the ring at the end of a piece of platinum wire, and grasp the wire in the Bunsen's holder, so that it may be held steadily just within the border of the Bunsen flame. The sodium carbonate (which is preferable to sodium chloride as being less volatile) will yield an intense yellow light for a considerable time. Place a blue flower, such as a lobelia or larkspur, or a few crystals of copper sulphate, on a sheet of white paper and hold them near the light. They will appear quite black, for the reason above explained; while if the air-holes of the burner are stopped so as to get the usual luminous gas-flame, their blue colour is plainly seen.

2. If the rays fall on a surface which can reflect them, and them only (i.e. what we call a 'yellow' surface), that surface appears undistinguishable in colour from a white surface placed near; since although the white surface could reflect other rays, there are in the sodium light no others to fall upon it.

(*a*) Place a piece of strongly yellow turmeric-paper or a few crystals of potassium chromate on a sheet of white paper, and hold it in the yellow light, all other light being excluded. The yellow substances will appear of the same colour as the paper; yellow rays only are falling upon both, and hence both appear yellow. On obtaining, however, a white flame by stopping the air-holes of the burner, the paper reflects all the white light which falls on it, while the turmeric and the potassium chromate reflect only the yellow.

(*b*) Form a scarlet stripe or pattern upon a sheet of white paper with mercury iodide[1]. When it is held in the yellow light it will almost disappear. The reason is that mercury iodide reflects a large number of yellow rays as well as red; there are none of the latter in the sodium light, hence the substance can only appear yellow, and the paper also. A scarlet geranium shows a similar effect, and if several stripes of different brilliant colours are painted on paper they all appear of different shades of yellow or black, according to the amount of yellow rays which they can reflect.

3. AMMONIUM.

[Formula of radicle, H_4N.
Weight of radicle, 18 hydrogen-atoms.]

The mode of preparation and properties of ammonia have been given already, p. 123; and it has been noticed that it combines directly with hydrogen salts to form substances which all agree in containing a radicle composed of 4 atoms of hydrogen united with 1 atom of nitrogen, which is called AMMONIUM. Some of these compounds are mentioned in the list given below.

This radicle, although it can be transferred from one compound to another (see, for instance, the experiment given on page 75), has never with certainty been isolated, owing to the extreme readiness with which it breaks up into ammonia and hydrogen. But the salts containing it so closely resemble those of potassium and sodium that it is at 'least possible that, if it is ever isolated, it may appear as a metal. The following experiment is thought to shew that it even resembles a metal in forming an amalgam with mercury.

[1] This has been already prepared, pp. 74, 79. It should be mixed with a little gum and applied with a camel's hair brush or a tuft of cotton wool.

Preparation of Ammonium-amalgam.

This is most readily formed by acting upon ammonium chloride with sodium-amalgam; a double decomposition taking place which results in the ammonium being transferred to the mercury [1].

Make a strong solution of ammonium chloride by dissolving 7 grms. of the powdered salt in 20 c.c. of water, noticing the cold produced by the solution of the salt. Support the test tube containing the sodium amalgam (already made, p. 358) in a beaker standing on a plate, and pour into it the solution of ammonium chloride. The amalgam will immediately begin to increase in bulk, swelling into a gray porous mass which will probably half fill the tube. Shake it out upon a dish; notice its pasty consistence, and also that when no longer immersed in the solution of ammonium chloride it begins to decompose, shrinking in bulk, giving off ammonia and hydrogen, and finally leaving nothing but a globule of mercury. The presence of ammonia will be proved by its smell and its alkaline action upon a piece of reddened litmus paper held just over the decomposing mass.

Tests for compounds of Ammonium.

[Typical examples,—Ammonium hydrate [(H₄N) HO], p. 128.

$$\text{Ammonium hydrate } [(H_4N)\,HO], \text{ p. } 128.$$

,, chloride [(H₄N) Cl], p. 192.
,, nitrate [(H₄N) NO₃], p. 135.
,, sulphide [(H₄N)₂ S].
,, carbonate [(H₄N)₂ CO].

The ordinary laboratory solution of ammonium chloride may be used.]

*1. They volatilise when heated, usually without decomposition.

Put a small bit of ammonium carbonate into a test-tube and heat it rather strongly. Observe that as it becomes hot it

[1] $Hg\,Na + H_4\,N\,Cl = Na\,Cl + Hg\,(H_4\,N).$

rolls about the test-tube when shaken, without apparent friction; owing to the formation of a layer of vapour which prevents it from actually touching the glass. It will volatilise entirely, without melting, forming a white sublimate in the upper part of the tube.

The volatility of ammonium chloride has been proved already, p. 192.

It has also been shown, in preparing nitrogen protoxide, p. 146, that ammonium nitrate is decomposed when heated and does not volatilise unchanged, thus forming an exception to the general rule.

2. With platinum perchloride they give a yellow crystalline precipitate.

This has been noticed already, under PLATINUM, p. 302. The precipitate closely resembles that formed by potassium salts under similar circumstances, and hence potassium must be proved to be absent before the test can be applied to detect ammonium.

*3. Heated with quicklime they give off ammonia.

Place a little powdered calcium oxide in a small beaker, add just enough water to cover it, and then one or two drops of solution of ammonium chloride. Cover the mouth of the beaker with a watch-glass, on the convex (and lowermost) side of which has been spread a moist slip of reddened litmus-paper and a similar slip of turmeric-paper. Heat the beaker very gently on the sand-bath. In a short time ammonia will be evolved by the action of the calcium hydrate on the ammonium salt, and will restore the blue colour to the litmus-paper and redden the turmeric-paper. When this is seen to be the case, remove the watch-glass, notice the odour of ammonia in the beaker, and also test for it by dipping into the beaker (without touching the sides) a glass rod moistened with strong hydrogen chloride or, better (since hydrogen chloride itself fumes perceptibly), strong hydrogen acetate.

Additional Experiments.

Formation of other radicles derived from ammonium.

Each of the four atoms of hydrogen in the molecule of ammonium can be replaced by an equivalent amount of some other radicle, and thus a large number of new radicles are obtainable. The formation of one of these, cuprammonium, has been noticed already, p. 271, and several similar radicles can be formed by the action of mercury salts.

1. **Dimercurammonium chloride** (Hg H₂N Cl).

Put some solution of ammonium hydrate into a test-tube and add a few drops of solution of mercury perchloride. A white precipitate will be formed, which is a compound of chlorine with dimercurammonium, a radicle derived from ammonium by the replacement of two atoms of hydrogen by one atom of (diatomic) mercury, as the formula shows.

2. **Tetramercurammonium iodide** [Hg₂NI] (**Nessler's test for ammonium**).

This substance is formed when mercury iodide acts upon ammonium salts in presence of an excess of potassium hydrate, and its formation affords an extremely delicate test for the presence of ammonium.

In the first place a solution of mercury iodide in potassium iodide must be prepared.

Take 5 c.c. of solution of potassium iodide (the ordinary laboratory solution), and add to it, drop by drop, solution of mercury perchloride, shaking after each addition. The precipitate of mercury iodide will at first be readily redissolved. Continue the addition of mercury perchloride until the last drop produces a slight permanent precipitate, insoluble even on being shaken : then add an equal volume of solution of potassium hydrate (the ordinary laboratory solution) and filter [1]. The solution thus made is called 'Nessler's test solution,' from the chemist who discovered it, and the following experiments will illustrate its application.

(*a*) Dilute two or three drops of solution of ammonium chloride with 10 c.c. of water and add a few drops of the Nessler's

---

[1] The excess of potassium hydrate is not enough to act materially upon the filter.

solution just made. A reddish precipitate will be formed, consisting of tetramercurammonium iodide[1]; the radicle being formed by the replacement of all the four atoms of hydrogen in a molecule of ammonium by two atoms of (diatomic) mercury.

(*b*) To shew the delicacy of the test, put one drop of solution of ammonium chloride into a large test-tube and add 50 c.c. of water. Pour away the whole of this solution, and add 50 c.c. of water to the small portion still adhering to the tube. If the tube be placed on a sheet of white paper, and a few drops of Nessler's solution added, a distinct yellow colour will appear, though mere traces only of ammonium chloride can be present.

(*c*) Test about 50 c.c. of common rain water with Nessler's solution. It will almost invariably become yellow or orange, owing to the ammonium salts which the rain dissolves in falling through the air.

4. HYDROGEN.

[Symbol of atom, H.

The hydrogen atom is taken as the unit of atomic weight.]

The preparation and properties of hydrogen have been already given, p. 104.

Properties of compounds of Hydrogen.

[Typical examples,—Hydrogen chloride (HCl), p. 190.
,,   sulphate (H_2SO_4), p. 228.
,,   phosphate (H_3PO_4), p. 235.]

The so-called 'acids' are, as has been already noticed, p. 103, salts in which hydrogen is present as the basic or electropositive radicle.

*1. They redden blue litmus-paper, but have no action on turmeric-paper.

---

[1] $(H_4N)Cl + 2HgI_2 + 4KHO = Hg_2NI + KCl + 3KI + 4H_2O.$

This property has been frequently observed already (see, for instance, pp. 191, 229) and it belongs to all the hydrogen salts which are likely to be met with in the course of elementary work. Hydrogen borate, however, while it has the usual acid reaction on blue litmus-paper, turns turmeric-paper brownish red (p. 241).

*2. They decompose sodium carbonate and other carbonates, with evolution of carbon dioxide.

This has been illustrated already, p. 162.

3. They are, in many cases, decomposed by metals such as zinc, with evolution of hydrogen gas, and formation of a neutral salt.

The action of hydrogen chloride, for instance, on zinc has been examined already, p. 71.

Hydrogen salts will be easily recognised in the course of an analysis by tests 1 and 2. But it must be remembered that many double salts containing another radicle as well as hydrogen are known (for example, sodium-and-hydrogen tartrate), and hence another metal should always be examined for, unless the substance has the characteristic properties, such as volatility, of the hydrogen salt of the non-metallic radicle which has been detected.

PART II.

SECTION I.

Explanation of the Analytical Course.

EVERY portion of matter, considered chemically, is either a single substance or an aggregate of two or more single substances. By a single substance, we mean a substance from which no part, having different properties from the rest, can be separated by any physical means, such as pounding and sifting, or the action of solvents, or distillation, or diffusion. Water, and salt, and hydrogen, and indeed all the various substances whose properties you have been studying, are single substances[1]. Other portions of matter, consisting of two or more single substances aggregated together, such as a mixture of salts, or a piece of rock, or any part of a plant or of an animal, are called mixtures.

According to this twofold division of natural objects, analysis is divided into the analysis of single substances, and the analysis of mixtures. Further, the analysis may in each case be either qualitative or quantitative, the former having for its object the identification of a single substance or of the ingredients of a mixture, the latter enquiring in what proportions the elements of a single substance, or the various single substances in a mixture, coexist.

Of the four divisions of analysis thus indicated, the present

---

[1] The term 'chemical substance' is often used in this sense, but has the disadvantage of suggesting a peculiar class of substances instead of expressing the simple fact.

volume treats only of the first—the qualitative analysis of a single substance.

Two further limitations should be named in order exactly to define the scope of the analytical course which follows. First, you will have observed that the preceding exercises deal only with the commoner or more abundant elements. About half the elements at present known are so rare that it is found expedient to exclude them from consideration in framing the analytical course. Secondly, of the compound radicles formed from these elements some are continually met with among natural or artificial products, while others occur but seldom. Here, again, it has been necessary to make an arbitrary limitation, and to select from a host of such radicles those of a few of the commonest substances. The problem, then, which the following analytical course supplies the means of solving should be stated thus :—Given a single substance consisting of one or more of the radicles already treated of, to find out what it is.

The great majority of the substances which present themselves for analysis consist of a combination of a simple radicle, belonging to the class of metals, with a simple or compound radicle belonging wholly or chiefly to the non-metallic class; a few contain more than one radicle of either class. For example, sodium chloride, $Na\,Cl$, consists of the simple metallic radicle Na and the simple non-metallic radicle Cl; silver nitrate, $Ag\,N\,O_3$, of the simple metallic radicle Ag, and the compound non-metallic radicle $N\,O_3$; sodium and potassium tartrate $K\,Na\,C_4\,H_4\,O_6$, sodium ammonium and hydrogen phosphate $Na\,(H_4N)\,H\,P\,O_4$, magnesium phosphate and fluoride $Mg_2\,P\,O_4\,F$, exemplify the union of more than two radicles. Conformably with this fact as to the constitution of substances, the course of analysis consists of two principal parts, the one determining the metallic radicle or radicles, the other the non-metallic radicle or radicles of which a substance is formed.

It is by no means necessary actually to obtain the simplest forms of matter (the so-called ' elements ') from a substance in order to ascertain its composition. It is sufficient to cause the formation of compounds which, from their colour, solubility, &c., and the conditions under which they are formed, may be recognised as substances containing particular elements. For instance, it was observed (p. 217) that when iron sulphide (a compound of iron and sulphur) is acted upon by an acid such as hydrogen sulphate, a gas is evolved which blackens paper moistened with a solution of lead acetate. The formation of this black compound of lead, under the circumstances there given, is accepted as a proof of the presence of sulphur in the substance.

In many cases a group of two or more elements appears to be more easily detached from a substance than the individual elements; the substance, like a crystal, cleaving more easily in some directions than in others. An illustration of this will be found on p. 154. The substance, marble, from which we have considerable difficulty in obtaining carbon, oxygen, and calcium individually, we found to yield by very simple treatment a gas, the properties of which were examined. Having once ascertained that this gas is composed of carbon and oxygen in certain proportions, we are satisfied, in the examination of other substances, if we can obtain the same gas from them by the action of hydrogen chloride. Substances which yield the gas when thus treated, we place under one head and call them 'carbonates.' Similarly, in examining marble for its other constituent, calcium, we do not attempt to isolate the element; we stop short of this point, and are satisfied with obtaining a compound with well-marked properties, from which it has once for all been proved possible to obtain calcium. Substances from which we can obtain this compound, we class together as 'calcium salts.'

We find, in fact, that certain groups of elements can be transferred from one compound to another, and the analysis of a substance is effected when it has been resolved into groups, the composition of which is known.

The acquaintance which has been gained in Part I with the properties of the various radicles and their compounds furnishes a means of distinguishing any one from the rest and identifying it. But it remains to be seen in what manner this end can be most surely· and quickly accomplished. It would be possible by trying in any order a sufficient number of the experiments described in the section upon the substance under examination, to discover at last its true nature; but much time would be lost upon experiments which yielded no information, the presence of other radicles might interfere with the tests for any particular one, and the analysis would not unfrequently end in inextricable confusion. We ought to start, then, with a clear perception of the relative value of the tests we are going to employ: we must select them on definite principles, and observe a definite order in applying them.

No better course can be followed in an analysis than the strictly logical one of

1st. Finding out the genus to which the radicle belongs.

2ndly. Ascertaining the species.

3rdly. Identifying the individual.

In this way we are enabled to concentrate our search on fewer and fewer subjects, dismissing altogether the other radicles from our attention, until finally we are led to the conviction that one particular radicle must be present.

Distribution of the Radicles into Groups for the purposes of analysis.

In following out the above principle, the first step will be to take a number of solutions each containing one of the radicles under consideration, and to try the effect of some one particular test upon each. We should find that in some cases the radicle was separated by precipitation, while in the rest no apparent effect was produced. We should place all those radicles which could be separated thus, in a group by themselves. We should then try the effect of another test on each of the remaining solutions, and those radicles which could be separated by means of it would constitute another group. Proceeding thus, we could sort all the radicles into different groups, which we might distinguish by the name of the test by which all the members of the group could be separated.

The process thus indicated has been gone through with great care by chemists, and the following experiments may be made to illustrate it.

A. Non-metallic Radicles.

Place five test-tubes in the stand, and put into them, respectively, eight or ten drops of solutions of

1.	2.	3.	4.	5.
SODIUM CARBONATE.	AMMONIUM OXALATE.	AMMONIUM CHLORIDE.	POTASSIUM IODIDE.	POTASSIUM NITRATE.

[The ordinary laboratory solutions may be used.[1]]

Add to each about 20 c.c. of water, and write the contents of each tube upon the strip of slate in front of it (or place a label on the tube itself).

1. Barium Chloride Test.

Pour a little of the solution of sodium carbonate into another tube and add a few drops of solution of **barium chloride**. A white precipitate will be formed, and if the barium chloride was added in excess the whole of the carbonate radicle might be separated as barium carbonate.

Test portions of each of the other solutions in a similar way. The general result is, that barium chloride produces a precipitate in the solutions of sodium carbonate and ammonium oxalate, but not in the solutions of ammonium chloride, potassium iodide, or potassium nitrate. And by examining solutions containing other non-metallic radicles in the same way, it would be found that they could be divided·into two groups, viz.

1. Those which are precipitated from neutral solutions by barium chloride.

2. Those which are not thus precipitated.

[See the Table on p. 377.]

---

[1] These salts are so selected that the presence of the other radicle (sodium, ammonium, &c.) has no influence on the formation of a precipitate.

The practical inference, then, is,—If on testing a neutral solution (containing an unknown single salt of an alkali-metal) with barium chloride, a precipitate occurs, one of the radicles in Group I must be present.

2. Silver Nitrate Test.

We next proceed to see whether, out of all the remaining radicles, some may be separated from the rest on a similar principle.

Take fresh portions of the solutions of ammonium chloride, potassium iodide, and potassium nitrate, and test each with a few drops of solution of silver nitrate. Precipitates will be obtained in the case of the chloride and iodide, but none in the solution containing the nitrate.

In this way we may divide all the radicles which are not precipitated by barium chloride into two groups, viz.

1. Those which are precipitated from neutral solutions by silver nitrate.

2. Those which are not thus precipitated.

[See the Table on p. 377.]

Hence, by applying only two tests we are enabled, in the further course of analysis, to limit our search to the members of that group to which the radicle has thus been proved to belong.

Formation of sub-groups.

The next step will be to try the action of other reagents on the members of each group, in order to be able to carry the principle of division still further, and to arrange them in smaller sub-groups. For instance,—

Take the portions of the solution of ammonium chloride and potassium iodide which contain the precipitates produced by the addition of silver nitrate, add to each some solution of ammonia, and shake the mixture. The precipitate produced in the solution of the chloride will be dissolved, while that which was produced in the solution of the iodide will remain undissolved. And if the same experiment was repeated with all the members of Group II, it would be found that they may be arranged in two divisions.

A.	B.
Radicles, the silver compounds of which are soluble in ammonia.	Radicles, the silver compounds of which are insoluble in ammonia.
CYANIDES.	IODIDES.
CHLORIDES.	SULPHIDES.
BROMIDES (sparingly).	
HYPOSULPHITES.	
HYPOPHOSPHITES.	
ACETATES.	

Characteristic and Confirmatory Tests.

Finally, when we have thus divided the radicles into conveniently small groups, we proceed to apply tests which give characteristic reactions with the individual members of each group. For instance,—

Take a fresh portion of the solution of potassium iodide, add a few drops of **carbon disulphide**, and one drop of **solution of chlorine**. The solution will become yellow, and after agitation the carbon disulphide will collect at the bottom of the tube in a globule of a violet colour. This, as we have already seen (p. 210), is a reaction which distinguishes an iodide from all other substances, a bromide being the only one which gives a result at all similar. We might here rest content with having obtained one reaction which is characteristic of the presence of one particular substance. But it is highly expedient to accumulate evidence, so as to leave no possible room for doubt as to the constitution of the substance under examination; and other confirmatory tests should always be tried.

Thus, in the case of the solution of potassium iodide, a precipitate has been obtained with silver nitrate, proving that the radicle belongs to Group II: this precipitate is insoluble in ammonia, and the radicle therefore is included in division B; the yellow colour of the precipitate and reaction with chlorine water and carbon disulphide indicate the presence of an iodide.

Assuming, then, that an iodide·is present, we try the action of other tests which are known to give characteristic reactions with an iodide. For example,—

Take a fresh portion of the solution of potassium iodide, add to it a few drops of solution of **starch**, and then a drop of **solution of chlorine**. The solution will acquire an intense blue colour. This· reaction is absolutely characteristic of an iodide, and confirms the previous results.

B. The Metallic Radicles.

Place in test-tubes about 1 c.c. of each of the following solutions (the ordinary laboratory solutions may be used), and dilute with 20 c.c. of water. Write the contents of each on the slate strip.

1. SILVER NITRATE.	2. COPPER SULPHATE.	3. IRON PROTOSULPHATE.	4. BARIUM CHLORIDE.

5. MAGNESIUM SULPHATE.	6. AMMONIUM CHLORIDE.

1. Hydrogen Chloride Test.

Test portions of each of the above solutions with a few drops of dilute **hydrogen chloride**. A white precipitate will be formed in the solution of silver nitrate, but not in any of the others.

We place, then, in one group substances (such as silver salts) which are precipitated from solutions by hydrogen chloride.

[See the Table on p. 378.]

2. Hydrogen Sulphide Test.

Take the portions of each of the solutions to which hydrogen chloride was added in the last experiment without producing a precipitate, and test each with some solution of **hydrogen**

sulphide. A precipitate will be produced in the solution of the mercury salt, but not in any of the others.

We associate in another group substances (such as copper salts) which are precipitated from solutions containing hydrogen chloride by **hydrogen sulphide.**

3. Ammonium Sulphide Test.

Take fresh portions of the solutions of iron protosulphate, barium chloride, magnesium sulphate, and ammonium chloride, and test each with solution of ammonium sulphide. A precipitate will be produced in the solution of the iron salt, but not in any of the others.

We form, therefore, into a third group substances, such as iron salts, which are not precipitated from solutions containing free acid by **hydrogen sulphide,** but are precipitated from solutions by **ammonium sulphide.**

4. Ammonium Carbonate Test.

Take the portions of the three solutions (barium chloride, magnesium sulphate, and ammonium chloride) to which ammonium sulphide was added in the last experiment without producing a precipitate, and test each with solution of **ammonium carbonate.** A precipitate will be produced in the solution of barium chloride, but not in either of the others.

We constitute, therefore, another group of those substances, such as barium salts, which are not precipitated from their solutions by either **hydrogen sulphide** or ·**ammonium sulphide,** but which are precipitated by **ammonium carbonate** from solutions containing ammonium salts.

5. Sodium Phosphate Test.

Take the solutions of magnesium sulphate and ammonium chloride used in the previous experiments (and which therefore contain ammonium sulphide and carbonate) and test each with solution of **sodium-and-hydrogen phosphate**. A precipitate will be formed in the solution of magnesium sulphate, but not in the solution of ammonium chloride.

We place, then, in a group substances, of which magnesium salts are the only known examples, which are not precipitated from solutions by **hydrogen sulphide, ammonium sulphide,** or **ammonium carbonate** (if the solutions already contain ammonium salts), but which are precipitated by **sodium phosphate** from solutions containing ammonium salts.

The 6th and last group consists of those substances, such as ammonium, which are not precipitated from solutions by any of the reagents above mentioned, and which remain as a residue, in the course of an analysis, when all the other substances have been removed from the solution under examination.

By repeating the above experiments with each of the metallic radicles treated of in p. 249 seq., we should be enabled to place it in some one of the above six groups, and we might construct a Table similar to that which is given on p. 378.

Summary.

The general course, then, to be pursued in a qualitative analysis of an unknown substance for one radicle, is—

Firstly, to ascertain the group to which the radicle belongs.

Secondly, to ascertain the subdivision of the group to which it belongs.

Thirdly, to ascertain which member of the subdivision is present.

Lastly, to confirm the results, by applying tests which give characteristic reactions with the radicle, the presence of which has been indicated.

In practice it is found advisable to examine a substance first for the metallic radicle it contains, chiefly for the reason that there are more cases in which ignorance of the metal present would interfere with the examination for the non-metallic radicle, than *vice versâ*.

TABLE

SHOWING THE DISTRIBUTION OF THE MORE IMPORTANT NON-METALLIC RADICLES INTO GROUPS FOR THE PURPOSES OF ANALYSIS.

These Radicles are divided into

I. Those which are separated from neutral solutions by **Barium Chloride.**
CARBONATES
TARTRATES (partially)
OXALATES
FLUORIDES (partially)
SILICOFLUORIDES
SULPHITES
SULPHATES
PHOSPHATES
BORATES (partially)
SILICATES
CHROMATES.

Those which are not separated from neutral solutions by **Barium Chloride.**

II. Those which are separated from neutral solutions by **Silver Nitrate.**
CYANIDES
CHLORIDES
BROMIDES
IODIDES
SULPHIDES
HYPOSULPHITES
HYPOPHOSPHITES
ACETATES (partially).

III. Those which are not separated from neutral solutions by **Silver Nitrate.**
NITRATES
CHLORATES.

TABLE

SHOWING THE DISTRIBUTION OF THE MORE IMPORTANT METALS INTO GROUPS FOR THE PURPOSES OF ANALYSIS.

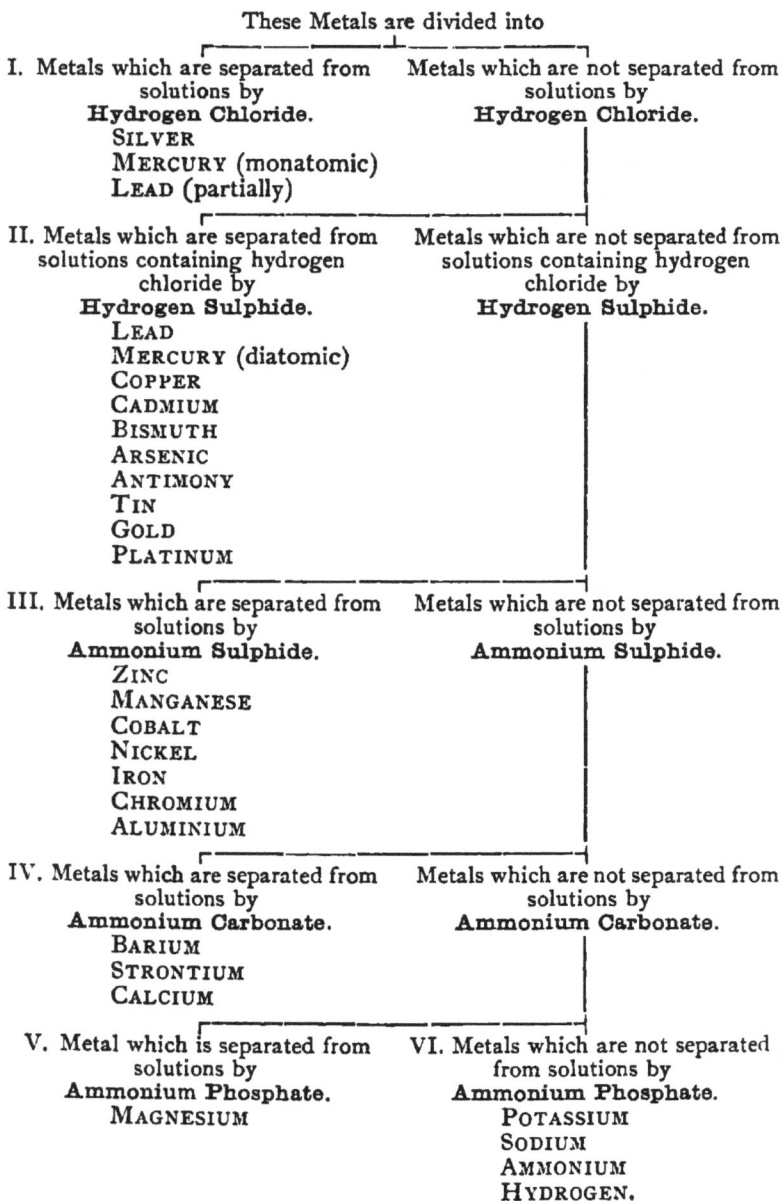

These Metals are divided into

I. Metals which are separated from solutions by **Hydrogen Chloride.** SILVER MERCURY (monatomic) LEAD (partially)	Metals which are not separated from solutions by **Hydrogen Chloride.**
II. Metals which are separated from solutions containing hydrogen chloride by **Hydrogen Sulphide.** LEAD MERCURY (diatomic) COPPER CADMIUM BISMUTH ARSENIC ANTIMONY TIN GOLD PLATINUM	Metals which are not separated from solutions containing hydrogen chloride by **Hydrogen Sulphide.**
III. Metals which are separated from solutions by **Ammonium Sulphide.** ZINC MANGANESE COBALT NICKEL IRON CHROMIUM ALUMINIUM	Metals which are not separated from solutions by **Ammonium Sulphide.**
IV. Metals which are separated from solutions by **Ammonium Carbonate.** BARIUM STRONTIUM CALCIUM	Metals which are not separated from solutions by **Ammonium Carbonate.**
V. Metal which is separated from solutions by **Ammonium Phosphate.** MAGNESIUM	**VI.** Metals which are not separated from solutions by **Ammonium Phosphate.** POTASSIUM SODIUM AMMONIUM HYDROGEN.

SECTION II.

PRELIMINARY EXAMINATION OF THE SUBSTANCE.

General Rules.

1. *Write out a full account of all that is done; the conditions of each experiment, the results obtained, the inferences drawn from these results.*

This should be done not more for the sake of the teacher or examiner than for the advantage of the student himself. The former has, in general, no way of ascertaining what has been done except from the contents of the note-book; the latter will find no means of gaining accuracy of thought, clearness of expression, and insight into analytical methods, more effectual than constant practice in writing out accounts of work done.

Excellent models for such descriptions may be found in the original papers on substances either newly discovered or examined for the first time, contributed by our best chemists to scientific magazines. The object of every such paper is to make clear to others the views of the author on the nature of the substance he has examined, and his reasons for adopting these views; to give the experiments which have been tried, and the inferences drawn from them. This is no less the object of the student commencing analysis. Every substance presents to him an original problem, which should be approached in the same spirit, and worked out with the same care as the most elaborate research of a chemist like Faraday or Dumas.

A meagre, tabular view of work done, full of chemical symbols introduced without reference to their quantitative chemical meaning, but solely in order to save a few seconds of time and a few millimetres of space, is not sufficient. The essential points in each experiment, and the inferences deducible from it, should be put down in plain logical English as soon as possible after the experiment has been made. Many moments of spare time, while a precipitate is being washed or an evaporation is going on, will be found available for writing ; while, if all description is deferred until the analysis is completed, there is great danger that some steps will escape the memory[1].

2. *Do not jump to conclusions too quickly, or desert the regular course of testing, unless some absolutely characteristic reaction (such as the formation of liquid globules of mercury, or of violet vapours of iodine) is obtained.*

3. *Learn to distinguish between slight precipitates, changes of colour, &c., due to impurities in the salt, and reactions which indicate the nature of the salt itself.*

Comparatively few single substances are met with in a state of absolute chemical purity, and reactions which are due to admixture of traces of other substances must be carefully distinguished from those which are due to the principal substance. In cases where the subject of examination is a solid, it will be easy to infer roughly what amount of a precipitate should be obtained from the known quantity of it which has been taken to form the solution ; and when a liquid is being examined, the amount of residue obtained by evaporating a known volume of it, or the bulk of any precipitate previously obtained, will be an indication of its strength. If, then, the addition of a reagent merely causes a turbidity in a solution which is known to be strong, it would be inexcusable thoughtlessness to take such a result as indicating the general constitution of the substance.

For instance, ordinary sodium carbonate invariably contains

---

[1] An attempt is made at the end of this Part (Sect. VI) to give an example of the way in which an account of the analysis of a single salt should be written out, and to this the student should refer before he begins his analytical work.

a little sulphate, and will, therefore, when tested with barium chloride, give a white precipitate which is not wholly soluble in hydrogen chloride. But the amount of the insoluble residue will be obviously disproportionate to the amount of the salt known to be present in the solution, and hence, while the fact is observed and recorded, the examination for another radicle is to be proceeded with.

Again, ordinary zinc generally contains a little lead and iron; and hence in applying the hydrogen sulphide test the solution may become more or less dark; also in applying the ammonium sulphide test, the precipitate may not be quite white, owing to the admixture of a trace of iron sulphide.

Such substances are always most safely analysed by the methods given for mixtures; but careful observation of the character and quantity of the precipitate ought generally to prevent the student from going astray.

4. *Do not apply tests recklessly. Before adding a reagent, consider what substances are already known to be present in that which is being examined, and in what way the use of too much or too little of the reagent may affect the result.*

If, for instance, hydrogen chloride has been used to dissolve the substance, it would be an absurd error to test for a chloride in that solution.

Similarly, when in the course of analysis ammonium salts have been added to the solution, it is easy, but not conclusive, to detect subsequently the presence of ammonium in the liquid.

Some indication is usually given of the amount of the reagent which should be used. It is, in general, easier to rectify the mistake of adding too little than that of adding too much.

[No analytical Tables (with the exception of those which show the distribution of the radicles into groups) are included in this volume; not because such tables are considered of no value, but from a conviction that they will be best made by the student himself, in the same way as a summary of the contents of a book, or a *précis* of a document, rather as a result of intelligent work than as an introduction to it.]

A. *The Substance for Examination is a Solid.*

1. General characters of the Substance.

1 Examine and note down the physical characteristics of the substance; its colour, shape, hardness, metallic or non-metallic appearance, &c. If it is brittle and in a single lump, break it into small fragments in a mortar, or by wrapping it in a fold of clean paper and striking it with a hammer on an anvil. Reduce the greater part of these fragments to a very fine powder in the mortar, which should not be of glass unless the substance is a friable salt[1]. The powdered substance may be conveniently kept in a small dry test-tube, fitted with a cork.

2. Solution of a Substance which has not a Metallic appearance.

[If it is a metal, pass on to **7**.]

2 Solution in water. Place about 0.5 grm. of the powder (as much as will lie on the broad end of a spatula) in a test-tube, pour on it about 20 c.c. of **water**, and heat the liquid to boiling, shaking it occasionally[2]. The tube may be supported on the sand-bath resting in a ring of the retort-stand, and while it is being gradually heated, you may proceed to other experiments (p. 387).

A *The substance is entirely dissolved.*

Filter the solution, if necessary, from any suspended particles of dirt, and examine it according to **13** (p. 393).

---

[1] For pulverising minerals, an agate mortar should, if possible, be used.

[2] Some acetates, tartrates, &c., are decomposed when boiled with water, a basic salt being formed: but this will dissolve in an acid. If, however, on boiling the substance alters in appearance, leaving a flocculent residue, it will be well to see whether it will not dissolve better in cold water.

B   *The substance is not dissolved, or at least not wholly*[1].

Allow the particles to subside, decant the liquid carefully, and boil the residue with a fresh quantity of **water**. If it *dissolves*, add this solution to the first portion, and proceed to **13**. If it *does not dissolve*, pass on to the next paragraph.

3   Solution in dilute hydrogen chloride.   Pour off the greater part of the liquid and add to the residue (or to a fresh portion of the original substance moistened with a little water) about eight or ten drops of concentrated **hydrogen chloride**, and if it does not dissolve, heat the mixture to boiling. If there is an *evolution of gas* observe its character[2].

   (a) *It is colourless, and without distinctive smell*. It is carbon dioxide, and shows that the substance is a CARBONATE. Confirm this by lowering into the tube the long branch of an elbow tube having a drop of **lime-water** in it, and gently sucking a little of the gas through it (p. 162). If the lime-water *becomes cloudy*, the gas is certainly carbon dioxide.

   (β) *It has the smell of rotten eggs*. It is hydrogen sulphide, and shows the presence of a SULPHIDE. Confirm this by moistening a bit of blotting-paper with solution of lead acetate, and holding it within the tube. If it turns brown or black, the gas is certainly hydrogen sulphide.

   (γ) *It has a sharp, penetrating smell, like that of burning sulphur*. [This must not be confused with the somewhat similar, but less pungent, smell of vapour of hydrogen chloride itself[2].] It is sulphur dioxide, and shows the presence of a SULPHITE or HYPOSULPHITE (if the latter, the liquid will be milky, owing to separation of sulphur). To test for these, boil a little of the original substance with solution of sodium carbonate, filter, and test the filtrate (which will contain the radicle combined with sodium) as directed in **15 B**.

[1] The best way of ascertaining whether any of the substance has been dissolved is to filter off a few drops of the liquid and evaporate it to dryness in a watch-glass. If a residue is left, something must have been dissolved by the water. But since nearly all carbonates are insoluble, it will generally be sufficient to add to the clear liquid, filtered or decanted, a few drops of solution of sodium carbonate. If no precipitate occurs, the substance is almost certainly insoluble, unless it is a salt of arsenic or an alkali-metal.

[2] In case of doubt, warm gently a small portion of the substance with strong hydrogen sulphate.

(δ) *It has a faint odour like essence of almonds.* It is hydrogen cyanide, and shows that the substance is a CYANIDE (or FERRO-CYANIDE). To test for these with certainty, boil a little of the original substance with solution of potassium hydrate, filter, and test the filtrate (which will contain the radicle combined with potassium) as directed in 15 D.

(ε) *It is greenish yellow, has the smell of chlorine, and bleaches moist litmus paper held in the tube.* It is chlorine, and points to the presence of a PEROXIDE, CHROMATE (or basic NITRATE).

(ζ) *It is orange, and has the characteristic smell of nitrogen trioxide.* The substance is a NITRITE.

A *The substance is dissolved by the hydrogen chloride.* Examine the solution as directed in 17.

B *The substance is not dissolved by the hydrogen chloride.* Pass on to the next paragraph.

4 **Solution in concentrated hydrogen chloride.** Take a fresh portion of the substance, add 6 or 7 c.c. of concentrated **hydrogen chloride**, and heat it to boiling, noting the character of any gas which may be evolved[1], as above (3).

A *The substance is dissolved.*

Dilute the solution with twice its bulk of water, and examine it as directed in 17. [If any precipitate is produced on dilution, add strong **hydrogen chloride**, drop by drop, and warm the liquid until it disappears[2].]

B *The substance appears to be acted upon, but there is a residue left, different in appearance to the original salt.*

Allow the residue to subside, pour off the solution and examine it after dilution with twice its bulk of water, as directed in 17. If the residue does not subside quickly, add

---

[1] Concentrated hydrogen chloride will often, in the case of sulphides, chromates, and cyanides, produce an evolution of gas, where the dilute acid has failed to do so.

[2] It may be due to the presence of BISMUTH or ANTIMONY, since their chlorides are decomposed by water; possibly to SILVER CHLORIDE, since this substance is soluble in concentrated hydrogen chloride, but not in the dilute acid; possibly also to LEAD, since lead chloride is much less soluble in cold water than in hot, and the addition of water will have cooled the solution. In this last case the precipitate will redissolve when the liquid is warmed, and will be reprecipitated in needle-like crystals as it again becomes cool.

to the solution about the same volume of water (that it may not act on the filter), filter it, and proceed to **17**.

The residue may consist of—

(*a*) SULPHUR. If so, it will be yellowish white, and remain long in suspension, being more easily filtered off after long boiling. It indicates the presence of a PERSULPHIDE, in which case hydrogen sulphide will be evolved (see **3** β); or of a HYPO-SULPHITE, when sulphur dioxide will be evolved (**3** ζ).

(β) SILICON HYDRATE, which is a white gelatinous substance, and will have been formed from the decomposition of a SILICATE. Wash the residue on a filter, dry it at a gentle heat, and heat some of it in a bead of microcosmic salt, as directed, p. 245. If it is silica it will float undissolved in the bead, and the original substance may be inferred with certainty to be a silicate.

(γ) The residue may also be a chloride of one of the metals in Group I. If so, it will be a dense white substance, and will turn black when a drop of ammonium sulphide is added. The original substance should be dissolved in hydrogen nitrate, as directed below (**θ**).

C   *The substance is unacted upon by concentrated hydrogen chloride.* Pass on to the next paragraph.

5

Solution in aqua regia.

Add three or four drops of strong **hydrogen nitrate** to the solution containing hydrogen chloride, and boil the liquid[1].

The substance is dissolved.

Boil the liquid until no more chlorine is evolved, dilute with water, and examine the solution as directed in **17**.

B   *The substance is not dissolved.* Pass on to the next paragraph.

6   Solution in hydrogen nitrate.

Boil a fresh portion of the substance with dilute **hydrogen nitrate**, and, if it does not dissolve, with strong **hydrogen nitrate**.

If *a clear solution is obtained*, dilute it with water (if the strong acid has been used), and test it as directed in **16**.

If *the substance remains undissolved*, it must be examined as directed in Sect. V.

---

[1] This should be done in a draught-cupboard, since chlorine is evolved.

3. Solution of a substance which has a metallic appearance.

[If, besides having a bright lustre, it is malleable, it is certainly a metal.

If it is brittle, it may be one of such minerals as galena, antimonite, iron pyrites, graphite, &c.]

7 Solution of a metal in hydrogen nitrate. Place a few small fragments in a test-tube, add a few drops of water and then about 5 c.c. of concentrated **hydrogen nitrate**, and (if no action begins) heat it to boiling.

A *The substance dissolves.*

Dilute the solution with three or four times its volume of water, and examine it as directed in **16**.

B *The substance does not dissolve, but is converted into a white powder.*

Add 8 or 10 c.c. of water, and boil again.

If it *dissolves*, examine the solution as directed in **16**.

If it *does not dissolve*, pour the whole on a filter, wash the residue with one or two changes of water, and pour a little **ammonium sulphide** over it.

    (a) *It turns slightly yellow.* It is tin dioxide, and the metal is TIN. Confirm by heating the metal on charcoal, when it should give an easily fusible, malleable globule, with white incrustation.

    (β) *It turns orange.* It is antimony pentoxide, and the metal is ANTIMONY. Confirm by heating the metal on charcoal, when it should give a brittle globule, from which white fumes rise, with a white incrustation.

    (γ) *It turns black.* It is lead sulphate, and the original substance is galena, LEAD SULPHIDE[1]. Boil some of the substance (finely powdered) with strong **hydrogen chloride**, noting the evolution of hydrogen sulphide (**3** β); dilute the solution with water, and test it at once for lead (**18 B**).

C *The substance is unacted upon.* It is probably either GOLD or PLATINUM; gold if it is yellow, platinum if it is white. It will dissolve in aqua regia, and the solution should be examined as directed in p. 401, note.

[Graphite has a metallic appearance, and would be insoluble, even in aqua regia. See Sect. V, for the method of treating it.]

---

[1] Hydrogen nitrate gives up oxygen to this, forming a sulphate.

4. Blowpipe Examination.

[This should be begun while the solution of the substance is being made, in order to economise time. If no characteristic results are obtained at once, and if the substance has entered readily into solution, the blowpipe examination may be continued at intervals during the testing of the solution, e.g. while a filtration is going on, or a precipitate is subsiding, and the results will be applicable as confirmatory evidence.]

8   Substance heated in tube. Place a small quantity of the substance in an ignition tube, and heat it over the Bunsen's burner, at first very gently, then to redness, and afterwards before the blowpipe to a temperature as high as the tube will bear.

A   *The substance remains unaltered, even at a high temperature.* Pass on to **9 B.**

B   *The substance does not fuse, even when heated to redness, and gives off no gas or vapour, but changes colour.*

    (α) *From white to yellow, becoming white on cooling.* It is ZINC OXIDE, TIN DIOXIDE, or BISMUTH TRIOXIDE (probably the last, if it fuses at a strong red heat). Pass on to **9 B.**

    (β) *From red to black, turning red again on cooling.* It is probably IRON PEROXIDE. Pass on to **9 B.**

C   *The substance fuses at a very moderate heat, and possibly gives off water (of crystallisation).*

Wipe the inside of the tube perfectly dry with twisted slips of blotting-paper, and when no further moisture is condensed, raise the temperature, and observe which of the following results takes place.

D   *The substance volatilises entirely, or, at any rate, yields a distinct sublimate.* It is a salt of AMMONIUM, ARSENIC, or MERCURY; or possibly HYDROGEN OXALATE, or SULPHUR. Observe the character of the sublimate in the tube.

    (α) *It is white and amorphous.* The substance is an AMMONIUM salt, or a MERCURY salt (not, however, mercury iodide, sulphide or oxide). Pass on to **9 A.**

    (β) *It is white and crystalline.* The substance is probably ARSENIC TRIOXIDE. Pass on to **9 A.**

    (γ) *It is yellow, the original substance being red, and fusing to a nearly black liquid before volatilising. The sublimate becomes red when touched with a wire or glass rod.* It is MERCURY IODIDE[1].

(δ) *It is yellow and amorphous, the original substance being also yellow.* It is ARSENIC SULPHIDE. Pass on to 9 A.

(ε) *It forms brownish yellow, transparent drops, which do not solidify for some time.* It is SULPHUR[2].

To make quite sure that the sublimate consists of sulphur, cut off the closed end of the tube, and heat the sublimate gently, holding the tube in a slanting position. The sulphur will be oxidised in the current of air, and sulphur dioxide will be formed, which may be recognised by its smell and by its acid reaction on litmus-paper held at the upper end of the tube.

The substance becomes charred, i. e. *turns black, and leaves a carbonaceous residue, while vapours are evolved which have a strong smell like burnt paper.* It is an ACETATE, TARTRATE, OXALATE, or other organic salt[3]. Pass on to 9 B.

The substance does not become charred, but evolves a gas, with or without previous fusion. This gas is—

(a) *Colourless, and has the peculiar suffocating smell of sulphur dioxide, and reddens a piece of blue litmus-paper held in the mouth of the tube.* This indicates the presence of a HYPOSULPHITE or a SULPHATE decomposable by heat.

(β) *Colourless, and has the peculiar smell of cyanogen, resembling essence of almonds.*

Hold a lighted match to the mouth of the tube. If the gas burns with a pink flame, a CYANIDE is present.

(γ) *Colourless and odourless.*

Place a single drop of **lime-water** on a watch-glass, and hold

---

[1] This reaction is so characteristic of mercury iodide that the substance may be at once examined for mercury and iodine. For this purpose, a little of the substance should be decomposed by boiling it with some solution of potassium hydrate. A solution containing potassium-and-mercury iodide will be obtained, which should be acidified with hydrogen sulphate and tested for iodine, as directed in 24 B, p. 412. The residue, which consists of mercury oxide, should be washed by decantation, dissolved in dilute hydrogen chloride, and tested for mercury with tin protochloride (p. 264, Expt. 4).

[2] Persulphides (e.g. iron pyrites) and hyposulphites, give a sublimate of sulphur.

[3] Some other substances, e.g. COPPER CARBONATE, MANGANESE CARBONATE, turn black when heated, but do not evolve strong-smelling vapours. To obtain a further proof that the substance is organic, it may be heated again with addition of a small quantity of potassium nitrate, when, if it is organic, the blackness will disappear, with slight deflagration.

it close to the mouth of the tube (which should be as far as possible from the lamp-flame). If the drop becomes turbid, the gas is CARBON DIOXIDE, and proceeds from the decomposition of a CARBONATE or an OXALATE.

If carbon dioxide is not detected, drop a small splinter of charcoal from the charred end of a match into the tube, add a little more of the substance, and again heat the end of the tube. If the charcoal burns vividly, the gas is OXYGEN and the substance is a CHLORATE, NITRATE, NITRITE, CHROMATE, or PEROXIDE [1].

(δ) *Orange-coloured, and reddens litmus-paper.* The substance is probably a NITRATE or NITRITE; if so, a splinter of charcoal dropped into the tube will deflagrate as in the case of chlorates. Some few bromides and iodides are decomposed by heat alone, and give off orange vapours of bromine, which condense to an orange liquid, or violet vapours of iodine, which condense in the form of steel-gray flakes.

[Hypophosphites, when heated, give off hydrogen phosphide, which inflames spontaneously at the mouth of the tube.]

G *The substance fuses to a clear transparent liquid, when heated to redness.* It is probably a SALT OF AN ALKALI METAL, or BARIUM-, STRONTIUM-, or CALCIUM-CHLORIDE.

Pass on to the next paragraph.

9 The course next to be followed will depend upon the appearance of the substance and the results obtained by heating it in a tube.

A *The substance volatilised in the tube.*

Reduction in tube. Mix a small quantity of the original substance with twice as much **sodium carbonate**, introduce into an ignition tube enough of the mixture to nearly fill the bulb, and heat it slowly over the lamp. If any moisture condenses, wipe the tube dry with slips of blotting-paper.

(a) *A bright metallic sublimate is formed, which may be seen with a magnifier to consist of liquid metallic globules.* The substance is a MERCURY salt.

To make quite sure of the character of the sublimate, pass

---

[1] AMMONIUM CHROMATE, however, gives off, not oxygen, but water and nitrogen, and a light, bulky, olive-green residue of chromium oxide is left. AMMONIUM NITRATE, moreover, gives off nitrogen protoxide.

down into the tube a thin slip of wood, such as a match, and scrape together the sublimed metal. If it is mercury, the particles will run together into one large globule, which may be allowed to drop into a watch-glass.

(β) *The sublimate is a steel-gray lustrous mirror.* The substance is an ARSENIC salt.

Break off the tube close to the bulb, hold it horizontally, and heat the part containing the sublimate over a lamp with a very small flame. If the deposit is arsenic, it will readily volatilise and become oxidised, condensing in the cool parts of the tube in sparkling transparent crystals of arsenic trioxide.

If neither mercury nor arsenic is found, and if the sublimate is white and amorphous, the substance may be examined at once for ammonium, as directed in 22.

The substance *did not volatilise, and is coloured.* If so, it is very probable that it will impart colour to a borax bead.

Examination in borax bead. Fuse some **borax** in a loop of platinum wire, as directed in p. 92, and when a clear transparent bead has been obtained, bring into it a minute portion of the powdered substance and heat it first at the tip of the blowpipe flame, then in the oxidising flame, and lastly in the reducing flame. Observe the colour, if any, which is imparted to the bead.

(a) *It is blue, in both oxidising and reducing flames.* The substance is a COBALT salt.

(β) *It is greenish blue in the oxidising flame, and becomes almost colourless in the reducing flame. If much of the substance has been added, the bead becomes red and opaque in the reducing flame.* The substance is a COPPER salt.

(γ) *It is green, both in the oxidising and reducing flame.* The substance is a CHROMIUM salt, or a CHROMATE.

(δ) *It is orange-red in the oxidising flame, becoming light yellow as it cools. In the reducing flame it is orange while hot, and dull bottle-green when cold.* The substance is an IRON salt.

(ε) *It is amethyst-red in the oxidising flame, and becomes quite colourless in the reducing flame.* The substance is a MANGANESE salt.

(ζ) *It is brownish red in the oxidising flame, and becomes gray and turbid in the reducing flame.* The substance is a NICKEL salt.

If no characteristic colour is observed, pass on to the next paragraph.

C   *The substance is colourless, or if coloured has given no reaction with a borax bead, and is infusible, or difficultly fusible.*

Pass on to the next paragraph.

10 Reduction on charcoal.   Mix a small portion of it with an equal quantity of **potassium cyanide**, and heat it on charcoal, as directed in p. 94 [1].

A   *A white residue or incrustation is left on the charcoal.* [If this residue is infusible and *strongly luminous* while held in the blowpipe-flame, ALUMINIUM, MAGNESIUM, STRONTIUM, or CALCIUM is probably present.]

Moisten the residue with one drop of solution of **cobalt nitrate**, and heat it again strongly for a few seconds. If on cooling the residue has a *blue colour*, ALUMINIUM is probably present (see, however, note 1, p. 333); if it has a *faint pink tinge*, MAGNESIUM is present; if it is *green*, ZINC is present.

If no characteristic result is obtained pass on to **11**.

B   *Bright metallic globules are obtained.*

Continue the heat, with addition, if necessary, of a little more **potassium cyanide**, until the small globules have run together into one large one, and the flux has disappeared. When the metal is cool [2], detach it from the charcoal with a knife, and try its malleability, as directed in p. 194, Expt. 2, noting also if any incrustation has formed on the charcoal in or around the cavity [3].

(a) *The metallic globule is malleable, with brilliant lustre, and there is no incrustation on the charcoal.* The substance is a SILVER salt.

(β) *The globule is malleable, and a slight white incrustation is formed on the sides of the cavity.* The substance is a TIN salt.

---

[1] The substance should be held just within the tip of the blowpipe-flame, and the blast should be pretty strong. As we have such powerful reducing agents present as potassium cyanide and charcoal, we can afford to sacrifice some of the reducing power of the blowpipe-flame itself, for the sake of getting as high a temperature as possible.

[2] The cooling may be hastened by very cautiously dropping water upon the charcoal at a little distance from the cavity, not into the cavity itself.

[3] Care must be taken not to mistake the white ash, which charcoal always leaves when burnt, for an incrustation of metallic oxide. The former is light, almost downy, in appearance; the latter is more compact, and generally has a definite border.

(γ) *The globule is malleable and, soft, and a yellow incrustation is formed on the charcoal.* The substance is a LEAD salt.

(δ) *The globule is brittle, and white fumes ascend from it while hot; a white incrustation is formed on the charcoal.* The substance is an ANTIMONY salt.

(ϵ) *The globule is brittle, and a yellow incrustation is formed.* The substance is a BISMUTH salt [1].

C *No globule is obtained, but a reddish brown incrustation is formed on the charcoal.* The substance is a CADMIUM salt.

If no characteristic result has been obtained it may be worth while to examine the substance in a borax bead, as directed in **9 B.**

11 Colour im- Moisten the end of a *perfectly clean* platinum
parted to flame. wire with strong **hydrogen chloride**, dip the wire into the powdered substance, and hold it at the edge of the flame of a Bunsen's burner. Observe the colour, if any, imparted to the flame.

(a) It is *yellow*. The metallic radicle is SODIUM.

(β)	„ *lavender*.	„ „	POTASSIUM.
(γ)	„ *crimson*.	„ „	STRONTIUM.
(δ)	„ *orange-red*.	„ „	CALCIUM.
(ϵ)	„ *green*.	„ „	BARIUM or COPPER [2].

[It must be borne in mind that traces of sodium are almost invariably present in a substance, and that the intense yellow flame due to its presence will often mask other colours. This difficulty is especially felt in the case of potassium salts; the barium, strontium, and calcium salts are so much less volatile that they remain long after the traces of sodium have volatilised. A transient yellow tinge will almost always be imparted to the flame when the salt is first introduced into it; and when this has passed away, the other characteristic colours should be looked for. In order to detect potassium the flame should be looked at through a piece of deep blue glass. If sodium alone is present, the flame will be invisible or very pale

---

[1] The globule of bismuth, although in reality brittle, does not always crumble to pieces under the pestle or hammer; the particles may remain aggregated so as to appear not unlike a flattened plate of malleable metal. But the edges of the plate are always jagged, as may be readily seen with a magnifier, and a blow or two of the pestle is generally sufficient to break it into several pieces.

[2] Or, possibly, hydrogen borate may be present.

blue, the yellow rays being absorbed by the glass. If potassium is present, the flame when thus looked at will appear deep crimson. The same blue glass may be used to distinguish between the colours imparted by strontium and calcium. The flame, when looked at through it, if strontium is present appears crimson, if calcium is present appears greenish-gray.]

B. *The Substance for Examination is a Liquid.*

12 Observe and note down the colour and smell (if any [1]) of the liquid.

Place about 2 or 3 c.c. of the liquid in a watch-glass and set it on the sand-bath to evaporate to dryness, while other experiments with the solution are proceeded with. The residue, if any, should be examined by heating in an ignition-tube and before the blowpipe according to the directions already given, and if the nature of the liquid has been discovered before the residue is obtained, the results of the blowpipe examination will serve as confirmatory evidence.

13 Colour imparted to flame. Dip a clean platinum wire into the solution, and hold it at the border of the flame of a Bunsen's burner. If any characteristic colour is imparted to the flame, refer to **11**.

14 Reaction on litmus-paper. Take out a drop on a glass rod and place it upon a piece of **blue litmus-paper**. If no effect is produced, place another drop upon a piece of **reddened litmus-paper**.

A *The liquid is neutral, or nearly so, to test-paper.* Pass on to **15**.

B *The liquid reddens blue litmus-paper strongly.*

This may be due to the presence either of a HYDROGEN salt or of a double salt of HYDROGEN and a METAL, or lastly of certain metallic salts, e. g. COPPER or MERCURY salts, which, although containing no basic hydrogen, have an acid reaction.

[1] A cyanide, for instance, may often be detected by the odour of hydrogen cyanide which it gives off.

To distinguish between these, place a little of the liquid in another test-tube, add a drop of solution of **sodium carbonate** on the end of a glass rod, and stir the mixture. If no precipitate is produced, or the precipitate redissolves immediately, with effervescence, a HYDROGEN salt is present. If a permanent turbidity is produced, the acid reaction is due to the presence of a metallic salt, soluble in water. Pass on to **15**.

C   *The liquid turns reddened litmus-paper strongly blue.*

This may be due to the presence of a BORATE, PHOSPHATE, SILICATE, HYDRATE, CARBONATE, SULPHIDE, or CYANIDE of a member of Group VI, or of a HYDRATE or SULPHIDE of a member of Group IV, V, or VI. Pass on to the next paragraph.

15 Action of hy-   Pour a little of the solution into a test-tube, drogen nitrate and add dilute **hydrogen nitrate**, drop by on solutions. drop, until the liquid is strongly acid; then warm, but do not boil, the mixture. Observe whether any gas is evolved, or precipitate formed, or both.

A   *No effervescence, or evolution of a gas recognisable by its smell occurs; and no precipitate is produced*[1].

If the original solution was *neutral*, pass on to **16**.

If the original solution was *alkaline*, add a few drops of it to some solution of **ammonium chloride** in another test-tube, and warm it.

(a) *No precipitate is produced.*
The substance present is probably a phosphate, borate or arsenite of a metal in Group IV, V, or VI, or a hydrate of an alkali-metal. Pass on to **16**.

(β) *A white gelatinous precipitate is produced.*
The solution contains a silicate of an alkali-metal.
Treat 20 c.c. of the original solution as directed below, under **H**.

B   *A gas is evolved, recognised to be sulphur dioxide by its penetrating smell* (see note [2]), and

---

[1] If the original solution was yellow, and becomes red when the acid is added, a CHROMATE is present.
[2] This must not be confounded with the slight acid odour of hydrogen nitrate itself. If there is any doubt, add a few drops of strong hydrogen sulphate to another portion of the original solution, warm (but do not boil) and observe the odour.

(*a*) *No precipitate is produced.*

The solution contains a SULPHITE.

Confirm this by adding to the acidified solution a few drops of a solution containing blue starch iodide, made as directed on p. 226, expt. 3 *b* (or potassium permanganate). If *the colour is discharged*, a sulphite is present.

Examine the original solution for the metal by the usual course.

(*β*) *A white milky precipitate is produced.*

A HYPOSULPHITE is present. Confirm this by pouring a few drops of solution of silver nitrate into a watch-glass, adding a drop of the original solution on a glass rod, and stirring. If *a white precipitate, quickly turning yellow and finally black, is formed*, a hyposulphite is present.

Add excess of dilute hydrogen chloride to about 20 c.c. of the original solution, and evaporate to dryness in a dish: warm the residue with 20 c.c. of water, filter, and test the filtrate for a metal by the usual course.

C *A gas is evolved, recognised to be hydrogen sulphide by its offensive smell and its blackening action on a slip of paper moistened with lead acetate and held in the tube;* and

(*a*) *No precipitate is formed.*

The solution contains a SULPHIDE of one of the metals in Group IV, V, or VI. Treat about 20 c. c. of the original solution as directed just below, in *β*.

(*β*) *A white milky precipitate is formed.*

The solution contains a POLYSULPHIDE of one of the metals in Group IV, V, or VI.

Add excess of dilute hydrogen chloride to about 20 c.c. of the original solution; boil in a dish until all hydrogen sulphide is expelled; filter from deposited sulphur (if any) and examine the filtrate, which contains the metal as chloride, as directed in **20**.

(*γ*) *A coloured precipitate is formed.*

The solution contains a double sulphide of an alkali-metal with arsenic, antimony or tin.

Add a slight excess of hydrogen chloride to about 20 c.c. of the original solution, warm and filter; test the filtrate for an alkali-metal as directed in **22**. Wash the precipitate, dissolve it in strong hydrogen chloride with addition of a drop or two of hydrogen nitrate, and test the solution for arsenic, antimony, and tin as directed in **17**.

D   *A gas is evolved, recognised to be hydrogen cyanide by its smell, like essence of almonds,* and

(a) *No precipitate is formed.*

The solution contains a CYANIDE of an alkali-metal (or possibly a FERROCYANIDE or SULPHOCYANATE). To confirm this, add to some of the acidified solution (1) a few drops of solution of **iron protosulphate**[1], (2) a slight excess of solution of **potassium hydrate**, warm the mixture slightly and shake it thoroughly: lastly, add excess of dilute **hydrogen chloride**. If a *deep blue precipitate remains*, a cyanide is present. Boil about 10 c.c. of the original solution in a dish with excess of hydrogen chloride until all hydrogen cyanide is expelled, and test the solution for an alkali-metal, as directed in **22**.

(β) *A precipitate is formed.*

The solution contains a double CYANIDE of an alkali-metal with some other metal. Confirm the presence of a cyanide as directed just above (*a*).

Evaporate about 20 c.c. of the original solution to dryness, and boil the residue with 6 or 8 c.c. of strong hydrogen nitrate, which will decompose all the cyanides present. When all hydrogen cyanide has been expelled, dilute the solution with 30 c.c. of water, and examine it in the usual way for metals; remembering that, when one metal has been found, the group reagent must be added in excess to another portion of the solution, in order to separate it entirely. The liquid must then be filtered, and the filtrate tested for an alkali-metal.

E   *A gas is evolved, recognised to be chlorine by its smell and its bleaching action upon litmus-paper.*

The solution contains a CHLORIDE, HYPOCHLORITE, or CHLORATE.

These will be detected in following the usual course. Pass on to **16**.

F   *Orange vapours of nitrogen trioxide are evolved.* The substance is a NITRITE. As a confirmatory test, add a few drops of the solution to some solution of potassium iodide mixed with a little starch solution and acidified with hydrogen sulphate (see p. 141). If a nitrite is present, iodine will be at once liberated and the liquid will turn blue.

---

[1] If a blue precipitate is formed at once, a FERROCYANIDE is present.
If the liquid turns red by degrees, a SULPHOCYANATE is present.

Boil about 20 c.c. of the original solution with a little hydrogen chloride until all decomposition is over, and examine it for a metal by the usual course.

G   *A strong effervescence occurs, and the gas given off is inodorous.*

The solution contains a CARBONATE.

Confirm the presence of carbon dioxide by holding within the tube a glass elbow tube dipped in lime water, as already directed, p. 162.

Boil until the carbon dioxide is expelled, and test the solution for a metal, as directed in **16**.

H   *No gas is evolved, but a gelatinous precipitate is formed which does not dissolve in the excess of acid.*

The solution contains a SILICATE of an alkali-metal.

Put about 10 c.c. of the original solution into a dish, add a slight excess of strong **hydrogen chloride**, and evaporate the whole to complete dryness, continuing the heat, with occasional stirring, until the residue loses all gelatinous appearance. Warm it with a little water, grinding the whole together with a pestle; then filter. Test the solution for an alkali-metal as directed in **22**. Wash the residue of silicon dioxide on the filter, dry it, and fuse a little of it in a bead of microcosmic salt (p. 245). If it floats undissolved in the bead it is silicon dioxide, and the original substance is proved to be a silicate.

I   *No gas is evolved, but a precipitate is formed which redissolves in the excess of hydrogen nitrate.* The substance contains one of the metals which form soluble double oxides or hydrates with an alkali-metal [1].

Pass on to **16**, bearing in mind that an alkali-metal must be sought for, as well as one of the others above mentioned, and therefore that the latter must be separated by addition of the proper group-reagent in excess (as explained under **D** β above) and the filtrate examined for the alkali-metal.

---

[1] A list of these metals is given at the foot of the Table of Solubilities, Appendix E.

SECTION III.

EXAMINATION OF THE SOLUTION OF A SINGLE SUBSTANCE FOR A METAL.

[If the substance has been dissolved in hydrogen chloride, or aqua regia, pass on to 17, since silver salts and mercury protosalts cannot be present[1].]

1. Examination for the Metals included in Group I.

SILVER, MERCURY (protosalts), LEAD.

16 *Hydrogen chloride test.* Add **dilute hydrogen chloride**, drop by drop, to a portion of the cold solution. If a precipitate is obtained, add several drops of the strong acid, and apply heat, in order to ascertain if the precipitate is soluble in excess.

A *No precipitate, or a precipitate which dissolves in excess of the acid, is obtained[2].*

SILVER salts, MERCURY protosalts, and also LEAD salts, unless the solution is very dilute, are absent. Pass on to **17**.

B *A permanent white precipitate is formed.*

A metal belonging to Group I is present.

Allow the precipitate to subside, pour off as much as possible of the liquid, and add 2 or 3 c.c. of ammonia.

---

[1] The original substance may, however, be a mercury protosalt, which would be converted into a persalt by aqua regia.

[2] Such a precipitate will be probably due to the presence of a BISMUTH or ANTIMONY salt, since bismuth and antimony chlorides (which will be formed on addition of hydrogen chloride) are more easily decomposed by water than their other salts.

(a) *The precipitate is redissolved.*

SILVER is present. This may be also inferred if the precipitate is curdy and becomes gray on exposure to light. As a confirmatory test, add to another portion of the original solution (which, if acid, must be nearly neutralised by ammonia) a drop of solution of **potassium chromate,** which will cause a crimson precipitate if silver is present.

(β) *The precipitate turns black.*

A MERCURY protosalt is present. To confirm this, place a clean strip of copper in another portion of the original solution, acidified, if necessary, with dilute hydrogen chloride. If the copper is coated with a gray deposit, which becomes silvery-bright when rubbed with a cloth, and disappears when heated over a lamp, the presence of mercury may be considered certain.

(γ) *The precipitate remains unaltered in appearance.*

LEAD is present. To confirm this, test another portion of the original solution with **dilute hydrogen sulphate:** a white precipitate will be produced if lead is present.

[The precipitate may consist of silica, sulphur, or certain sulphides, if the original solution was alkaline. This, however, should have been already ascertained, see 15.]

2. Examination for the Metals included in Group II.

MERCURY (persalts), LEAD, COPPER, BISMUTH, CADMIUM, TIN, ARSENIC, ANTIMONY, GOLD, PLATINUM.

17 Hydrogen sulphide test. Add to the portion of the solution which has been acidified with hydrogen chloride, solution of **hydrogen sulphide** in excess. Observe whether the addition of a single drop of the reagent produces a precipitate, and if so, whether the precipitate is altered in appearance when more of the reagent is added (see note [1]). If no precipitate is produced at once, warm the solution.

[1] The precipitate produced in solutions of mercury persalts by the first drop of the solution of hydrogen sulphide is white, and quickly changes to yellow, brown, and finally black, on further addition of the reagent.

Salts of lead, if a large excess of hydrogen chloride is present, give a dull red precipitate of lead sulphochloride.

A   *No precipitate is formed.*   .,

None of the metals included in Group II are present. Pass on to **19.**

B   *A perfectly white, milky precipitate is produced.* This consists solely of SULPHUR, and is due to the decomposition of the hydrogen sulphide by some oxidising substance present in the solution, probably a CHROMATE or an IRON PERSALT. If the solution is red or yellow, and turns light green simultaneously with the precipitation of sulphur, one of the above two substances is certainly present; they will, however, be detected in the course of examination for Group III (**19**), to which you should pass on.

The decomposition of the hydrogen sulphide may also be due to the presence of CHLORINE (if aqua regia has been used to dissolve the substance), a HYPOCHLORITE, a CHLORATE, a NITRITE, or a SULPHITE. These will, even if not detected in the preliminary examination, be examined for in Sect. IV. Bear in mind the result, therefore, but pass on to **19**, as if the precipitate had not occurred.

C   *The precipitate is light yellow.*

CADMIUM, ARSENIC, or TIN (as persalt) is present.

Add to the fluid in which the precipitate is suspended, solution of potassium hydrate in excess, and warm the mixture.

> (*a*) *The precipitate dissolves.*
>
> TIN (as persalt) or ARSENIC is present. To distinguish between them, add a few drops of **ammonia** to a portion of the original solution. If *a white precipitate is produced*, TIN is present; the reduction of the salt before the blowpipe (*malleable bead, white incrustation*) will afford confirmatory evidence. If *no precipitate is obtained*, ARSENIC is present; the production of a bright metallic sublimate when the substance is heated in an ignition tube with **sodium carbonate (9 A**) will remove any doubt.
>
> If arsenic is found, the substance may be an ARSENITE or ARSENATE; if the latter, the precipitate caused by hydrogen sulphide will have formed only on boiling the liquid. Refer to p. 284, Expt. 2, and p. 286, Expts. 2, 3, for other means of distinguishing between them.
>
> (*β*) *The precipitate does not dissolve.*

CADMIUM is present. Confirm this inference by adding ammonia to some of the original solution, which should produce a white precipitate, soluble in excess, and precipitated as yellow sulphide on addition of hydrogen sulphide to the ammoniacal solution (or the blowpipe test, p. 277, may be applied).

D   *The precipitate is orange-coloured.*

ANTIMONY is present. Confirm this by laying a piece of platinum foil in a dish, putting on it a bit of zinc, pouring over it 5 or 6 c.c. of dilute hydrogen chloride, and adding a few drops of the original solution. If a black deposit is formed on the platinum, the presence of antimony is certain.

E   *The precipitate is dark brown or black.*

MERCURY, LEAD, COPPER, BISMUTH, TIN (protosalt), GOLD, or PLATINUM is present [1].

Add to the liquid in which the precipitate is suspended, excess of solution of **potassium hydrate** and heat the mixture.

If *the precipitate is dissolved*, TIN, as a protosalt, is present (or possibly GOLD or PLATINUM). In confirmation, add a drop of the original solution to some solution of **mercury perchloride**, which latter will be reduced to protochloride and form a white precipitate [2].

If *the precipitate does not dissolve*, pass on to the next paragraph.

18   The remaining metals must be successively tested for, as follows.

A   Place a clean slip of **copper** in a portion of the original solution, acidified, if necessary, with dilute hydrogen chloride. The formation of *a gray metallic film* on the copper proves that MERCURY is present (see **16 B β**). The changes in colour of the precipitate produced by addition of successive portions of hydrogen sulphide to the original solution (see p. 399, note), will have also indicated the presence of mercury, and the result of

----

[1] If the solution has a blue or green colour, copper may be at once tested for as directed in **B**.

[2] It is scarcely probable that gold and platinum will be met with; but if tin is not detected by the above tests, separate portions of the original solution should be tested (*a*) for GOLD, with solution of **iron protosulphate** (p. 300), (*b*) for PLATINUM, with solution of **ammonium chloride** (p. 302).

heating the dry substance with sodium carbonate in a tube (9 A *a*) will afford absolute certainty.

B   Add to a portion of the original solution **ammonia** in excess.

(*a*) *A precipitate is at first formed which readily dissolves in excess of ammonia, forming a deep blue solution.* COPPER is present. To remove any doubt, test another portion of the original solution with solution of **potassium ferrocyanide**, which will give a reddish brown precipitate of copper ferrocyanide.

(β) *A white precipitate is produced which does not dissolve in excess of the ammonia.* LEAD or BISMUTH is present. To distinguish between them, pour off a portion of the liquid in which the precipitate is suspended into another tube, and add an equal volume of dilute **hydrogen sulphate**.

If a white residue remains insoluble, LEAD is present. The reduction of the substance on charcoal before the blowpipe, p. 267 (*malleable globule, yellow incrustation*), will serve as a confirmatory test.

If the precipitate dissolves, BISMUTH is present. To confirm this, add to the rest of the liquid (containing the precipitate formed by ammonia) dilute **hydrogen chloride**, drop by drop, until the precipitate is *just* redissolved; then fill up the test-tube with water. *If the fluid becomes milky*, the presence of BISMUTH is confirmed. As a further proof, a portion of the original substance may be mixed with some **sodium carbonate** and reduced on charcoal before the blowpipe (*brittle globule, yellow incrustation*).

3. Examination for the Metals included in Group III.

IRON, COBALT, NICKEL, MANGANESE, CHROMIUM, ALUMINIUM, ZINC.

19   Ammonium   Add to a portion of the original solution (1) sulphide test.   one-fourth of its volume of solution of **ammonium chloride**[1], (2) solution of **ammonia** until the liquid smells of it strongly, and lastly (3), whether ammonia has produced any precipitate or not, add three or four drops of solution of **ammonium sulphide**, and warm the mixture.

---

[1] The ammonium chloride is added to prevent the possible precipitation of magnesium as hydrate by the ammonia.

A *No precipitate is produced.*

None of the metals included in Group III are present. Pass on to **20**.

B *A black precipitate is produced.*

IRON, NICKEL, or COBALT is present.

[Before applying the next test, consider whether the original substance (or the residue from the original solution) became charred when heated in a tube (**8 E**). Some organic substances, e. g. tartrates, citrates, sugar, prevent the precipitation of the hydrates by potassium hydrate. If, therefore, an organic substance is present, it will be safest to filter and wash the precipitate produced by ammonium sulphide, and then to dissolve it in hydrogen chloride (adding, if necessary, but not otherwise, a little hydrogen nitrate), and boil until all hydrogen sulphide is expelled. This solution should then be used instead of the original solution for the following test.]

Add to a portion of the original solution some solution of **potassium hydrate.**

(α) *A dull green precipitate is formed, which becomes nearly black, and finally reddish brown when shaken up for a short time in the tube, in order to expose it to the air.*

An IRON protosalt is present. In confirmation, test another portion of the original solution with solution of **potassium ferricyanide,** which should produce a deep blue precipitate (p. 310, Expt. 5).

(β) *A light green precipitate is produced, which does not alter in colour when exposed to the air or when the solution is boiled.*

NICKEL is present. The colour imparted to a borax bead by a portion of the substance will be the best confirmatory test (**9 B ζ**).

(γ) *A light blue precipitate is formed, which slowly turns green in the air, and becomes reddish brown when the solution is boiled.*

COBALT is present. The blue colour imparted to a borax bead by a little of the substance will afford absolute certainty (**9 B a**).

(δ) *A reddish brown precipitate is formed.*

An IRON persalt is present. If so, hydrogen sulphide will have produced a precipitate of sulphur (**17 B**). Confirm this by testing another portion of the original solution with solution of **potassium ferrocyanide,** which will produce a deep blue precipitate of Prussian blue.

C   *A precipitate is produced which is not black.*

[Before applying the next test, consider whether the original
substance (or the residue from the original solution) *became
charred* when heated in a tube (**8 E**). If it did, read the preliminary remark in the last paragraph (**B**) and proceed as there
directed.]

Add to a portion of the original solution, some solution of
potassium hydrate, drop by drop, observing whether a precipitate is produced, and if so, whether it dissolves in excess.

(*a*) *No precipitate is produced.*

If the original solution is red or yellow, and lost its colour
when the ammonium sulphide was added just now, a CHROMATE
is present. This will be confirmed in the usual course of testing for acids (Sect. IV). Pass on to **20**.

If the original solution was colourless, there is reason to test
for a SILICATE, if this has not been already done, as directed
in **15 H.**

(*β*) *A gray precipitate is produced, insoluble in excess, and turning
brown when shaken up with air.*

MANGANESE is present. To confirm this, heat some of the
original substance in a borax bead (**9 B ε**).

(*γ*) *A dull bluish green precipitate is produced, soluble in excess, forming a green solution.*

CHROMIUM is present. The colour imparted to a borax bead
(**9 B γ**) will confirm this.

(*δ*) *A white precipitate is produced, soluble in excess.*

ZINC or ALUMINIUM is present.

To distinguish between them, pour off some of the solution
(containing excess of potassium hydrate) into another tube and
add solution of **hydrogen sulphide.**

(*aa*) If this *produces a white precipitate*, ZINC is present. This
should be confirmed by application of the blowpipe test, **10 A**
(incrustation on charcoal, yellow while hot, white when cold;
turning green on ignition with cobalt nitrate).

(*ββ*) If *no precipitate is formed*, ALUMINIUM is present. To
confirm this add some of the remainder of the solution (in
potassium hydrate) to some solution of **ammonium chloride**
in another test tube, and warm the mixture. The formation
of a white gelatinous precipitate will confirm the presence of
ALUMINIUM [1].

[1] Solution of potassium hydrate often contains a silicate and aluminate,

(ε) *A white precipitate is produced, insoluble in excess of potash, unaltered in air.*

The substance is an OXALATE, PHOSPHATE, FLUORIDE, BORATE, or SILICATE; the other radicle being BARIUM, STRONTIUM, CALCIUM, or MAGNESIUM[1]. If so, it will have been found soluble in hydrogen chloride, not in water.

Add to a portion of the original (acid) solution an equal volume of solution of **calcium sulphate**.

(αα) *An immediate white precipitate is formed.*

BARIUM is present. The green colour imparted to flame (11) will serve to confirm this.

(ββ) *A precipitate is produced, but only after the lapse of three or four minutes.*

STRONTIUM is present. The crimson colour imparted to flame (11) will serve to confirm this.

(γγ) *No precipitate is produced even after the lapse of eight or ten minutes.*

Add to a portion of the original solution a few drops of **hydrogen sulphate** and then 5 or 6 c.c. of alcohol.

If *a white precipitate is formed*, CALCIUM is present. The orange-red colour imparted to flame (11) will serve to confirm this.

If *no precipitate is formed*, MAGNESIUM must be present. To obtain positive evidence some of the original substance must be decomposed in the following way.

Mix a portion of the original substance with about twice its bulk of **sodium carbonate**, and fuse it before the gas blowpipe on a piece of platinum foil turned up at the edges so as to form a cup (see p. 8). When the mass is thoroughly fused, allow it to cool, then place the whole in a beaker, add a little water, and boil for a minute or two.

which would give a precipitate under the above conditions. To avoid any mistake on this account, take two equal portions of the solution of potassium hydrate, add to one a few drops of the solution you are analysing, then add to both half the volume of solution of ammonium chloride and heat them equally. If the potassium hydrate is impure, a precipitate will be formed in both cases; but if the solution you are analysing contains aluminium, then there will be more precipitate in that tube to which it has been added.

[1] These salts are soluble in acids, but are reprecipitated, unaltered in composition, when the solution is neutralised.

Hence either they must be examined for by tests which can be applied to acid solutions, or they must be decomposed by other means.

By this treatment the salt is decomposed, the metal forming an insoluble carbonate, while the other radicle combines with sodium, and passes into solution. (While this is going on you may examine other portions of the original substance for an oxalate, a phosphate, and a fluoride, as directed in **25 B, C,** and **E.**)

Filter the solution containing excess of sodium carbonate, wash the residue on the filter several times with warm water, then pour over it a little **dilute hydrogen chloride,** and pour the fluid which runs through again on the filter, until the whole of the precipitate is dissolved. The solution, which contains the metal as chloride, should be boiled for a minute, to free it from carbon dioxide, then mixed with a slight excess of ammonia, filtered from any precipitate (of undecomposed salt) formed, and tested for magnesium as directed in **21.**

4. Examination for the Metals included in Group IV.

BARIUM, STRONTIUM, CALCIUM.

20 Ammonium carbonate test. To the portion of the solution in which ammonium sulphide has produced no precipitate, and which also contains ammonium chloride and ammonia, add a few drops of solution of **ammonium carbonate,** and apply heat.

A *No precipitate is formed.*

None of the metals included in Group IV are present. Pass on to **21.**

B *A white precipitate is formed.*

BARIUM, STRONTIUM, or CALCIUM is present.

Add to a fresh portion of the original solution an equal volume of solution of **calcium sulphate.**

(a) *A white precipitate is produced immediately.*

BARIUM is present. As a confirmatory test, add to another portion of the solution some **hydrogen and silicon fluoride,** which will produce a crystalline precipitate if barium is present. Observe also the colour imparted by the original substance to flame (**11**).

(β) *No immediate precipitate is formed.*

STRONTIUM or CALCIUM is present.

Allow the solution to stand for four or five minutes, and then, if no turbidity appears, warm it.

(γ) *A precipitate is formed after the lapse of some time.*

STRONTIUM is present. The carmine-red colour imparted to flame will be a conclusive proof.

(δ) *No precipitate is formed, even after the solution has been allowed to stand and warmed.*

CALCIUM is present. In confirmation, add to a fresh portion of the original solution enough **ammonia** to render it alkaline, and then a drop of solution of **ammonium oxalate**, which should produce a white precipitate[1]. Observe also if the substance imparts to flame a brick-red colour, which appears greenish gray when looked at through a piece of blue glass.

5. Examination for the Metal included in Group V.

MAGNESIUM.

21 Sodium phos-   To the portion of the solution in which am-
phate test.   monium carbonate produced no precipitate, and
which also contains other ammonium salts, add a drop of solution of **sodium phosphate**, and, if no precipitate is produced, rub the sides of the tube gently with a glass rod.

A   *No precipitate is formed.*

MAGNESIUM is absent. Pass on to **22**.

B   *A crystalline precipitate is formed.*

MAGNESIUM is present. In confirmation, ignite some of the original substance on charcoal, moisten the white residue with a drop of cobalt nitrate, and again ignite strongly. If the mass is light pink when cool, the presence of magnesium is confirmed (p. 351).

---

[1] It must be borne in mind that this result is only characteristic of the presence of a calcium salt under the circumstances here mentioned; since a barium or strontium salt would, if present, yield a similar precipitate.

6. Examination for the Metals included in Group VI.

POTASSIUM, SODIUM, AMMONIUM, HYDROGEN.

22 To a fresh portion of the original solution placed in a small beaker, add a little powdered **calcium hydrate** or **oxide**, cover the beaker with a watch-glass on which a slip of moist **reddened litmus-paper** has been placed (p. 363), and warm the mixture very gently for half a minute.

A *The litmus-paper is turned blue.*

AMMONIUM is present. In confirmation, observe the odour of the gas, and see whether white fumes are formed when a glass rod dipped in **hydrogen chloride** (or **acetate**) is held within the beaker. The original substance will also have been found wholly volatile.

B *The litmus-paper is unaltered.*

POTASSIUM, SODIUM, or HYDROGEN is present.

Add to another portion of the original solution, rendered slightly acid, if necessary, by hydrogen chloride, a drop or two of solution of **platinum perchloride**, and, if no precipitate is formed, evaporate the solution to dryness in a watch-glass and treat the residue with dilute alcohol (p. 356).

(a) *A yellow crystalline precipitate, or insoluble residue, is obtained.*
POTASSIUM is present. To remove any doubt, test another portion of the solution (concentrated, if necessary) with solution of **sodium-and-hydrogen tartrate** (p. 356), and also observe if the substance imparts a lavender colour to flame, appearing crimson when looked at through blue glass (p. 357).

(β) *No precipitate or insoluble residue is obtained.*
SODIUM or HYDROGEN is present.

To test for the former metal, place a little of the original substance in a loop of platinum wire, and hold it in the flame of a Bunsen's burner. If an intense yellow colour, lasting for some time, is imparted to the flame, which is nearly or quite invisible when looked at through a piece of blue glass, the substance is a SODIUM salt.

[It must be remembered that mere traces of sodium (which is nearly always present as an impurity in other substances) will colour the flame yellow. But such traces are soon volatilised, and hence if the substance continues for some minutes to impart an intense yellow to the flame, it may be certainly inferred to be a sodium salt.]

If sodium is not found, and no other metal has been detected in the previous stages of the analysis, the basic radicle is probably HYDROGEN alone.

The presence of a hydrogen salt will have been already inferred in the preliminary examination (14 B). Proceed with the examination for the other radicle (Sect. IV); and when it is found, consider whether the original substance possesses the properties, such as volatility, of its hydrogen salt.

SECTION IV.

EXAMINATION OF A SINGLE SUBSTANCE FOR A NON-METALLIC RADICLE.

23   Consider, in the first place, what radicles may be inferred to be present from the solubility or insolubility of the substance in water, and the results of the preliminary examination.

(*a*) If it is soluble in **water**, the radicle sought must be one of those which form compounds soluble in water with the metal which is known to be present.

(*b*) If it is insoluble in water, but soluble in **acids**, the radicle must be one of those which form compounds soluble in acids only, with the metal which is known to be present.

The Table in Appendix E will serve as a guide, but there are so many degrees of solubility that it is not always safe to infer the presence or absence of a radicle on this ground alone.

The following radicles will have been detected with certainty in the preliminary examination, and will not be further referred to in the course.

CARBONATE (**3 *a*, 15 G**).   SILICATE (**4 B, 15 A, H**).
SULPHIDE (**3 β, 15 C**).   NITRITE (**3 ζ, 15 F**).
SULPHITE (**3 γ, 15 B**).   HYPOPHOSPHITE (**8 F**).
HYPOSULPHITE (**3 γ, 15 B**).

If none of the above have been found, refer to the results obtained by heating some of the original substance (or some of the residue obtained by evaporating the solution) in an ignition-tube (**8**).

A   *The substance became charred.*

It is a TARTRATE, or ACETATE (or possibly an OXALATE [1]).

---

[1] These (unless quite pure) become, in many cases, more or less charred when heated.

Boil some of the original substance with some solution of **sodium carbonate**; or, if it was a liquid, add excess of **sodium carbonate** to it; filter, if necessary, and add a few drops of solution of **silver nitrate** to the clear liquid, sufficient to produce a precipitate of silver carbonate; lastly, boil it for half an minute.

(*a*) *The precipitate turns black.*
The substance is a TARTRATE.

(β) *The precipitate remains unaltered.*
The substance is an ACETATE (or OXALATE). To confirm this, add a little pure **alcohol** to the original substance, or solution, and then a few drops of strong **hydrogen sulphate,** and warm gently. If the fragrant smell of ethyl acetate (acetic ether) is perceived, an acetate is certainly present [1].

If it is not found, test for an oxalate as directed in **25 D**.

[CITRATES, though not of very common occurrence, may be here examined for, if an acetate is not found, before passing on to **25 D**. Add to some of the original solution about 1 c.c. of solution of **ammonium chloride,** then enough **ammonia** to render it alkaline, and lastly some solution of **calcium chloride.**

If *a precipitate forms at once,* an OXALATE is present. Pass on to **25 D**.

If *no precipitate is formed,* boil the liquid for half a minute. If *a crystalline precipitate gradually forms,* a CITRATE is present.]

B *The substance did not become charred.*

Add to a portion of the original solution (which, if acid, must be made slightly alkaline by ammonia: any precipitate thus formed, being filtered off) a few drops of solution of barium chloride (or barium nitrate, if lead, silver or mercury as protosalt is present [2]).

(*a*) *No precipitate is formed.* Pass on to **24.**

(β) *A yellow precipitate is formed.*

---

[1] If there is any doubt, mix a little hydrogen acetate with alcohol and hydrogen sulphate in another tube, and warm it; compare the smell of the vapours from this with that of the vapours from the substance under examination.

[2] A little barium nitrate for this purpose may be readily prepared by dissolving a small quantity of barium oxide, hydrate, or carbonate, in dilute hydrogen nitrate.

The substance is a CHROMATE. This will have been suspected already, and may be confirmed by the colour imparted to a borax bead (green in both flames).

(γ) *A white precipitate is formed.*

Add to this an excess of dilute **hydrogen nitrate**. If *it dissolves*, pass on to the next paragraph (24). If *it does not dissolve*, a SULPHATE is present. Confirm by testing the original solution (which must not contain hydrogen chloride), acidified with hydrogen nitrate, with solution of **lead acetate**, which will cause a white precipitate if a sulphate is present [1].

[SILICOFLUORIDES also give a precipitate with barium chloride, insoluble in acids. Hence if no precipitate is obtained with lead acetate, it will be well to test the substance for a fluoride as directed in **25 E**, and, if it is present, to place over the leaden cup a slip of glass on the under side of which a drop of water has been deposited. If a white film is formed in and around this, silicon fluoride is being evolved as well as hydrogen fluoride, and the substance is a silicofluoride.]

24 Add to some of the original solution (which must have been made with water or hydrogen nitrate) sufficient dilute **hydrogen nitrate** to render it strongly acid, if not already so, and test it with solution of **silver nitrate**.

A *No precipitate is formed.* Pass on to **25**.

B *A white or light yellow precipitate is formed.*

The solution contains a BROMIDE, IODIDE, CHLORIDE or CYANIDE.

Add to some of the original solution, acidified with hydrogen nitrate, two or three drops (not more) of **solution of chlorine**.

(a) *The liquid turns yellow.*

A BROMIDE or IODIDE is present.

Pour off some of the yellow solution into another tube and add some fresh **solution of starch**.

If *the liquid turns deep blue*, an IODIDE is present.

If *the liquid does not change colour*, a BROMIDE is present. Confirm by adding to the remainder of the yellow liquid (to which more chlorine may be added) some **carbon disulphide**, and shaking it up. If the globule which sinks to the bottom is orange-coloured, the original substance is a BROMIDE.

[1] Another very good confirmatory test for a sulphate is, to fuse the substance on charcoal with a little **sodium carbonate** (see p. 231), and examine the fused mass for a sulphide.

- (β) *The liquid does not turn yellow.*

A CHLORIDE or CYANIDE is present.

Put some of the original substance, or solution (in water or hydrogen nitrate) into a test tube, add a little **manganese dioxide** and then 2 or 3 c.c. of strong **hydrogen sulphate**, and warm it gently.

If *a greenish yellow gas, which bleaches moist litmus-paper, is given off*, the substance is a CHLORIDE (see, however, note [1]).

If *vapour smelling like essence of almonds is given off*, the substance is a CYANIDE. This should, as a rule, have been detected already (3 δ, 15 D). Confirm by the test given in 15 D *a*.

[A ferrocyanide or ferricyanide, might also evolve hydrogen cyanide when thus treated. If a ferrocyanide is present, the addition of **iron protosulphate** will give a pale blue precipitate, rapidly becoming dark blue.

If a ferricyanide is present, the precipitate produced by silver nitrate will be orange, and the original solution will give a dark blue precipitate at once on addition of **iron protosulphate,** but no precipitate when tested with **iron perchloride.**]

25 The remaining radicles should be tested for in separate portions of the original substance or solution, in the following order,—

A Add to a portion of the original solution (which must not have been made with hydrogen nitrate) several drops of **strong hydrogen sulphate** [if lead, barium, or strontium is present, a precipitate will, of course, be formed, which must be filtered off] then add one drop of solution of **indigo sulphate**, and heat the liquid to boiling.

If *the blue colour of the indigo is destroyed,* a NITRATE or CHLORATE is present. To distinguish between them, place a portion of the solid salt, or of the residue obtained on evaporation of the solution in a test-tube, and add a little **strong hydrogen sulphate.**

---

[1] A HYPOCHLORITE would give a similar reaction. If there is reason to suspect the presence of this from the results obtained in 15 E and 17 B, add to some of the original solution, rendered alkaline with potassium hydrate, a few drops of solution of lead acetate, and boil the mixture. If the precipitate *turns brown*, a hypochlorite is present. If not, the substance is a chloride.

(a) *The substance turns yellow, and a yellow gas is given off, posses-sing the odour of a chlorine oxide.*

The substance is a CHLORATE.

(β) *The substance does not turn yellow, or evolve gas.*

It is a NITRATE.

To confirm this, allow the solution to become quite cold, then pour down the side of the tube a little solution of iron proto-sulphate. If a dark brown ring is formed (either immedi-ately or in a minute or two) at the junction of the fluids, a NITRATE is present (p. 137, Expt. 4).

B   Add to a portion of the original substance moistened with water, or of the solution, some **strong hydrogen sulphate**, then put in a little **manganese dioxide**, and warm slightly (do not heat to boiling).

If *a gas is given off with effervescence* [1], which *causes cloudiness in a drop of lime water held in the tube* (p. 162), it is carbon dioxide, and the substance is an OXALATE [2].

C   Place some solution of **ammonium molybdate** in a tube, add a few drops of the original solution (acidified, if necessary, with hydrogen nitrate) and warm it gently (but do not boil it).

If *a yellow precipitate is formed*, a PHOSPHATE is present.

D   Place a little of the original solution in a watch-glass, render it slightly acid (if not already so) by dilute **hydrogen chlo-ride**, dip into the liquid one half of a strip of **turmeric-paper**, and dry it by a gentle heat.

If *the dipped portion becomes red, and turns bluish black when a drop of solution of sodium carbonate is placed upon it*, a BORATE is present. To obtain further proof, put a little of the substance (or solution) into a small dish, add some **alcohol** and several drops of strong **hydrogen sulphate**; stir them together, and set fire to the mixture. If a borate is present, the flame will be edged with green.

E   Put some of the original substance (or of the residue obtained by evaporating the solution) into a capsule of lead or platinum ;

---

[1] This is best recognised by its sound, the open end of the tube being held close to the ear.

[2] Carbonates, tartrates, and citrates must have been previously proved to be absent.

add a few drops of strong **hydrogen sulphate,** and stir the whole together. Cover the capsule with a glass plate coated with wax (see p. 211), and having some letters traced on it, as there directed ; and place the capsule on some hot sand for four or five minutes.

If, on removing the wax, *the letters are permanently etched into the glass,* the substance is a FLUORIDE.

26 Separation of   If the above tests have given no satisfactory
the metal.   results, it will be advisable to separate the metal and to obtain the other radicle in combination with sodium or ammonium.

The means to be employed for effecting this will depend upon the metal present.

A   The metallic radicle belongs to Group I, II, or III.

Boil some of the substance with 5 or 6 c.c. of solution of **ammonium sulphide** for a few minutes : allow any residue to subside, decant the clear solution into a porcelain dish, and evaporate it to dryness. Warm the residue with some water, filter the solution, which (unless the original substance was an ARSENIC, ANTIMONY, or TIN salt) will contain only the non-metallic radicle associated with ammonium, and examine it as directed in **23.**

[If ARSENIC, ANTIMONY, or TIN have been found to be present the solution will contain ammonium sulpharsenate, sulphantimonate, or sulphostannate. In this case, add to the solution a slight excess of dilute **hydrogen nitrate,** which will throw down the metal as sulphide, filter the solution, and examine it as directed in **23,** remembering that a NITRATE is now present.]

B   The metallic radicle belongs to Group IV or V.

Boil some of the substance with solution of **sodium car-bonate** for a few minutes, filter the solution, and test it as directed in **23,** decomposing the excess of sodium carbonate with dilute hydrogen chloride or nitrate, as required.

If no non-metallic radicle can be detected by the above course of examination, the substance is probably an OXIDE. Observe whether it has the characteristic properties of the oxide of the metal which has already been found.

SECTION V.

ANALYSIS OF A SINGLE SUBSTANCE WHICH IS INSOLUBLE IN WATER AND ACIDS.

It will be seen, by a reference to the Table of Solubilities (Appendix E), that the salt must be one of a very limited number of substances. The following list includes all the insoluble substances which are commonly met with in inorganic analysis.

(LEAD SULPHATE).	BARIUM SULPHATE.
SILVER CHLORIDE.	STRONTIUM SULPHATE.
SILVER BROMIDE.	(CALCIUM FLUORIDE).
SILVER IODIDE.	ALUMINIUM-AND-SODIUM FLU-
TIN DIOXIDE.	ORIDE.
ANTIMONY TRIOXIDE.	BARIUM SILICOFLUORIDE.
ANTIMONY PENTOXIDE.	SILICON DIOXIDE.
IRON-AND-CHROMIUM OXIDE.	SILICATES of various kinds.
(CHROMIUM SESQUIOXIDE).	SULPHUR.
(ALUMINIUM SESQUIOXIDE).	CARBON.

[The substances enclosed in brackets dissolve with great difficulty in acids.]

27 Refer to the results obtained by heating the substance in an ignition-tube (**8**).

A *The substance was unaltered. and is dark-coloured.* It may be iron-and-chromium oxide (chrome iron stone), chromium sesquioxide, carbon, or one of certain silicates.

Powder a little of it and mix it intimately with about twice as much **potassium nitrate**; put some of the mixture into an ignition-tube and heat it, at first gently, afterwards as strongly as possible.

(*a*) *The substance deflagrates.*

It is CARBON (charcoal or graphite; if the latter, it will have been acted on only with difficulty, and will have a metallic lustre). To confirm this, dip the bulb, while hot, into a little water in a dish, when it will crack off; pour off the solution into a test tube and examine it for a carbonate as directed in p. 162.

β) *The substance is acted on with formation of a yellow mass.* •

It is IRON and CHROMIUM OXIDE, or CHROMIUM SESQUIOXIDE (if the latter, it will be green). Detach the yellow mass by dipping the hot bulb into water, as directed just above; powder it, and boil it with water; then filter it and examine the filtrate for a chromate as directed in **23 B.** If any brown residue remains on the filter, it is probably iron oxide, and should be dissolved in a little hydrogen chloride, and the solution tested for iron with **potassium ferrocyanide.**

(γ) *No yellow mass is obtained.*

The substance is probably a SILICATE. Pass on to **28.**

B   *The substance was unaltered, or only fused, and is white or light yellow.*

Mix a little of it with **potassium cyanide,** and heat it strongly on charcoal, as directed in **10,** if this has not been done already.

(α) *A metallic globule is obtained.* The substance is one of the above compounds of SILVER, LEAD, TIN, or ANTIMONY.

Refer to **10 B** to decide which metal is present from the character of the globule and incrustation; and, if there is any doubt, boil the globule with hydrogen nitrate as directed in **7,** and examine the solution for the above metals.

If LEAD is found, boil some of the original substance with solution of **sodium carbonate,** and test the solution for a sulphate.

If SILVER is found, mix a little of the original substance with about three times as much **sodium carbonate,** and fuse the mixture in an ignition tube, or on a bit of porcelain: powder the fused mass and boil it with water, filter, and examine the filtrate for a bromide and iodide as directed in **24 A,** and, if neither are found, for a chloride as directed in **24 B β.**

(β) *No metallic globule is obtained.*

In this case it will be best to decompose the substance by fusion with sodium carbonate, as directed in the next paragraph.

28 Reduce some of the substance to an impalpable powder (using an agate mortar, if possible), and mix it with about four times its weight of **sodium carbonate** (or of a mixture of sodium-and-potassium-carbonates, which fuses at a much lower temperature than sodium carbonate alone). Place the mixture in a platinum capsule or on a piece of platinum foil turned up at the edges (p. 8), and heat it strongly over the gas blowpipe, until all action is over and the mass is in a state of calm fusion. After allowing it to cool, detach it from the capsule, and grind it with a little water in a mortar; then transfer the whole to a beaker, and boil it for a few minutes with about 20 c.c. of water.

[While this is going on, a fluoride may be tested for by heating a little of the substance (in a platinum vessel) with strong hydrogen sulphate, and observing whether vapours are evolved which etch glass (p. 211). If a fluoride is found to be present, continue heating the substance with hydrogen sulphate for five or six minutes, then allow the mass to cool, add a little water, mixing it thoroughly with the substance, and boil the mixture in a test-tube. Test the filtered solution for aluminium (**19 C**) and calcium (**19 C** γγ).]

By this treatment the metals are separated as insoluble carbonates or oxides, while the non-metallic radicles are obtained in solution as sodium salts.

Pour the whole on a filter, and test the filtrate (which, of course, contains excess of sodium carbonate) by the usual course for a silicate, sulphate, and fluoride.

Wash the residue, dissolve it in dilute hydrogen chloride (pouring the warm acid several times through it, on the filter) and test it for a metal by the usual course.

SECTION VI.

EXAMPLE OF THE ANALYSIS OF A SINGLE SUBSTANCE.

(The substance taken was manganese chloride.)

The single substance for examination was a light pink solid, not crystalline, which soon became moist when exposed to the air, and was readily powdered in a mortar.

A. Preliminary Examination.

1. A portion of the substance, heated in a test-tube with some water, readily dissolved, forming a nearly colourless solution which was neutral to test-paper.

2. A little of the solid substance was heated gently in an ignition-tube. It gave off traces of water, and when heated nearly to redness it fused into a brownish liquid which solidified, on cooling, into a pink crystalline mass. No vapours were given off, and no charring of the substance was observed.

Hence no organic radicle is present.

3. Since the substance was coloured, a little of it was heated in a borax bead before the blowpipe. It gave a transparent bead which was amethyst-red in the oxidising flame and colourless in the reducing flame.

Therefore manganese is present.

B. Examination of the Solution for a Metal.

1. A portion of the solution was tested with a few drops of dilute hydrogen chloride.

No ppte. was formed.

Therefore silver, mercury (as protosalt), and probably lead, are absent.

E e 2

2. To the same portion of the solution (containing hydrogen chloride) was added an excess of solution of·hydrogen sulphide, and the mixture was warmed.

No ppte. was obtained.

Therefore no metal belonging to Group II is present.

3. Another portion of the original solution was mixed with some solution of ammonium chloride, rendered alkaline with ammonia (which caused a ppte.), and tested with a drop of solution of ammonium sulphide.

A ppte was formed, which was not black.

Therefore the metal belongs to Group III, but is not iron, cobalt, or nickel.

4. Some of the original solution was tested with potassium hydrate.

A gray ppte. was formed, which became brown on being shaken up.

.Therefore the metal present is Manganese.

This was confirmed by the colour imparted to the borax bead.

C. Examination for the non-metallic Radicle.

Since the substance is a manganese salt readily soluble in water, the following radicles may be present :—Nitrate, Acetate, Chloride, Iodide, Bromide, Sulphate, Chromate.

It is not an acetate, since it did not blacken when strongly heated.

It is not a chromate, since the solution was nearly colourless, and gave no ppte. of sulphur when warmed with hydrogen sulphide.

It is probably not a nitrate, since no perceptible decomposition took place when it was heated.

1. A portion of the original solution was tested with a drop of solution of barium chloride.

A slight turbidity was produced, which did not disappear when excess of hydrogen nitrate was added ; but as the ppte. was quite disproportionate to the quantity of salt known to be in solution, it proved that a sulphate was present only as an impurity.

2. Another portion of the original solution was acidified with hydrogen nitrate and tested with solution of silver nitrate.

A white flocculent ppte. was formed.

> Therefore the substance is an iodide, bromide, or chloride.

3. Some of the original solution was acidified with hydrogen nitrate and tested with solution of chlorine.

No yellow colour appeared.

> Therefore the substance is not an iodide or bromide, but probably a chloride.

To confirm this a little of the solid substance was mixed with some manganese dioxide, and warmed in a test-tube with some strong hydrogen sulphate. A gas was given off which was proved to be chlorine by its odour, its yellowish green colour, and its bleaching action on a piece of litmus-paper held in the tube.

> The non-metallic radicle is, therefore, Chlorine; and the substance is Manganese Chloride.

APPENDIX A.

SUGGESTIONS FOR THE CONSTRUCTION OF CHEMICAL APPARATUS.

THE set of apparatus described at the beginning of this book is, necessarily, a rather expensive one; and although it includes nothing which would not be found in a regular laboratory, yet a student who is working by himself may not find it possible or advisable to spend so much on the materials for his experiments.

The object of the present chapter is to supply a few hints for the construction of simpler and cheaper forms of apparatus, such as may be made by any one possessed of a little mechanical skill and a few carpenter's tools. The student should never rest satisfied with clumsy makeshifts; he should endeavour to do the best he can with the materials he has at hand.

[The numbers refer to those in the list of apparatus at the beginning of the book.]

2. Measures. A method of making these has been illustrated in p. 47. Larger measures than the one there mentioned can, of course, be made in the same way; a tall, narrow, wide-mouthed bottle or jar[1]. being counterpoised in the balance, and 10, 20, &c. grms. of water weighed into it. Or, successive portions of 10 c.c. of water may be measured into the jar from the small test-tube measure already made. The graduation-marks may be either made with a file or, more neatly, by covering the whole of the outside of the jar with wax before graduating it, scratching the requisite marks and figures through the wax with a pointed bit of hard wood or of wire, and exposing it to the vapours of hydrogen fluoride prepared in the usual way (p. 211). A long narrow trough of sheet lead will serve to contain the fluor spar and acid.

4. Weights. Duplicates of any of the smaller weights to which

---

[1] A common flower-vase may often be obtained which answers the purpose well.

you have access, may be readily made out of strips of sheet brass or lead, cut at first decidedly heavier than the intended weight, and then carefully reduced, first by cutting off small strips with a pair of shears or scissors, and lastly by filing, until they exactly balance the standard weight.

When a strip has been cut off and the weight is still too heavy, you may ascertain whether it will be safe to cut off another similar strip by putting the piece cut off in the same scale as the standard weight, and observing whether the weight you are making is still too heavy; if it is, another strip equal in size to the last may be cut off. If the piece of metal is inadvertently made too light, it may still, of course, serve for a duplicate of the next smaller weight. Figures or marks showing its value should be scratched or punched upon the metal *before* it is quite reduced to the proper weight.

If no standard weights are at hand to be copied (but only in such a case), a set of small weights such as fractions of the gramme may be made as follows:—

Obtain a piece of rather thin brass wire about 12 or 15 cm. long (10 cm. of brass wire, No. 18 wire gauge, 1.22 mm. diameter, weigh almost exactly 1 grm.), and reduce it to the weight of 1 grm., by cutting off pieces as above directed. Then straighten it and cut it *exactly* in half (using cutting-pliers or scissors, not a file). This will give two pieces each weighing 5 decigrammes; keep one of these as a weight, bending it into a 5-sided figure, thus, to indicate its value. Divide the other piece accurately into five equal parts. This may be done by drawing a line equal to it in length on a piece of paper, and stepping along this line with a pair of compasses; a few trials will give the proper length of one division, ⌐1 2 3 4 5⌐. Lay down the wire close to the line on the paper, and mark each division on it with ink; then cut it at the 1st and 3rd divisions. You have now a piece weighing 1 decigramme, and two others each weighing 2 decigrammes. Bend each of the two latter in the middle to a right angle, thus, ⌐, to indicate its value, but leave the 1 decigramme straight. You will then have all the necessary decigramme weights, 5, 2, 2, 1; and centigramme weights may be made in the same way, a piece of thinner wire being taken.

5. Pneumatic Troughs. An earthenware pan, a large biscuit tin, or even a wooden box (covered with three or four coats of good paint) will answer the purpose. The shelf may be made of a sheet of tin plate or zinc, about 14 or 15 cm. wide, and sufficiently long to reach across the trough. A portion on each side

about 1 cm. broad should be bent, down, as shewn in Fig. 66, to
give stiffness to the shelf, and a hole about 2 cm. in diameter should
be cut in the centre. The size of the hole should first be traced
out with a pair of compasses, and the metal may then be cut with
a chisel or with the blade of a strong knife struck with a hammer,
the plate being laid on a piece of board. To suspend this shelf in
its place in the trough the ends of the strip of tin should be turned
up and bent over the rim of the trough. Or, four pieces of strong
brass wire may be passed through holes made at the four corners
of the shelf, and bent up over the rim. Instead of tin plate the
shelf may be made of wood (well painted), having one or two
pieces of lead nailed on its under side to sink it, and suspended
by wire hooks.

Instead of a shelf reaching across the trough, a 'bee-hive' shelf,
Fig. 65, may be used. This is intended to stand on the bottom
of the trough, the delivery tube being introduced into the side-

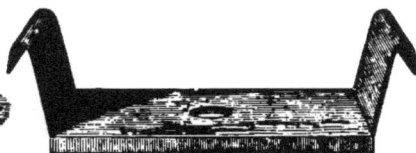

Fig. 65.   Fig. 66.

opening, and the gas jar being placed over the central aperture.
Such a shelf may be purchased for a shilling, or less; but it may
be easily made out of a common gallipot, or flower-pot saucer.
The hole in the centre may be pierced by gradually chipping away
the material with the point of an old three-square file, care being
taken that the blows of the hammer are light. The notch at the
side may be cut out by chipping or crumbling away the earthenware with a pair of pliers. There is, of course, some chance of
breaking the pot, but if it is placed on a folded cloth, and if the
chipping is carefully and patiently done, the risk is slight.

Finally, an extempore shelf may be made by placing two bricks
side by side in the trough, a sufficient interval being left between
them to admit the delivery tube.

6, 7. Supports for Flasks, Tubes, &c. A very useful tripod-
stand for supporting sand-baths, &c., may be made by taking three
pieces of stout iron wire or thin rod, about 45 cm. long, bending
them into this form ⊓, and binding them together by folds of thin
iron wire, so as to form the stand represented in Fig. 67.

For small crucibles, a triangle support is extremely useful, Fig. 68.
It is made by taking three bits of tobacco-pipe stem about 6 cm. in

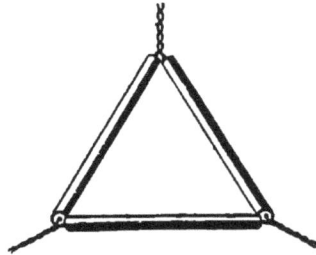

Fig. 67.   Fig. 68.

length, passing through each a piece of iron wire of such a length

Fig. 69.   Fig. 70.

that about 4 cm. projects at each end of the pipe stem, and then
twisting together these projecting pieces of wire. It is intended to

be laid on the ring of a retort-stand or on the tripod-stand above mentioned[1].

A simple form of **holder** is represented in Fig. 69, which may be constructed by the student himself. The base (12 × 24 cm.) and the upright (5 × 30 cm.) are made of deal about 2.5 cm. in thickness. The upright is mortised and glued into the base, and has a slit about 5 mm. wide cut in it to a distance of 20 cm. from the top. The arm *a* is of harder wood, such as birch or beech, and is 3 cm. square and 7 cm. long. The screw *b* is one of the common brass thumb-screws sold by ironmongers; a hole is bored in the arm *a* to receive it, and its nut is inserted in a small mortise cut in the arm about 2 cm. from the end. The screw passes through a collar (for which a small brass or iron pulley answers well) on the other side of the upright, and by tightening it the arm may be secured at any height. At the other end of the arm a saw-cut is made in which are placed two strips of stout tin or brass plate about 12 × 3 cm., firmly secured by the two screws shewn in the figure. The outer ends of these strips are curved so as to grasp a tube or a retort, and their grip is tightened by slipping outwards the ring *c*. This latter is a common curtain ring 2.5 cm. in diameter carefully flattened into the shape of an elongated ellipse.

8, 9. Gas-burners. It will be best to obtain at starting a fish-tail burner attached to a firm base, such as is represented in Fig. 5, (p. 5), since a blow-pipe burner and a Bunsen's burner can be easily fitted to it.

A serviceable, though not ornamental, form of base is shown in Fig. 70. It consists of a block of wood, about 8 cm. square, on which is fixed, by means of two screws, a 'three-eighths angle burner' with fish-tail nipple, which may be procured from any gas-fitter. A stop-cock need not necessarily be fitted to it, since the stop-cock of the supply pipe will regulate the current of gas.

A **Blowpipe Burner**, of the form of Fig. 5 *a* (p. 5), may be made of a piece of brass tubing about 1 cm. in diameter and 5 cm. in length, having one extremity cut obliquely and flattened in a vice. The flattening should not be carried so far as to close the opening, but a narrow slit about 1 mm. in width should be left. This tube may then be fitted over the fish-tail nipple by an india-rubber connector.

---

[1] Such a triangle is also very useful for supporting a funnel, containing a filter, over a beaker. For this purpose a piece of thin board, about 6 × 12 cm., with a hole in the middle, will answer equally well.

A Bunsen's Burner may be made, as shewn in the figure, by placing over the fish-tail nipple a tube of glass or brass open at both ends, about 1.5 cm. in diameter and 12 or 14 cm. in length, secured in its right position by two wedges of cork or soft wood. Plenty of room should be left for air to enter at the lower end, and the tube should be raised or lowered a little, until a non-luminous flame is obtained. A tube of brass is, of course, the best, but a piece of 'composition' (lead and tin) gas tubing will answer. Instead of leaving air-space at the bottom of the tube, it may be fitted close round the burner, and two large notches may be filed, one on each side, at such a height that, when the tube is fixed in its place, the top of the fish-tail jet may be 1 mm. below the highest part of the notch.

A test-tube, the closed extremity of which has been cut off, makes a very useful burner; the only drawback being that the flame is slightly coloured by the sodium of the glass. This may be obviated by fitting over the top of the tube a short metallic ferrule, made by wrapping a strip of thin brass or tin plate, about 3 cm. broad, round a cylindrical rod a little smaller than the burner itself. The strip should be of such a length that its ends overlap each other, and the joint need not then be soldered. The elasticity of the metal will keep it in its place, when it is fitted over the end of the glass tube burner.

A still simpler form of burner may be made as follows. The gas-jet is formed by flattening one extremity of a short elbow tube, as in making the blowpipe burner, until a narrow slit remains. Over this is fitted a test-tube, as above described, by wedges of cork, and the elbow tube is passed through a notch cut in a large cork. This latter is fitted over the beak of an inverted funnel which serves as the base.

A Ring-burner, which will serve the purpose of an Argand burner, is represented in Fig. 71. It is nothing more than a flat, hollow, brass door or cupboard-handle, about 4 cm. in diameter, round which eight or ten small nicks have been cut with a file, through which the gas issues. The hole into which the axis of the latch fitted should be enlarged with a rat-tail file and with a broach (for which latter the tapering end of a screw-driver will serve), until the cylindrical part of the fish-tail nipple will just enter it.

Fig. 71.

12. **A Sand-bath** may be made by turning up the edges of a piece of tin plate about 16 cm. square, so as to form

a shallow box 2 cm. in depth. The shallow tin 'pastry-pans' sold by ironmongers answer very well.

A **Water Bath**, which is often convenient and sometimes necessary for evaporations at a gentle heat, may be made of a saucepan or a short tin canister, in the lid of which a hole is cut rather smaller in diameter than the evaporating dish. The bath should be about half filled with water, which must be kept gently boiling over a lamp: the dish containing the solution to be evaporated being placed in the opening of the lid, so as to be exposed to the heat of the steam.

A beaker will answer the purpose fairly, the dish being placed upon it, and slightly raised at one part by interposing a bit of wire or a splinter of wood, so as to allow free exit for the steam.

14. A Mouth Blowpipe, made of glass, is represented in Fig. 72. A good sound cork is selected, a hole is bored two-thirds of the way through it, and another hole is then bored in the side to meet

Fig. 72.

the first. Into this latter hole is fitted a short jet of hard glass tubing, and into the other hole is fitted a piece of glass tube about 20 cm. in length. A bulb may be blown on this latter tube near its lower end, to serve as a reservoir for condensing the moisture of the breath.

17. For a **Cork-borer** the tube of a steel pen-holder will answer pretty well, but it is hardly strong enough to last very long. Holes may be made in corks by means of the rat-tail file alone, the latter being worked into the cork like a gimlet. As, however, the file does not cut through, but pushes aside the particles of cork, there is some risk of splitting the cork.

21. A Deflagrating Cup may be made of the bowl of a tobacco-pipe, partially filled up with plaster of Paris (for the method of using this, see p. 348), in order to reduce its depth. A piece of copper wire should be twisted round the bowl, and the end bent up and passed through a large cork, or bung. Instead of a tobacco-pipe bowl, a small block of chalk or of plaster of Paris, hollowed into a cup, may be used.

22. A Deflagrating Jar, which answers every purpose, may be made by cutting off the bottom of a large wide-mouthed glass bottle. This is best done by leading a crack round it by means of a piece of heated glass (p. 42). The crack should be begun at the edge, the red-hot glass being pressed against it for a short time, and then a drop of water applied. There is generally some little difficulty in inducing the crack to follow the heated glass evenly, owing to the irregular thickness of the glass near the bottom of the bottle, and the crack should never be left to itself for any time, lest it should extend itself in a wrong direction. A bottle may often be found which is thinnest where it should be strongest, viz. at the angle formed by the sides and the bottom. In this case the bottom may generally be detached evenly by a sharp blow of an iron rod introduced through the neck of the bottle.

The edges of the glass should be roughened with a file, and it is advisable, though not necessary, to grind them level upon a flat paving stone with the assistance of a little fine sharp sand, or emery, and water. The jar should be rubbed on the stone with circular strokes, not backwards and forwards, or the edges are likely to chip away.

24. Very useful **Gas Jars** may also be made out of bottles; the necks being cut off in the manner above described. For most purposes, however, tall wide-mouthed bottles will answer without alteration. The small bottles, about 14×5 cm., in which quinine is sold, are particularly suitable for experiments; but gases should not be exploded in them.

27. Instead of the stoneware **Trays for Gas Jars,** common saucers may be used.

29. Florence Flasks will answer almost every purpose for which flasks are required. Their necks are rather narrow for a cork in which two tubes have to be fitted; the latter must be of small diameter, and the holes in the cork made with extra care. For preparing gases which are evolved without the necessity of applying heat, such as hydrogen and carbon dioxide, wide-mouthed bottles will answer quite as well as flasks.

36. Thistle Funnels. In place of these, ordinary funnels, attached to long pieces of glass tubing by india-rubber connectors, may be used. The connectors, however, will be slowly acted on if strong acids are brought in contact with them.

43. A simple form of **Spirit Lamp** is represented in Fig. 73.
A cork is fitted to a wide-mouthed bottle, and a bit of glass tubing
about 6 mm. in diameter and 8 cm. in length,
passed through a hole bored in the cork, serves
as the wick-holder. A small nick should be
cut in the side of the cork to admit air, otherwise the spirit will not rise readily. A short
wide test-tube is fitted over the cork as a cap,
to prevent evaporation of the spirit when the
lamp is not in use.

48 Flasks, and Retorts, the necks of
which are broken, may be converted into
evaporating dishes by leading a crack down
from the broken part to within 5 or 6 cm.
from the bottom, and then continuing it horizontally round the flask. A capsule about 5
cm. deep will thus be obtained, which will
serve as well as, or even better than, a porcelain dish for evaporations, but cannot be
used for ignitions. A clean iron spoon, not tinned, will often serve
for the latter purpose.

Fig 73.

50. A Drying Tube may be made out of a piece of glass tubing
about 1.5 cm. in diameter and 15 or 20 cm. in length: a test-tube,
if of stout glass, will answer the purpose, the closed end being cut
off. The ends should be fused in the lamp, and turned outwards
(see p. 44), for the sake of strength. They should be closed with
corks into which short pieces of glass tube should be fitted. If
a larger drying tube is required, a paraffin-lamp-glass will answer
well. If it is desired to use fragments of pumice soaked in hydrogen
sulphate in order to dry a gas, the above form of drying tube cannot be used, as the acid which drains away would soon destroy the
cork. A U-shaped tube must be obtained, or a tube such as is
shown in Fig. 74 may be made without much difficulty. A test-
tube of thick glass is selected, about 2 × 20 cm. A portion of the
glass about 3 cm. from the closed end is moderately heated all round,
and then a pointed blowpipe-flame is directed upon one spot. The
softened portion of the glass is forced inwards, by touching it with a
piece of iron wire, so as to form a projection in the interior, reaching
nearly to the centre. Two other similar projections are made by
heating and pushing inwards other portions of the circumference,
at the same distance from the closed end of the tube. Finally, the
tube is annealed with especial care, the projections and the adjacent

portions of the tube being heated again until the glass just begins to soften, and then cooled very slowly. A cork is now adapted to

the tube, and fitted with two elbow tubes of small diameter, the one having each of its branches about 6 or 8 cm. in length, the other having one branch about 6 cm. in length, and the other branch sufficiently long to reach, when the cork is fitted into its place, nearly to the bottom of the test-tube, passing between two of the projections.

In order to fill the tube, the cork should be loosened and raised just so far as to admit of its being turned half round and then rested on the rim of the tube. There will now be sufficient room left by the side of the cork to introduce, first four or five pieces of pumice rather larger than peas, which will rest upon the ledge formed by the projections, and then a quantity of smaller bits of pumice, until the tube is nearly full. Finally, a little concentrated hydrogen sulphate should be poured

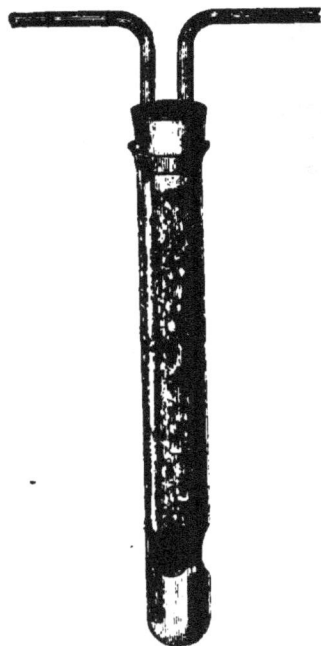

Fig. 74.

in through a funnel, and the cork turned round again and gently pressed into its place. The acid will, as it descends, moisten the pumice, and any excess will..collect at the bottom of the tube, and may be withdrawn by blowing through the short elbow tube, which will cause the acid to rise through the long tube.

One advantage of this form of drying tube is, that the pumice can be easily drenched with fresh acid, and the excess of acid withdrawn, without removing the pumice from the tube. Moreover, any number of these tubes can be arranged in an ordinary test-tube-stand, if required, or clustered together and held by india-rubber bands. A short wide-mouthed bottle forms a good support for a single tube, the latter being passed through a large hole in the cork.

69. Gas Blowpipes. The best form of portable blower is undoubtedly that lately brought out by Mr. Fletcher, of Warrington, which is represented in Fig. 75. In this, a piece of sheet india-

rubber is stretched over a round disc containing a valve, screwed
upon the top of the feeder; this sheet, when expanded to a hemisphere by blowing in air, forms the reservoir, and by its elasticity
supplies the requisite pressure.

A similar blower may be made out of a common pair of bellows;
the nozzle being stopped up, and two or three holes made in the
centre of the upper board and covered with a flap of leather, nailed
along one edge only, so as to act as a valve. Upon the upper board
of the bellows should be fitted (by screws and glue) a thick wooden
ring, having a groove cut in its rim as if for a pulley, but close to
the upper edge. Just below this groove a hole should be bored in
the direction of a radius of the ring, to form an outlet for the
compressed air; a bit of brass tube should be fitted tightly into
it, with which india-rubber tubing may be connected, as in the
figure. A circular piece of vulcanised sheet india-rubber[1] about

Fig. 75.

1 mm. thick should be laid on the wooden ring, and its edge
secured air-tight in the groove by folds of wire or string. It will
be advisable to place a net, or a calico bag, over the reservoir,
and secure it also in the groove, so as to prevent the india rubber
expanding beyond a hemisphere and bursting. The bellows should
be screwed to a base-board, one end being slightly tilted up by
fixing a piece of wood under it, so as to allow air to enter the
lower valve. A couple of $\subset$-shaped springs of thick brass or
iron wire should be fitted, one on each side of the bellows, in
order to keep the upper board raised except when pressed upon
by the foot.

---

[1] If there is any difficulty in procuring india-rubber sheet, a large bladder
may be cut in half, and one hemisphere of it may be secured by its edge in
the groove as above described. In order to obtain the pressure, a couple
of large india-rubber rings may be stretched across it, at right angles to one
another, between nails or hooks driven in on opposite sides of the wooden
ring: these will keep it flat until air is blown into it.

The blowpipe for use with this blower may be made out of a brass **T**-piece fitted with an internal glass jet, as described just below.

A very cheap and simple gas blowpipe is represented in Fig. 76.

Fig. 76.

The air reservoir is a bladder, round which a large and strong india-rubber ring is placed to give it elasticity[1]. . The air is supplied at intervals from the lungs through the tube *a*, at the other end of

[1] A bladder, when allowed to dry, loses most of its pliability. To keep it moist and fit for use, it should be rubbed over with some glycerine, diluted with an equal bulk of water (a few drops of carbolic acid being added, to prevent putrefaction), a little of the liquid being also poured into the interior.

Another method of preserving a bladder is to allow it to dry in a distended state, and then to rub over it a rag dipped in sweet oil, a little oil being also poured inside. If the bladder is now twisted and worked in all directions between the hands, it will become permanently pliable, and will not have the unpleasant smell of a moist bladder.

which is a valve to prevent the return of the air. This valve is represented in full size at *b*, and is thus constructed. A cork about 1.5 cm. in diameter is selected and a hole bored through it, into which the tube *a* is fitted so that its extremity may be just level with the surface of the cork. Across the end of the cork is laid a strip of oiled silk (oilskin) about 1 cm. or rather less in breadth, and its ends are turned down on each side of the cork and secured by an india-rubber ring, or folds of thread. The mouth of the bladder is closed by a cork about 2 cm. in diameter through which two holes are pierced: into one the tube *a* is fitted; into the other a short tube *c*, bent to an obtuse angle and terminating in a jet. Over this jet is fitted by a cork a common 'three-eighths T-piece' (which may be obtained from any gas-fitter); and to the side-branch of this latter is adapted a short piece of glass (or, better, brass) tubing, to be connected with the gas supply. The valve having been fitted to the tube *a*, the cork is placed in the neck of the bladder, and tightly secured by several folds of string; the whole may then be supported in a Bunsen's holder, and the gas turned on and lighted at the open end of the T-piece. On blowing into the tube *a* the bladder will become distended with air, which will be forced out through the jet by the elasticity of the india-rubber ring. A very good and fairly uniform brush-flame may be thus obtained, quite hot enough for most glass-blowing operations. Moreover, with a small air-jet, a low pressure of air, and a limited supply of gas, a well-defined pointed flame may be formed, which answers well for analytical blowpipe operations. The air-jet should not, in any case, be made larger than 1 mm. in diameter, otherwise the labour of supplying air from the lungs becomes considerable. An obvious expedient for avoiding this labour is, to connect the tube *a* by a long piece of india-rubber tubing with the nozzle of a common pair of bellows; the latter being fixed in a frame or box and placed on the floor. The upper board of the bellows may be kept raised by a spring, or by an india-rubber band, and pressed down by the foot.

The gas should not be left burning at the end of the T-piece longer than is required, since the latter will get very hot, and the cork may be loosened. Indeed it is decidedly better to have a brass jet screwed in; but this will add to the expense. The materials for such a blowpipe as is above described will not cost more than 1*s.* 6*d.*

74. A Retort and Receiver for distillations on a small scale are represented in Fig. 77. The retort is a very small flask, or a test-tube having its lower end expanded into a bulb, and connected by an india-rubber joint with a tube which has been bent to an acute angle and drawn out (but not very fine) just beyond the bend. The drawn-out portion passes through the cork nearly to the bottom of the receiver, which is also a test-tube. The cork should have several nicks in it, to afford escape for uncondensable vapours, and the receiver may be placed in an evaporating dish filled with cold water, or snow. Nitrogen tetroxide may be readily condensed in such an apparatus.

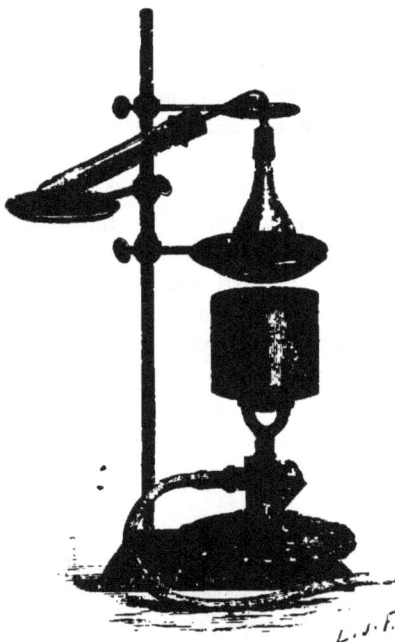

Fig. 77.

The best form of **condenser** is that known as 'Liebig's Condenser,' shewn in Fig. 78.

It consists of a long straight tube surrounded by a wider tube; in the space between the two a stream of cold water is kept constantly flowing.

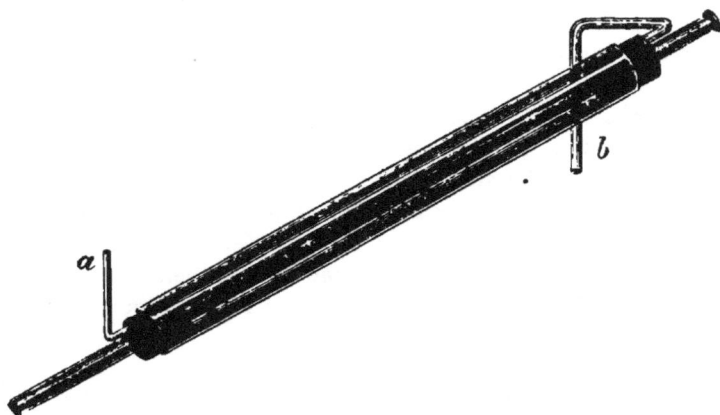

Fig. 78.

Such a condenser may be easily made out of a moderator lamp-

F f 2

glass, having corks fitted tightly 'to its ends. In the middle of each cork a hole should be bored, into which should be fitted a piece of glass tubing about 1 cm. in diameter, or rather less. One end of this should project 3 or 4 cm. beyond the cork, and the beak of the retort should be connected with it by a ferule of india-rubber tubing secured with string[1]. The other end of the tube may project 10 or 12 cm. beyond the cork, and enter the receiver.

Into the corks should also be fitted small tubes bent as shewn in the engraving. The end of the lower one, *a*, should be connected by an india-rubber tube, with pinch-cock, to a large jug or bottle of cold water, placed on a higher level (the bottle with siphon arrangement shewn in Fig. 81, will do very well). Under the end of the other tube, *b*, a jug should be placed, to catch the waste warm water, or the water may be led off at once into the sink.

78. A Screw Pinch-cock may be made by cutting out two strips of tin plate, about 6 × 1.5 cm., turning over a narrow portion

Fig. 79.

on each side, to give additional stiffness, boring a hole at each end (which may be done with a bradawl driven through by a hammer), and connecting the two strips by a couple of the small nuts and screws used for holding papers or letters together, which may be procured from a stationer's. Fig. 79 shows a pinch-cock thus made. Strips of hard wood, such as beech or box, may be used instead of the strips of tin.

Gas Apparatus. A convenient apparatus for obtaining a constant stream of a gas, such as hydrogen sulphide, which may be generated without the application of heat, is represented in Fig. 80. A paraffin-lamp-glass (one of those which bulge out near the middle) is obtained, and a cork is fitted to its lower extremity. In this cork five or six holes are bored, and before being finally fitted into its place it is thoroughly saturated with wax or paraffin, to

---

[1] If strong acids are to be distilled, it is better to make the beak of the retort enter the condensing tube a little way; the end of the latter being bordered and expanded by the method explained in p. 44. The beak of the retort may be drawn out slightly until, when a ferule of india-rubber tubing is fitted over it, it will just fit tightly into its place.

protect it from the action of acids[1]. Some lumps of iron sulphide are placed in the glass, and a cork with a short elbow tube is fitted into the upper end of the glass, a piece of india-rubber tubing, carrying a screw pinch-cock, being adapted to the outer end of the elbow tube. The glass is now lowered into a tall wide-mouthed bottle or jar, and the latter is filled nearly to the neck with common hydrogen chloride, previously diluted with twice its bulk of water, and allowed to cool. When the pinch-cock is opened, the acid will rise through the holes in the cork, and will act upon the iron sulphide, causing an evolution of hydrogen sulphide, which may be led into a flask or tube, as required. When the pinch-cock is closed, the gas collects in the glass, and finally drives the acid back into the outer bottle; the evolution of gas ceasing when the acid is withdrawn from contact with the iron sulphide. The lamp-glass should be retained in its place in the bottle by being passed through a bung fitted into the neck of the latter, and having one or two nicks cut in it to admit air. The same form of apparatus will, of course, serve for other gases, such as hydrogen or carbon dioxide.

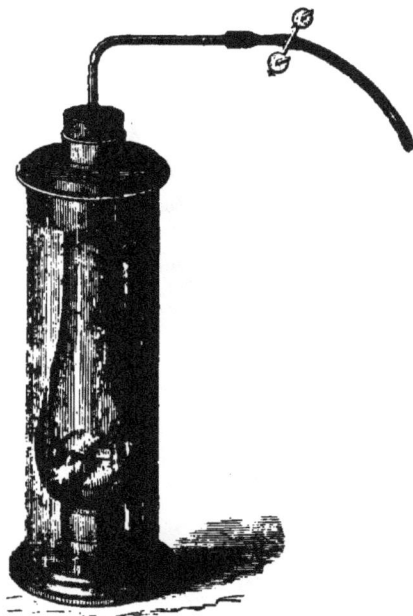

Fig. 80.

A simple form of **gas-holder**, which may be easily constructed and will be found very useful, is represented in Fig. 81. The bottles used are those which are known by the name of 'Winchester quarts,' and which may be procured from any chemist. A cork is adapted to one of the bottles and fitted with two elbow tubes, the one having both its branches about 8 or 10 cm. long, the other having one branch long enough to reach to the bottom

---

[1] This may be done by placing it for several minutes in a dish containing some paraffin or wax heated a little above its melting-point over a lamp.

of the bottle. An elbow tube similar to this latter is fitted to the other bottle by a cork which does not fit tightly into the neck.

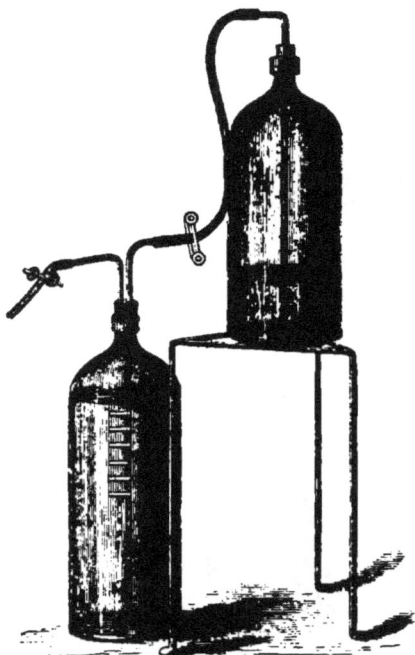

Fig. 81.

The two tubes last mentioned are connected by a piece of india-rubber tubing about 40 or 50 cm. in length, over which is slipped a screw pinch-cock. A piece of india-rubber tubing, also carrying a pinch-cock, is connected with the short elbow tube. All the connections must be made tight by folds of string or wire.

In order to use the apparatus, the first bottle is completely filled with water, the cork with the two elbow tubes being temporarily withdrawn. The short elbow tube is then connected with the delivery tube of the gas-generating flask, and gas is passed into the bottle, both pinch-cocks being opened. The gas displaces the water in the bottle, which passes over into the other bottle; the latter being so placed that there may never be more than a slight difference in level between the water in the two bottles, but that the gas may always pass in under some pressure. When the first bottle is full of gas, both pinch-cocks should be closed, and the generating flask should be at once disconnected from the apparatus. If, now, a constant stream of gas is required, e.g. for burning at a jet, the second bottle (now full of water) should be placed on a stand or shelf at such a height that its bottom may be a little above the level of the neck of the other bottle. If the pinch-cock on the tube connecting the two bottles is now opened, water will flow from the upper into the lower one, and will exert a pressure on the gas contained in the latter. It will then only be necessary to open the other pinch-cock cautiously, in order to obtain a regular stream of gas.

APPENDIX B.

I.

IT is intended that these solutions shall be made of such a strength that volumes which bear a simple relation to each other may contain weights of the different substances which correspond to the weights indicated by their chemical formulæ, i.e. to the weights of their molecules.

There are very obvious advantages in adopting such a system in place of the usual one of dissolving 1 part of the substance in 10 or 15 parts of water without strict reference to its molecular weight. It is clear that on the latter will depend the amount of chemical work which a given weight of the substance can do; and it would be wrong, or at least wasteful, to take for a solution the same weight of a sodium salt as of the corresponding potassium salt, when much less of the former than of the latter is required to effect a given amount of chemical change.

By carrying out the above principle the reactions between measured volumes of solutions become almost as simple as those between measured volumes of gases. For example, suppose that a little ammonium acetate is required for an experiment. The reaction which occurs between ammonium hydrate and hydrogen acetate is expressed by the following equation:—

$$(H_4 N) H O + C_2 H_4 O_2 = (H_4 N) C_2 H_3 O_2 + H_2 O.$$

From this we learn that one molecule of each substance is required for the combination; thus, 35 hydrogen-atom-weights of ammonium hydrate, and 60 hydrogen-atom-weights of hydrogen acetate, or, at any rate, weights bearing that proportion to each other must be taken; for instance, 35 centigrammes of ammonium hydrate, and 60 centigrammes of hydrogen acetate. If, then, the ordinary laboratory solutions of these substances are so made that 5 c.c. of them contain the above weights, the ammonium acetate can be made at once by simply mixing *equal volumes* of the solutions.

It has been found most convenient to take the centigramme as the practical unit of weight, corresponding to the hydrogen-atom-weight, which is, of course, a small but definite fraction of it; and the weights indicated by the formulæ printed in italics are to be understood to mean centigrammes, instead of hydrogen-atom-weights, as usual. Thus $(H_4N)_2CO_3$ expresses a weight of 96 centigrammes of ammonium carbonate, and '$(H_4N)_2CO_3$ in 5 c.c.' means that 5 c.c. of the solution contain 96 centigrammes of the salt.

[Reagents, even though procured, as they ought to be, from a trustworthy chemist, are liable to contain impurities; and before use they should be invariably tested with care, and rejected at once (at least for analytical purposes) if found impure[1].

The more usual impurities are mentioned in connection with each substance, together with simple tests for detecting them.

In the following directions, although 100 c.c. is mentioned in all cases, yet it will be well to prepare five or ten times this quantity of some solutions, such as dilute hydrogen sulphate, which are constantly required.]

1. Ammonium Carbonate ($(H_4N)_2CO_3$ in 5 c.c.). Dissolve 17 grms. of the pure white sublimed salt in 75 c.c. of distilled water, add 15 c.c. of solution of ammonium hydrate (sp. gr. 0.96), place the solution in a measure, and add more water until the whole measures 100 c.c.

The salt often contains ammonium chloride and sulphate, and also iron chloride. If either of the two latter are present the salt is unfit for use, and another sample should be obtained.

(*a*) Heat a small piece of it on perfectly clean platinum foil over a Bunsen's burner. It should volatilise entirely.

(*b*) Add to a portion of the solution in a test-tube a drop of ammonium sulphide. If the solution becomes dark, or if a black precipitate is formed, iron is probably present.

(*c*) Add to another portion of the solution pure dilute hydrogen nitrate, drop by drop, until no further effervescence occurs and a drop of the liquid placed on blue litmus-paper turns it decidedly red: then add a drop of solution of barium chloride. If a cloudiness is produced, a sulphate is probably present.

---

[1] In most cases, where the quantity purchased is small, it will probably be best to exchange it for a pure sample, and not to attempt to purify it yourself.

(*d*) Acidify another portion with hydrogen nitrate, as in *c*, and add a drop of solution of silver nitrate. If a cloudiness is produced, a chloride is probably present.

2. Ammonium Chloride (($H_4 N$) Cl in 5 c.c.). Dissolve 10.7 grms. of the pure salt in 80 c.c. of water, and dilute with water to 100 c.c.

The salt may contain ammonium sulphate and iron chloride. It should be tested in the same manner as the carbonate (with the omission of Experiment *d*), and rejected, except for making ammonia, if found impure. Both these impurities may be removed by recrystallisation.

3. Ammonium Hydrate (($H_4 N$) HO in 5 c.c.). Dilute 50 c.c. of solution of ammonia (sp. gr. 0.96, containing 7 grms. of ammonium hydrate in 100 c.c.) with water to 100 c.c.

If the concentrated solution (sp. gr. 0.88) is at hand, 20 c.c. of it may be diluted with water to 100 c.c.[1].

The solution should leave no residue on evaporation at a temperature not much above 100°, and should give little or no precipitate on addition of an equal volume of lime water.

4. Ammonium Molybdate (($H_4 N$) $H Mo O_4$ in 30 c.c.). Dissolve 6 grms. of the salt in 20 c.c. of solution of ammonium hydrate (sp. gr. 0.96) by a gentle heat; pour the solution slowly, with constant stirring, into 60 c.c. of dilute hydrogen nitrate (equal volumes of the strong acid and water): dilute it to 100 c.c. and allow it to stand for a day before filtering it, in order that any traces of ammonium phospho-molybdate may separate.

The solution should contain so large an excess of acid, that little or no precipitate (of molybdic acid) forms when it is heated to boiling.

5. Ammonium Oxalate (($H_4 N$)$_2$ $C_2 O_4$ in 50 c.c.). Dissolve 2.84 grms. of the salt in 60 c.c. of water, and dilute to 100 c.c.

6. Ammonium Sulphide (($H_4 N$)$_2$ S in 5 c.c.). The solution usually sold varies in strength, but may generally be diluted with three times its bulk of water for use.

It may be easily prepared by saturating 50 c.c. of solution of ammonium hydrate (sp. gr. 0.96) with hydrogen sulphide (p. 221), and then adding 50 c.c. of the same solution of ammonium hydrate.

---

[1] Since the strength of solution of ammonia is liable to vary, owing to the volatility of the substance, it will be best to see whether the solution is of approximately the right strength by putting 10 c.c. of it into a test-tube, adding a drop or two of solution of litmus, and seeing whether 5 c.c. of dilute hydrogen sulphate (20) are sufficient to neutralise it. It should be made rather too strong than too weak, since it loses strength on keeping.

The solution is colourless at first, but soon becomes yellow, owing to the formation of ammonium polysulphides. It still, however, remains fit for use in analysis for some time.

The solution is liable to contain a sulphate, and possibly a calcium salt. Test one portion with a drop of solution of barium chloride, and another portion with solution of ammonium oxalate. No precipitate should be produced in either case.

7. **Barium Chloride** ($Ba\ Cl_2$ in 50 c.c.). Dissolve 4.9 grms. of the crystallised salt ($Ba\ Cl_2$, H_2O) in 80 c.c. of water, and dilute to 100 c.c.

The purity of the salt may be tested as follows. Place about 5 c.c. of the solution in a test-tube, heat it nearly to boiling, and add a slight excess of dilute hydrogen sulphate. While the precipitate of barium sulphate is subsiding, get ready a filter (of Swedish paper), wash it two or three times with warm water, then filter off the barium sulphate and evaporate the filtrate to dryness on a clean watch-glass. No solid residue should be left.

· This test will not, however, serve to show whether lead is present, as it frequently is. Its presence may be suspected if the solution becomes dark in colour when mixed with some solution of hydrogen sulphide.

8. **Calcium Chloride** ($Ca\ Cl_2$ in 10 c.c.). Dissolve 11.1 grms. of the dry salt (or 22 grms. of the crystallised salt) in 80 c.c. of water, and dilute to 100 c.c.

This reagent is, however, required but seldom, and may be dispensed with.

The salt often contains iron perchloride, from which it may be purified by crystallisation from a small quantity of water. The solution should be quite neutral to test-paper, and should not become dark coloured, or give a precipitate when a drop of solution of ammonium sulphide is added.

9. **Calcium Hydrate** ($Ca\ H_2O_2$ in 540 c.c.). For the method of making this solution, see p. 58.

10. **Calcium Sulphate** ($Ca\ SO_4$ in 500 c.c.). This solution should be made in the same manner as that of calcium hydrate, about 5 grms. of the best plaster of Paris being shaken up with 200 c.c. of water, and allowed to stand for a day or two, with occasional shaking, before being filtered.

11. **Chlorine, solution of.** For preparing this, an apparatus similar to that used for preparing solution of hydrogen sulphide (p. 221) should be set up. The small wash-bottle is necessary in this case, to retain hydrogen chloride which may distil over: the

other bottles need not be so large as those used for hydrogen sulphide, as less of the chlorine water is likely to be wanted, and it does not keep well.

Prepare the gas as already directed (p. 185), and pass it into the water until the latter assumes a perceptible yellow colour. It will be better to stop short of the point of saturation, as a comparatively weak solution of chlorine is all that is required for analytical purposes. It should be kept in a stoppered bottle, protected carefully from the light, which causes its decomposition[1]. A good method of protecting it is, to paste over the bottle two or three thicknesses of brown paper, through which, when it is dry, two narrow slits should be cut on opposite sides of the bottle, in order that the quantity of solution contained in the bottle may be seen.

12. Cobalt Nitrate ($Co (N O_3)_2$ in 20 c.c.). Dissolve 7.25 grms. of the crystallised salt in 40 c.c. of water, and dilute to 50 c.c.

13. Copper Sulphate ($Cu S O_4$ in 25 c.c.). Dissolve 10.7 grms. of the crystallised salt ($Cu S O_4 5 H_2 O$) in 80 c.c. of water, and dilute to 100 c.c.

14. Hydrogen Acetate ($H C_2 H_3 O_2$ in 5 c.c.). Place 70 c.c. of the acid sold as "Acid. Acet. Fort." (sp. gr. = 1.04) in a measure, and dilute with water to 100 c.c. [The best and purest acid is the "Glacial Acetic Acid," which is solid at 10° C. (50° F.): if this is at hand, 18 c.c. of it may be put into a measure and diluted with water to 100 c.c.] It will be well to try whether 5 c.c. of this approximately neutralises 5 c.c. of solution of ammonium hydrate (No. 3). It often contains a chloride and a sulphate, but traces of these will not interfere with its use in analysis.

(*a*) Evaporate a portion of the acid on a watch-glass. It should leave no residue.

(*b*) To another portion add three or four drops of dilute hydrogen nitrate, and then a drop of solution of silver nitrate. If a cloudiness is produced, a chloride is present.

(*c*) To another portion add a drop of solution of barium chloride. If a cloudiness is produced, a sulphate is present.

15. Hydrogen Chloride ($H Cl$ in 1 c.c.). The acid of which the sp. gr. is 1.158 is very nearly of the right strength. It is liable to contain a sulphate, and also iron and arsenic, and must, for analytical purposes, be obtained free from these impurities.

---

[1] The bottles of deep yellow glass, used in photography, answer very well for this purpose.

(*a*) Dilute a portion with three or four times its volume of water, and test for a sulphate by solution of barium chloride.

(*b*) The presence of iron is generally indicated by the yellow colour of the acid. Add to a portion a drop of hydrogen nitrate, and boil the mixture for a few seconds; then add an equal volume of water, and allow it to cool. If the liquid turns red on addition of a drop of potassium sulphocyanate, iron is present.

(*c*) Arsenic must be tested for by Reinsch's or Marsh's method (see pp. 283, 289).

16. Hydrogen Chloride, dilute (*H Cl* in 5 c.c.). Place 20 c.c. of the strong acid in a measure, and dilute with water to 100 c.c.

17. Hydrogen Nitrate (*H N O*, in 1 c.c.). The acid of which the sp. gr. is 1.3 is very nearly of the right strength. It is liable to contain lower nitrogen oxides, and also a sulphate and a chloride.

(*a*) It should leave no residue when evaporated on a watch-glass.

(*b*) It should be colourless. If it is yellow, lower nitrogen oxides are present. These, however, are usually found in the acid when it has been exposed to light, and do not in general interfere with its use in analysis.

(*c*) When separate portions are diluted with water and tested with solution of barium chloride and silver nitrate, no cloudiness should be produced in either case.

18. Hydrogen Nitrate, dilute. (*H N O*, in 5 c.c.). Place 20 c.c. of the strong acid in a measure, and dilute it with water to 100 c.c.

19. Hydrogen Sulphate (*H, S O,* in 0.55 c.c.). The acid of which the sp. gr. is 1.845 is very nearly of the right strength. It may contain lead and arsenic, and also a nitrate, and must, for analysis, be obtained free from these impurities.

(*a*) Evaporate a few drops on a porcelain dish, taking care not to inhale the dense white fumes of the acid. No solid residue should be left.

(*b*) Dilute a few drops of the acid with 2 or 3 c.c. of water. If the mixture becomes turbid, lead is present.

(*c*) Place a little of the acid in a test-tube and pour, very slowly, down the side of the tube some solution of iron protosulphate, so that the lighter fluid may rest upon, not mix with, the heavy acid. If, after a time, a brown stratum appears at the junction of the two fluids, a nitrate is present.

(*d*) Arsenic should be tested for by Marsh's method (p. 289).

20. Hydrogen Sulphate, dilute (*H, SO,* in 5 c.c.). Measure out 60 c.c. of water into a beaker and add by degrees (see p. 132) 11 c.c of the strong acid, stirring the mixture continually with a glass rod. When the liquid is cool, pour it into the measure, and dilute to 100 c.c.

21. Hydrogen Sulphide, solution of. Directions for making this are given on p. 220.

22. Iron Perchloride ($Fe_2 Cl_6$ in 50 c.c.). Dissolve 2.25 grms. of fine iron-wire in about 30 c.c. of dilute hydrogen chloride, heating the mixture in a small flask. When all the iron is dissolved, pour the solution into a porcelain dish, and add strong hydrogen nitrate, a few drops at a time, as long as the addition of a drop produces a transient brown colour in the liquid. The excess of acid may then be driven off by evaporating it to a small bulk (not to complete dryness). Finally, dilute it with water to 100 c.c. and filter it.

23. Iron Protosulphate ($Fe S O_4$ in 50 c.c.) Dissolve 5.5 grms. of the pure green crystals in 60 c.c. of cold water, with addition of about 5 c.c. of dilute hydrogen sulphate, and dilute with water to 100 c.c. Owing to the great tendency of iron protosalts to absorb oxygen and pass into persalts, the solution cannot be preserved unaltered for any length of time, and will contain an increasing quantity of iron persulphate the longer it is kept. This tendency is lessened by the addition of acid as above recommended.

24. Lead Acetate ($Pb (C_2 H_3 O_2)_2$ in 50 c.c.). Dissolve 6.5 grms. of the salt in 80 c.c. of water, with the addition of one or two drops of hydrogen acetate, and dilute to 100 c.c.

25. Litmus, solution of. This may be bought ready made; but it is best to buy a little of the solid, and boil 2 or 3 grms. of it (previously powdered) with 100 c.c. of water, and filter the deep blue solution obtained. This may be kept for a long time unaltered, if two or three drops of carbon disulphide are added.

26. Magnesium Sulphate ($Mg S O_4$ in 50 c.c.). Dissolve 5 grms. of the salt in 80 c.c. of water, and dilute to 100 c.c.

27. Mercury Perchloride ($Hg Cl_2$ in 100 c.c.). Dissolve 3.2 grms. of the crystallised salt in 80 c.c. of water, and dilute to 100 c.c.

28. Mercury Protonitrate ($Hg_2 (NO_3)_2$ in 100 c.c.). Dissolve 5.6 grms. of the crystallised salt in 80 c.c. of water (adding a few drops of dilute hydrogen nitrate) and dilute to 100 c.c.

If the solid salt is not at hand, put 3 or 4 grms. of mercury into a test-tube, pour on it 2 c.c. of strong hydrogen nitrate diluted with 1 c.c. of water, and allow the action to go on for several hours: then pour the whole into a measure and dilute to 100 c.c., adding a few drops of hydrogen nitrate to prevent the formation of any basic salt. Keep the solution in a bottle containing 2 or 3 grms. of metallic mercury.

29. Platinum Perchloride ($Pt\,Cl_4$ in 50 c.c.). A comparatively small quantity of this substance will be required, and it will on the whole be best to purchase it in the state of solution (3.2 grms. of the salt dissolved in 50 c.c. of water). The method of obtaining it from platinum scraps and residues (which, however, often contain iridium, see note, p. 302) is given in the next section of this Appendix.

30. Potassium Chromate (K_2CrO_4 in 50 c.c.). Dissolve 4 grms. of the salt[1] in 80 c.c. of water, and dilute to 100 c.c.

31. Potassium Ferricyanide ($K_3Fe(CN)_6$ in 50 c.c.). Dissolve 6.6 grms. of the salt in 80 c.c. of water, and dilute to 100 c.c. The solution, especially if it has been kept for a month or two, is liable to contain traces of potassium ferrocyanide. It should give no blue precipitate, but only a brown colouration when tested with a drop of a solution of pure iron perchloride.

32. Potassium Ferrocyanide ($K_4Fe(CN)_6$ in 50 c.c.). Dissolve 8.4 grms. of the salt in 80 c.c. of water, and dilute to 100 c.c. Potassium ferricyanide is sometimes present as an impurity: it may be detected by adding to a portion of the solution a drop of solution of silver nitrate. If a reddish-brown and not a white precipitate is found, potassium ferricyanide is present. The salt may easily be purified by recrystallisation.

33. Potassium Hydrate (KHO in 5 c.c.). Dissolve 11 grms. of the substance (which is sold in sticks) in 80 c.c. of water, closing the mouth of the flask with a cork, to prevent absorption of carbon dioxide from the air[2]. When the solution is cool, pour it into the measure, and dilute to 100 c.c.; then transfer it without loss of time to a stoppered bottle, which should not be made of flint glass. It is difficult to obtain pure potassium hydrate, and its solution attacks the glass of the bottles in which it is kept, and takes up carbon dioxide from the air. A fresh solution should therefore be made from time to time; and for some experiments it will be best to dissolve a small lump of the substance in water and use it at once.

If any deposit forms in the bottle, it should be separated by decantation, not filtration.

---

[1] Instead of this, 6 grms. of potassium dichromate may be used.
[2] Be careful, however, that no considerable vacuum is formed in the flask, by the contraction of the air on cooling; otherwise the pressure of the outside air may force the bottom inwards.

The usual impurities are—a carbonate, an aluminate, a silicate, a chloride, and a sulphate, of which the last two are not important. Traces, at any rate, of the three first-mentioned impurities will be found in a solution which has been kept some time : the carbonate derived from the carbon dioxide in the air, the aluminate and silicate from the glass of the bottles.

(*a*) Add to a portion of the solution pure dilute hydrogen nitrate, until a drop of the liquid placed on blue litmus-paper colours it decidedly red. No effervescence indicative of the presence of a carbonate should be produced.

(*b*) Test one portion of the solution which you have acidified with hydrogen nitrate, for a sulphate by adding a drop of solution of barium chloride.

(*c*) Test another portion of the same solution for a chloride, by adding a drop of solution of silver nitrate.

(*d*) Acidify a portion of the solution of potassium hydrate with dilute hydrogen chloride, then add solution of ammonia until the liquid is alkaline to test-paper, and boil the solution. If a flocculent, semi-transparent precipitate is formed after a little time, an aluminate or silicate is present.

34. Potassium Iodide (KI in 50 c.c.). Dissolve 3.3 grms. of the salt in 80 c.c. of water, and dilute to 100 c.c.

The salt usually contains a little potassium iodate. It should be bought in the form of colourless, well-defined cubic crystals. The solution should not turn yellow at once when one or two drops of pure dilute hydrogen sulphate are added.

35. Potassium Sulphocyanate ($KCNS$ in 50 c.c.). Dissolve 2 grms. of the salt in 80 c.c. of water, and dilute to 100 c.c.

36. Silver Nitrate ($AgNO_3$ in 50 c.c.). Dissolve 3.4 grms. of the salt in 80 c.c. of water, and dilute to 100 c.c.

37. Sodium Carbonate (Na_2CO_3 in 10 c.c.). Dissolve 28.6 grms. of the crystallised salt (Na_2CO_3 , $10 H_2O$) in 80 c.c. of water, and dilute to 100 c.c. [Instead of this, 10 grms. of the anhydrous salt may be taken.]

The salt is liable to contain sodium sulphate and chloride, and should, for analytical purposes, be free from these impurities.

Add to a portion of the solution dilute hydrogen nitrate as long as an effervescence occurs, warm the liquid until the carbon dioxide is completely expelled, and divide it into two portions. Test one portion for a sulphate by adding a drop of solution of barium chloride, and the other portion for a chloride by adding a drop of solution of silver nitrate. No precipitate should be produced in either case.

38. Sodium-and-Hydrogen Phosphate ($Na_2 H P O_4$ in 50 c.c.). Dissolve 7.2 grms. of the salt in 50 c.c. of water, and dilute to 100 c.c.

The salt is liable to contain a chloride and a carbonate. These impurities should be tested for in the manner described under the head of Potassium Hydrate, p. 447.

39. Sodium-and-Hydrogen Tartrate ($Na H C_4 H_4 O_6$ in 50 c.c.). Dissolve 3.8 grms. of the salt in 80 c.c., and dilute to 100 c.c.

The solution does not keep for any length of time, a growth of fungus soon appearing in it. When a little of it does not give an almost immediate crystalline precipitate on addition of one drop of solution of potassium hydrate (see p. 174), it should be thrown away.

40. Starch, solution of. Heat 100 c.c. of water to boiling in a flask or beaker. Weigh out 1 grm. of pure white starch, and mix it thoroughly with 4 or 5 c.c. of cold water in a test-tube. When the water in the flask boils, take away the lamp, and, as soon as the boiling has ceased, pour the starch *little by little* into the water, shaking after each addition. The starch granules swell and burst, their transparent contents being scattered through the liquid, giving the appearance of having dissolved in it. Replace the lamp, and boil the liquid again for half a minute, stirring or shaking it all the time : then allow it to cool slightly and filter it[1], but do not use it for testing until it is quite cool.

The solution does not keep well, the starch being gradually converted into dextrine, which gives a brownish violet colour with free iodine, instead of the characteristic deep blue formed by starch.

The addition, however, of about one-twentieth of its volume of alcohol (or of a few drops of ammonia) seems to preserve it in good condition for at least six months.

41. Tin Protochloride ($Sn Cl_2$ in 50 c.c.). Boil 3 grms. of tin-foil with 10 c.c. of strong hydrogen chloride until the metal is *nearly* all dissolved; then dilute the solution with water to 50 c.c. and filter it into a bottle in which some pieces of granulated tin have been placed.

An excess of the metal should always be kept in the bottle to prevent the formation of any perchloride, and if the solution becomes turbid, 2 or 3 c.c. of strong hydrogen chloride should be added.

---

[1] This, although a rather tedious operation (since the pores of the filter soon get clogged) and not necessary for general work, results in a decided gain in the sensitiveness of the test.

II.

RECOVERY OF SILVER, GOLD, AND PLATINUM FROM RESIDUES.

1. Treatment of Silver Residues.

It has been already recommended (p. 249) that all waste solutions containing silver should be poured into a stock-bottle kept for the purpose, in order that when a sufficient quantity has accumulated the metal may be recovered from the mixture. The first step will be the conversion of any soluble silver salts which may be present into soluble silver chloride, by the addition of a considerable excess of common hydrogen chloride. This, if added in sufficient quantity, will even throw down any silver chloride which may have dissolved in sodium hyposulphite. The mixture, after being warmed in a flask and shaken, should be filtered, and the precipitate thoroughly washed, first with dilute acid and then with water, and finally dried by placing the filter containing it in a porcelain dish on the sand-bath. When it is thoroughly dry, it must be shaken out of the filter, and mixed in a mortar with twice its weight of dried sodium carbonate and one-tenth its weight of potassium nitrate, a little borax being also added to act as a flux. While the mixture is being made, a common Cornish clay crucible, about 12 or 13 cm. in height, should be gradually heated by being placed mouth downwards on the top of an ordinary fire, if there is no wind-furnace at hand. When it is thoroughly warmed, it may be turned up and surrounded with hot coals. When it becomes red-hot, the mixture should be thrown into it, little by little, with an iron spoon, the fire being urged by a pair of bellows, so as to keep the crucible at a full red heat. The mixture will at first effervesce considerably, and should be stirred with an iron rod to prevent its frothing over, but it will finally fuse into a clear fluid, under which the silver will collect in a button. When this takes place, the crucible must be kept at a bright yellow heat for about five minutes longer, the contents being occasionally stirred to hasten the aggregation of the small particles of reduced silver

G g

disseminated through the mass. ·Finally, the crucible should be lifted from the fire by a pair of tongs, and the fluid mass poured out on a piece of slate or iron plate. When cold the silver may be separated from the flux by breaking off the latter with a hammer, and boiling the ingot with water to separate the last traces of slag.

It is a good plan to re-melt the ingot of silver in a small clean crucible, sprinkling a little borax over it. When thoroughly melted the metal should be poured out as before on a piece of slate, and an approximate estimate of its purity may be formed by observing the degree of 'spirting' which takes place as the silver solidifies, owing to the escape of absorbed oxygen. Unless the metal be very pure and especially free from copper and lead, this spirting will hardly take place at all, and the surface of the ingot will remain unruffled.

The lump of silver thus obtained may be either exchanged at a shop for the crystallised nitrate of silver, or it may be boiled for some time in a flask with hydrogen nitrate diluted with twice its volume of water, more of the acid being added when the action becomes slow[1]. The solution should be evaporated to complete dryness, in order to drive off the excess of acid, the residue should be redissolved in water, and the solution, after filtration, should be evaporated down and left to crystallise, the basin being covered with a glass plate or a sheet of paper to exclude dust. The crystals should be drained in a funnel and dried on filter-paper. The mother liquor, on further evaporation, will yield another crop of crystals, and the liquid drained from these may either be poured into the 'silver residues' bottle, or reserved for purposes in which absolute purity is not required.

This is decidedly the best method of reducing silver residues, since not only the chloride, but also the sulphide, iodide, &c., are reduced by fusion with sodium carbonate. If, however, the chloride only has to be dealt with, it may be reduced by hydrogen in the manner described, p. 250; the only disadvantages being that it is rather difficult to wash out the last traces of zinc salt from the spongy mass of silver, and also that organic impurities are liable to be carried down with the chloride, which form nitro-compounds when the metal is dissolved in hydrogen nitrate, and render the salt unfit for such delicate processes as photography.

---

[1] If a perfectly pure salt is wanted, it will be necessary to precipitate the silver again as chloride from the hydrogen nitrate solution, and to reduce it with pure sodium carbonate in the manner above described.

2. Treatment of Platinum Residues.

These residues will usually contain the metal either as perchloride or as ammonium- or potassium-chloroplatinate. In the first place, add about one-tenth of the volume of strong solution of ammonium chloride (to separate any platinum which may be in the solution), and evaporate the whole to dryness. Heat the residue to redness in a porcelain dish or a clean clay crucible, to get rid of any organic impurities, and to decompose the chloroplatinates. The resulting porous mass, to which any waste scraps of platinum foil or wire may be added, should be thoroughly washed with warm water to remove all soluble salts, and then boiled for some time with a mixture of three parts of strong hydrogen chloride and one part of hydrogen nitrate, in a long-necked flask, having a test tube full of cold water placed in the neck, to prevent loss of acid by evaporation.

This will be best done in the open air, or at any rate the chlorous vapours must be led into a chimney or through a window, and not allowed to escape into the laboratory. When the mixture has boiled for half an hour, the solution should be poured off, fresh acids added, and the boiling repeated. The solutions should then be mixed, and the whole evaporated at a gentle heat nearly, but not quite to dryness, a little hydrogen chloride being added at the last, to decompose any remaining hydrogen nitrate. The solution of crude platinum perchloride, obtained by warming the residue with water, should be filtered and mixed with a saturated solution of ammonium chloride, which should be added until no further precipitate is thrown down. This precipitate, of ammonium chloroplatinate, should then be collected in a filter, and carefully washed with water to which about half its volume of common methylated spirit should be added, in order to prevent the loss of any considerable quantity of the salt, since ammonium chloroplatinate is not wholly insoluble in water. Lastly, it should be dried on the filter, and strongly heated in a porcelain dish, until the whole of the salt is decomposed (p. 303). The spongy platinum thus obtained should be dissolved in aqua regia, and evaporated down, at first over the lamp to a small bulk, and then on a water-bath [1]

[1] In default of a regular water-bath for the above evaporations, the porcelain dish may be placed on the mouth of a beaker or saucepan rather smaller in diameter; the latter being previously half filled with water, which is kept gently boiling on a sand-bath.

to complete dryness. The highly· deliquescent residue should be transferred quickly to a well-stoppered bottle, or, better, dissolved at once in water for use.

A solution of convenient strength is made by weighing the spongy platinum before dissolving it in the acids, multiplying the weight in grammes by 25, and making up the solution to this number of cubic centimetres. This will give a solution 50 c.c. of which contains 3.19 grms. of platinum perchloride ($Pt\,Cl_4$ in 50 c.c.).

3. Treatment of Gold Residues.

These may be worked up in the same way as platinum residues, except that the solution in aqua regia must, after evaporation, be mixed with excess of a strong solution of hydrogen oxalate, and boiled for some time. The gold will thus be precipitated in the metallic state, and the yellow spongy mass obtained will readily dissolve in aqua regia. The solution should be carefully evaporated on the water-bath, and the residue dissolved in water. For each gramme of gold taken, 50 c.c. of solution should be made. The liquid will then contain 3.03 grms. of gold perchloride in 100 c.c. ($Au\,Cl_3$ in 100 c.c.).

APPENDIX C.

SHORT COURSES OF ANALYSIS FOR BEGINNERS.

I.

ANALYSIS

of a solution containing a salt which may be any one of the following: NITRATE, CARBONATE, ACETATE, TARTRATE, OXALATE. (The other radicle being potassium, sodium, or hydrogen.)

1 Test the action of the solution on **litmus paper**. Take out a drop on a clean glass rod, and put it on a strip of **blue litmus paper**. If it does not change the colour of the litmus, put another drop on **reddened litmus paper**.

A *The liquid does not decidedly alter the colour of either paper.* It is a solution of a neutral salt. Pass on to **2**.

B *The liquid reddens blue litmus paper strongly* (if only slightly, it may be considered as practically neutral). Then it is probably a hydrogen salt, i.e. an acid. Pass on to **2**.

C *The liquid turns reddened litmus paper strongly blue.* Then it contains a CARBONATE.

To test for this, pour about 5 c.c. of the solution into a test-tube, and add several drops of **dilute hydrogen sulphate**. If a CARBONATE is present there will be an effervescence, owing to liberation of carbon dioxide.

To make sure that the gas is really carbon dioxide, dip the long branch of an elbow tube in **lime water** (calcium hydrate), and lower it into the test-tube containing the gas, but not so far as to reach the solution, then suck a very little air through the tube. If the lime-water *becomes cloudy* (seen by holding it up to the light) the gas is certainly carbon dioxide, and the solution contains a CARBONATE.

2   Put about 5 c.c. of the original solution into a test-tube, add 1 c.c. of **strong hydrogen sulphate**, then one drop of solution of **indigo sulphate**, and boil the mixture.   Observe if there is any change of colour.

A   *The blue colour of the indigo remains unaltered.*   Pass on to **3**.

B   *The blue colour of the indigo disappears.*   Then the salt is a NITRATE.

To confirm this inference, put about 2 or 3 c.c. of the solution originally given for analysis into a test-tube, add an equal volume of **strong hydrogen sulphate**, and cool the mixture by holding it in a stream of water.   When cold, hold the tube slanting, and pour gently down the side (from a pipette) some solution of **iron proto-sulphate**, so that it may float upon the liquid in the tube : and let the tube remain in the stand undisturbed, for a minute.   If a *brown layer is formed where the two liquids meet*, the solution certainly contains a NITRATE.

3   Put about 6 or 8 drops of the original solution into a test-tube, add about 5 c.c. of water, and enough of solution of **sodium carbonate** to make the liquid alkaline to test paper ; then add 2 or 3 drops of solution of **silver nitrate**, which will produce a white precipitate, and boil the mixture.   Observe whether the precipitate changes colour.

A   *It is unaltered.*   Pass on to **4**.

B   *It turns black.*

Then the solution contains a TARTRATE.

To confirm this, pour some of the original solution into a test-tube, add one drop (not more) of solution of **potassium hydrate**, stir the mixture with a glass rod, and shake it violently.   If a *crystalline precipitate forms*, especially along the lines where the glass rod touched the sides of the tube, a TARTRATE is certainly present.

4   Put a little **manganese dioxide** into a test-tube, add about 1 c.c. of **strong hydrogen sulphate**, and then 2 or 3 drops of the original solution.   Observe whether any *effervescence* occurs (best heard by holding the open end of the tube close to the ear).

A   *No gas is given off.*   Pass on to **5**.

B   *A gas is given off, with effervescence.*

The solution contains an OXALATE.

To confirm this, add to some of the original solution sufficient **ammonium hydrate** to make it alkaline to test paper, and then add 2 or 3 c.c. of solution of **calcium sulphate**.   If a *white precipitate is produced*, an OXALATE is certainly present.

5 If no other radicle has been found, the solution must contain an ACETATE. Test for this as follows,—

Put about 2 or 3 c.c. of the original solution into a test-tube, add about 1 c.c. of **pure alcohol** and 1 c.c. of **strong hydrogen sulphate**, and warm the liquid gently.

If *vapours are given off which have the fragrant smell of acetic ether*, an ACETATE is present.

II.

ANALYSIS

of a Salt which may contain any one of the following radicles:—
CARBONATE, OXALATE, TARTRATE, FLUORIDE, CYANIDE, CHLORIDE, BROMIDE, IODIDE, ACETATE, NITRATE, CHLORATE.

(The other radicle being potassium, sodium, ammonium, or hydrogen.)

A. *The substance is a Solid.*

1 Examine and note down its appearance, e.g. colour, shape, crystalline form (if any).

Then powder a little of it (about as much as two peas) in a glass mortar, and put the powder in a test-tube; add about 20 c.c. of water, and warm it until the salt is dissolved.

While it is dissolving, heat a little of the solid salt in an ignition tube over a Bunsen's burner; observe which of the following effects are produced:—

A *The substance fuses, but gives off no gas, and is not otherwise altered.*

Then it is probably a FLUORIDE, CYANIDE, CHLORIDE, BROMIDE, or IODIDE. Pass on to **3**.

B *The substance fuses, and gives off a colourless gas.*

Drop in a small splinter of **charcoal** (from the charred end of a match) and again heat. If *the charcoal burns brightly*, the gas is oxygen, and the substance is a CHLORATE or a NITRATE. Pass on to **6 B**.

C *The substance becomes charred, i.e. turns black and gives off vapours which have a strong smell like burnt paper.*

Then it is either an ACETATE or a TARTRATE. Pass on to **3**.

D   *The substance volatilises entirely.* .,

Then it is either HYDROGEN OXALATE or an ammonium salt. Pass
on to **3**.

B.   *The substance is a Liquid.*

2   See whether it has any colour or smell (a cyanide would smell
like essence of almonds). Then take out a drop on a clean glass
rod, and place it on a bit of **blue litmus paper.** If it does not
change the colour of the litmus, put another drop on **reddened
litmus paper.**

A   *The liquid does not alter (or only alters slightly) either litmus paper.*
Then it contains a neutral salt. Pass on to **3**.

B   *The substance reddens blue litmus paper strongly.*
Then it is probably a hydrogen salt, i.e. an acid. Pass on to **3**.

C   *The liquid turns reddened litmus paper strongly blue.*
Then it is a CARBONATE or a CYANIDE.

Pour about 5 c.c. of the solution into a test-tube, and add several
drops of **dilute hydrogen sulphate.**

   a. There is a strong effervescence.
   Then the salt is a CARBONATE.
   Dip the long branch of an elbow tube in **lime water** (calcium
   hydrate) and lower it into the test-tube, but not so far as to
   reach the solution, and suck a very little air through the tube.
   If *the lime water becomes cloudy*, the substance is certainly a CAR-
   BONATE, and no further tests are required.

   b. There is little or no effervescence. The salt is probably a CY-
   ANIDE. Pass on to **4**.

3   **Barium chlo-**   Put about 5 c.c. of the original solution into
   ride test.   a test-tube (if it is acid, make it neutral by adding
      ammonia) and add 1 or 2 drops of solution of
barium chloride.

A   *No precipitate is formed.* Pass on to **4**.

B   *A white precipitate is formed.*
The salt is a TARTRATE, OXALATE, or FLUORIDE.

Put about 6 or 8 drops of the original solution into a test-tube,
add about 5 c.c. of water and then enough solution of **sodium carbonate** to make it alkaline to test paper: then add 2 or 3 drops
· of solution of **silver nitrate** (which will produce a white pptc.)
·and boil the liquid.

   (a.) The precipitate turns black on boiling.
   The substance is a TARTRATE.

(*b.*) *No blackening of the precipitate is observed.*

The salt must be an OXALATE or a FLUORIDE.

To decide which is present, put a little **manganese dioxide** into a test-tube, add about 1 c. c. of **strong hydrogen sulphate**, and then 2 or 3 drops of the original solution.

(*a.*) *Carbon dioxide is given off with effervescence.*

(Prove that the gas is really carbon dioxide by the lime water test, **2 C** *a.*).

The salt is an OXALATE.

(*b.*) *No gas is given off.*

The salt must be a FLUORIDE.

Confirm this as follows:—Pour a little of the original solution into a leaden cup and evaporate it to dryness. (If the original substance was a solid, some of this may be taken.) While the evaporation is going on, cover a glass plate with wax, and trace some letters on it with a pointed piece of wood. Pour a few drops of **strong hydrogen sulphate** on the residue in the cup: place over it the glass plate with the wax downwards, and warm gently for 5 minutes, taking care not to melt the wax.

If the lines are etched in the glass the substance is a FLUORIDE.

4 **Silver nitrate test.** Put 3 or 4 drops of the solution into a test-tube, add 6 or 8 c. c. of water and 2 drops of **dilute hydrogen nitrate**, then add 2 drops of solution of **silver nitrate.**

A *No precipitate is formed.* Pass on to 6.

B *A white or light yellow precipitate is formed.*

The salt is an IODIDE, BROMIDE, CHLORIDE, or CYANIDE.

5 Add to a portion of the original solution one or two drops (not more) of **solution of chlorine.**

A *The liquid turns yellow.*

A BROMIDE or IODIDE is present. Pour off one half of the yellow liquid into another tube, and add some freshly made **solution of starch.**

(*a.*) *The solution turns deep blue.* An IODIDE is present (no confirmatory test is needed).

(*b.*) *The solution remains unaltered.* A BROMIDE is present. Confirm this by adding to the rest of the yellow liquid (containing chlorine) about 1 c. c. of **carbon disulphide,** and shaking it up. If the globule of carbon disulphide which settles to the bottom is *orange-coloured,* the presence of a BROMIDE is certain.

B *The liquid does not turn yellow.*

The substance present must be a CHLORIDE or a CYANIDE.

To distinguish between them, put a little **manganese dioxide** into a test-tube, add about 2 c.c. of **strong hydrogen sulphate**, and then 6 or 8 drops of the original solution and warm gently (but do not boil the liquid).

(a) *A greenish gas is given off which has the smell of chlorine, and bleaches a strip of moist litmus-paper held in the tube.* The substance is a CHLORIDE.

(b) *A gas is given off which has the smell of essence of almonds.* The substance is a CYANIDE.

To confirm this, add to some of the original solution, first 6 or 8 drops of solution of **potassium hydrate**, and then 2 or 3 drops of solution of **iron protosulphate**, when a greenish precipitate will form. Shake the whole thoroughly for a few seconds, and warm it gently; then add 2 or 3 c.c. of dilute hydrogen chloride. If *a deep blue precipitate remains undissolved*, the presence of a CYANIDE is certain.

6

Indigo sulphate test. Put about 5 c. c. of the solution into a test-tube, add about 1 c. c. of **strong hydrogen sulphate**, then one drop of solution of **indigo sulphate**, and boil the mixture.

A *The blue colour of the indigo disappears.*
The salt is a CHLORATE or a NITRATE.

Put 5 or 6 drops of the original solution (or of the solid substance) into a test-tube, and add about 3 c.c. of **strong hydrogen sulphate**.

(a) *The liquid turns yellow, and gives off a yellow gas, when warmed, which smells like a chlorine oxide.*
The substance is a CHLORATE.

(b) *The liquid does not turn yellow.*
The substance is a NITRATE.

Allow the liquid to become quite cold; then holding the tube slanting, pour gently down the side from a pipette, some solution of **iron protosulphate**, and let the liquid stand for a few minutes. If *a brown ring is formed between the two liquids* a NITRATE is present.

B *The blue colour of the liquid is unaltered.*
The salt is an ACETATE.

To confirm this, pour about 2 c. c. of the original solution into a test-tube, add about 1 c.c. of pure **alcohol**, and 1 c.c. of **strong hydrogen sulphate** and warm the mixture.

If *vapours are given off which have the fragrant smell of acetic ether*, an ACETATE is present.

APPENDIX D.

I.

LAWS OF CHEMICAL COMBINATION.

It was only at the end of the last century that chemists, by accurate experiments and constant use of the balance, were enabled to establish certain laws which express the proportions by weight in which substances combine with, and act on, each other.

They are the following,—

I. Law of Constant Proportion.

A particular compound always consists of the same elements united in the same proportion.

Thus water, whether obtained from wells, or the clouds, or the sea, or formed in the laboratory in the course of experiments, invariably consists of the two elements, oxygen and hydrogen, united in the proportions of 16 parts by weight of oxygen to 2 parts by weight of hydrogen. If we endeavour to form water by combining oxygen and hydrogen in any other proportions, we find that the excess of one or the other remains uncombined.

Similarly, ammonia, whether formed in nature, or obtained by decomposing wool, silk, coal, ammonium chloride, &c., invariably contains nitrogen and hydrogen, united in the proportions of 14 parts by weight of nitrogen to 3 of hydrogen.

Two results follow from this law.

1. A few really well-made experiments are sufficient to settle once for all the composition of a substance.

If a substance is brought to a chemist, which has all the properties of water, he can feel sure what its composition is without taking the trouble actually to analyse it.

2. Any substance which is not quite constant in composition, is certainly not a chemical compound, but only a mixture. For

example, one reason for considering air to be a mixture is that the proportion of oxygen in it is slightly variable.

II. Law of Multiple Proportion.

When one body combines with another in more than one proportion, the higher proportions are multiples of the lowest.

A good example of this law is afforded by the series of compounds of nitrogen and oxygen, of which a list is subjoined.

COMPOUNDS OF NITROGEN AND OXYGEN.

Name.	Composition by weight. Nitrogen. Oxygen.	Acid formed by its combination with water.
Nitrogen Pentoxide	28 : 80	Hydrogen Nitrate.
„ Tetroxide	28 : 64	Hydrogen Nitrate and Nitrite.
„ Trioxide	28 : 48	Hydrogen Nitrite.
„ Dioxide	28 : 32	(None).
„ Protoxide	28 : 16	(None).

If we examine what quantity of oxygen is combined with the same quantity of nitrogen, say 28 parts, we find that in the compound which contains least oxygen it is 16 parts: in the next, 32 parts, or twice 16; in the next, 48 parts, or three times sixteen, and so on.

Two results follow from this law.

1. We can predict the composition of compounds which have not yet been formed. Thus, if a nitrogen oxide is ever discovered which contains rather more oxygen than the pentoxide (nitrogen 28 : oxygen 80), we may say with certainty that it will be composed of 28 parts of nitrogen united with 96 parts of oxygen.

2. If any substance is found to contain nitrogen and oxygen not in the proportions of 28 to 16 or a multiple of 16, we can say with certainty that it is not a chemical compound but a mixture.

Thus air contains nitrogen and oxygen, but the proportion of nitrogen to oxygen is as 28 : 8.36, which is not a simple multiple of 16. This is another reason for considering air to be a mixture.

III. Law of Reciprocal Proportion.

If two bodies, A, and B, each combine with a third body, C, they can only combine with each other in proportions which are measures or multiples of the proportions in which they each combine with C.

Thus, nitrogen and hydrogen each combine with oxygen, in the proportions shewn below.

	Nitrogen.		Oxygen.
Nitrogen Protoxide	28	:	16

	Hydrogen.		
Water	2	:	16

Nitrogen and hydrogen also combine with each other in the proportions shown below

	Nitrogen.		Hydrogen.
Ammonia............	28	:	6

in which we notice that 6 is a multiple of 2.

IV. Law of Compound Proportion.

The proportion in which a compound unites with anything else is the sum or a multiple of the sum of the proportions in which its elements are present in it.

Thus, taking two compounds, ammonia and hydrogen nitrate, the proportions by weight in which their elements are present in them are the following :

Ammonia.		Hydrogen Nitrate.	
Hydrogen	3 parts	Hydrogen	1 part
Nitrogen	14 ,,	Nitrogen	14 ,,
		Oxygen	48 ,,
Sum =	17	Sum =	63

Now it is found that ammonia only unites with hydrogen nitrate in the proportion of 17 parts of ammonia to 63 of hydrogen nitrate. If any other proportions are taken, there is found an excess of one or the other remaining uncombined.

These laws are simply the expression of facts, proved by actual experiment ; but it was soon felt that there might be some simple underlying fact which would account for them all, and the hypothesis put forward by Dalton in 1801 has now been almost universally accepted as true and sufficient.

The Atomic Theory[1].

According to Dalton, all kinds of matter are made up of distinct particles, respecting which the following assertions can be made.

1. They are incapable of being divided.

2. The particles of the same substance are similar to one another, and equal in weight.

3. The particles of different substances differ in weight.

These small indivisible particles are called 'Atoms' (ἄτομος, that which cannot be divided).

Dalton further said that the 'combining proportions' referred to in the above Laws simply express the relative weights of these atoms.

---

[1] A fuller account of this is given in the next section, to which this should be considered as introductory.

Thus, the n° 16, by which we always *can* and generally *must* express the proportion in which oxygen combines with other things, is considered to be the weight of the atom of oxygen as compared with the weight of the atom of hydrogen, which is found to be the lightest of all atoms.

In other words, the oxygen-atom is regarded as 16 times as heavy as the hydrogen-atom. Similarly, the nitrogen-atom is believed to be 14 times as heavy as the hydrogen-atom, and so on for other elements; each having its own definite combining proportion, which indicates the weight of its atom.

If it be asked how the relative proportions by weight in which things act on each other, as ascertained by our comparatively rough balances, can be taken to indicate the relative weights of their single atoms, which are far too small for us to weigh, the reply would be, that it is reasonable to suppose that in the large masses of atoms which we deal with in ordinary practice, each atom of the mass is acting and being acted on similarly to the rest: just as in a regiment, when marching, every individual moves similarly to and simultaneously with every other. We can hardly imagine, for example, when we mix a quantity of oxygen weighing 16 grms. with a quantity of hydrogen weighing 2 grms. and make them combine, that some of the atoms in each are affected and not others. We have every ground for believing that any chemical action between masses represents accurately the action which is going on between the individual atoms of the masses.

Now, if we allow that these atoms exist, and that all chemical changes consist in the shifting of them, like pieces on a chess-board, from one position to another, or from one group to another, all the observed facts of chemical combination can readily be explained.

I. The 1st law would be a necessary consequence, because if the atoms of each kind of matter have definite unalterable properties, different from those of the atoms of other kinds, then, whenever we find two or more specimens of matter having absolutely the same properties, they must be composed of the same number of atoms of the same kinds, united in the same way.

II. The 2nd law follows because in forming compounds containing larger proportions of a given element we must add a whole atom at a time, (and not half an atom or one-and-a-half atoms, since the atom is not divisible.)

Thus if the smallest quantity of nitrogen protoxide contains a single atom of oxygen weighing 16 hydrogen-atoms, we must add another whole atom, also weighing 16 hydrogen-atoms, to make the dioxide. Thus nitrogen dioxide will contain twice the weight of oxygen contained in the protoxide. Similarly nitrogen trioxide will contain three times the weight, and so on.

III. The 3rd Law follows because the atom is unalterable in weight, whatever may be the compound it exists in ; and therefore when an element unites with any other substance it must do so in the proportion which the weight of its atom indicates or in some multiple of that proportion, if more than one atom of it combines.

Thus if the weights of the atoms of hydrogen, nitrogen and oxygen are 1, 14, and 16 respectively, all their compounds, nitrogen protoxide, water, ammonia, &c. &c. must contain them in these proportions, or multiples of these proportions.

IV. The 4th Law also follows from the unalterability in weight of atoms. The weight of the particle of a compound which takes part in a chemical change is as certainly the sum of the weights of the atoms which compose it, as the weight of a bag of shot is the sum of the weights of the individual shot in it.

CHEMICAL SYMBOLS.

Chemists have agreed to represent the atoms of substances by symbols ; the first letter of the name of the element being generally taken to express its atom.

Thus, the atom of Oxygen is denoted by O.
,,   Hydrogen   ,,   H.
,,   Nitrogen   ,,   N.
,,   Carbon   ,,   C.

These symbols therefore represent definite weights of the respective elements.

Thus,

H represents the unit of atomic weight, i.e. the weight of the hydrogen-atom, whatever that may be.
O represents a weight of oxygen = 16 hydrogen-atoms.
N   ,,   ,,   nitrogen = 14   ,,   ,,
C   ,,   ,,   carbon = 12   ,,   ,,

The group of atoms which form the smallest particle of a compound which can exist in a free state is called its 'molecule'; and we can express the molecule of a compound by simply putting together the symbols of the atoms which compose it, just as we form a word by putting letters together. This group of symbols is called a Formula.

Thus the molecule of water contains 1 atom of oxygen and 2 atoms of hydrogen, and may therefore be expressed by the formula HHO. When however several similar atoms are present the symbol is only written once, and a small n⁰ is put on the right of it and a little below, to shew how many atoms are present.

Thus the usual formula for the molecule of water is H_2O.

On the same principle the formulae of the molecules of some of the substances already examined should be written out from the following data as to their composition.

Substance.	Composition.			Formula.
Ammonia .... {	Hydrogen 3 atoms.	Nitrogen 1 atom.		
Hydrogen Nitrate ... {	Hydrogen 1 atom.	Nitrogen 1 atom.	Oxygen 3 atoms.	
Potassium Nitrate ... {	Potassium 1 atom.	Nitrogen 1 atom.	Oxygen 3 atoms.	
Ammonium Nitrate . {	Hydrogen 4 atoms	Nitrogen 2 atoms	Oxygen 3 atoms.	
Or (to show that it contains the AMMONIUM and NITRATE radicles) .... }	(Hydrogen 4 atoms	Nitrogen 1 atom)	(Nitrogen 1 atom	Oxygen 3 atoms).
	Nitrogen	Oxygen		
Nitrogen Pentoxide ...	2 atoms.	5 atoms.		
,, Tetroxide ...	2 atoms.	4 atoms.		
,, Trioxide ...	2 atoms.	3 atoms.		
,, Dioxide	2 atoms.	2 atoms.		
,, Protoxide ...	2 atoms.	1 atom.		

H. G. M.

II.

ON CHEMICAL SYMBOLS.

The aim of these Exercises has been to present some of the facts of chemistry without entering upon questions of chemical theory. But symbols having been used to express the principal chemical changes which form the subject of the exercises, and the nature and strength of the various reagent solutions, a statement of the meaning of such symbols is subjoined, together with a Table of Atomic Weights.

In the beginning of Part II is an explanation of the term 'single substance.' A chemical change consists in the conversion of any quantity of one or more single substances into an equal quantity, by weight, of one or more different single substances. The remembrance of some of the most important facts relating to these changes is facilitated by the adoption of the following hypothesis.

It may be the case that a portion, a gramme for instance, of any single substance is an aggregate of a vast but finite number of little particles, each of which has exactly the same properties as any other and as the whole mass. These little particles are, in almost all cases, themselves divisible; but when they are divided, it is not into smaller portions of the same substance, but into distinct parts which are called atoms, as being incapable of further division. These atoms unite into fresh groups which are the molecules of substances different from the original substance; and it is in this way, according to the hypothesis, that a chemical change takes place.

A molecule may consist of any number of atoms, from one up to a very large number. The molecules of the elements are supposed to consist either of single atoms or of two or more similar atoms united together. Differences in the properties of different substances may be due to the differences in the nature, or number, or arrangement, of the atoms of which their molecules consist.

The weight of a molecule is the sum of the weights of its atoms; and, since different atoms have very different weights, and different molecules consist of very different numbers of atoms, the weight of one molecule may be several hundred times as great as that of another.

Chemists have agreed to represent the various atoms by letters, that assigned to each being generally the initial letter of the name of that element whose molecule is made up of such atoms. Thus H represents the hydrogen-atom and O the oxygen-atom. But as the names of several elements have the same initial letter, the requisite variety has been obtained by taking the Latin instead of the English name, or by using two letters. Thus K represents the potassium-atom (Kalium), Co the cobalt-atom, Sb the antimony-atom (Stibium). The union of two or more atoms to form a molecule or group of atoms is represented by placing their symbols together, thus Co O represents a molecule of cobalt oxide, which consists of a cobalt-atom united to an oxygen-atom. When similar atoms are united, a numeral is affixed to the symbol of the atom, to represent its repetition so many times; thus, the molecule of water is expressed by H$_2$O instead of H H O.

It will be plain that, according to the hypothesis stated above, the atoms are not actual portions of any known substance except in the case of those substances whose molecules consist of single atoms, where, consequently, the atom is also a molecule. The smallest particle, or molecule, of common salt is represented by the formula Na Cl: its atoms Na and Cl can be transferred to other molecules, but cannot be obtained as distinct substances; we know them only as constituents of a number of different molecules. And in the same way the molecule of hydrogen is represented by H_2, the letter H representing, not a minute portion of hydrogen, but a constituent of hydrogen, a hydrogen-atom, a portion of matter weighing half as much as the smallest particle of hydrogen which is transferable in chemical changes from one molecule to another.

The determination of the relative weights of different atoms depends upon (1) the accurate quantitative analysis of a number of single substances, (2) a decision as to the constitution of the molecules of these substances. For example, analysis shows that nine weights of water can be decomposed into eight weights of oxygen and one weight of hydrogen. Choosing the weight of the hydrogen atom, as being the smallest relative weight we have to do with, as the unit of our system, if we decide that a molecule of water consists of one hydrogen-atom combined with one oxygen-atom, the weight of the oxygen-atom must be 8. If we decide that a molecule of water consists of one hydrogen-atom combined with two oxygen atoms, we must take 4 to be the atomic weight of oxygen; or if we have reason, as in fact we have, to regard the molecule of water as consisting of two hydrogen-atoms and one oxygen-atom, we must call the weight of the oxygen-atom 16.

The reasons which guide us in deciding what is the constitution of the molecule of a substance may be stated summarily under two heads: (1) that constitution is most probably true which affords the simplest account of the changes which the molecule is known to undergo, and which best exhibits the analogy between the substance in question and other substances which it resembles; (2) that constitution is most probably true which makes the molecular weight of the substance bear to the molecular weights of other substances the same ratio which its vapour-density bears to theirs. As this latter guide to the molecular constitution of substances is only available in the case of substances which can be volatilized without decomposition at a moderate heat, the weights of those atoms have been most definitely fixed which enter into the molecules of such substances.

It has also been observed that the relative weights of the atoms are in most cases nearly in inverse proportion to the specific heats of the elements which they form. This observation furnishes us with another criterion of the true atomic weight.

The following Table of Atomic Weights has been arrived at in the manner indicated.

TABLE OF ATOMIC WEIGHTS.

H	represents a	Hydrogen-atom weighing		1
Li	,,	Lithium-atom	,,	7
B	,,	Boron-atom	,,	11
C	,,	Carbon-atom	,,	12
N	,,	Nitrogen-atom	,,	14
O	,,	Oxygen-atom	,,	16
F	,,	Fluorine-atom	,,	19
Na	,,	Sodium-atom	,,	23
Mg	,,	Magnesium-atom	,,	24
Al	,,	Aluminium-atom	,,	27.5
Si	,,	Silicon-atom	,,	28
P	,,	Phosphorus-atom	,,	31
S	,,	Sulphur-atom	,,	32
Cl	,,	Chlorine-atom	,,	35.5
K	,,	Potassium-atom	,,	39
Ca	,,	Calcium-atom	,,	40
Ti	,,	Titanium-atom	,,	50
V	,,	Vanadium-atom	,,	51.2
Cr	,,	Chromium-atom	,,	52.5
Mn	,,	Manganese-atom	,,	55
Fe	,,	Iron-atom	,,	56
Co	,,	Cobalt-atom	,,	59
Ni	,,	Nickel-atom	,,	59
Cu	,,	Copper-atom	,,	63.5
Zn	,,	Zinc-atom	,,	65
As	,,	Arsenic-atom	,,	75
Se	,,	Selenium-atom	,,	79
Br	,,	Bromine-atom	,,	80
Sr	,,	Strontium-atom	,,	87.5
Mo	,,	Molybdenum-atom	,,	92
Pd	,,	Palladium-atom	,,	106.5
Ag	,,	Silver-atom	,,	108
Cd	,,	Cadmium-atom	,,	112
Sn	,,	Tin-atom	,,	118
U	,,	Uranium-atom	,,	120
Sb	,,	Antimony-atom	,,	122
I	,,	Iodine-atom	,,	127
Ba	,,	Barium-atom	,,	137
W	,,	Tungsten-atom	,,	184
Au	,,	Gold-atom	,,	196.7
Pt	,,	Platinum-atom	,,	197
Hg	,,	Mercury-atom	,,	200
Tl	,,	Thallium-atom	,,	204
Pb	,,	Lead-atom	,,	207
Bi	,,	Bismuth-atom	,,	210

This Table, which includes all the atoms which are not of very rare occurrence, enables us to calculate the weight of any molecule whose formula is known to us,, relatively to the weight of a hydrogen-atom which is assumed as the unit. Thus, the formula of calcium carbonate being Ca C O,, the weight of its molecule is $(40 + 12 + (3 \times 16) =)$ 100; that is to say, a molecule of calcium carbonate weighs a hundred times as much as a hydrogen-atom.

A Table of Molecular Weights, which might be extended to all substances whose constitution has been determined, would have the following form.

H, represents a molecule, or 2 parts by weight, or 1 volume of Hydrogen.

N,	„	28	„	„	Nitrogen.
Hg	„	200	„	„	Mercury.
P,	„	124	„	„	Phosphorus.
N H,	„	17	„	„	Ammonia.
C, H,	„	28	„	„	Ethylene.
		&c. &c.			

In the equations by which chemical changes are represented, the sign = has not simply its algebraic sense of 'equals,' but should rather be read 'is converted into.' That the substances formed in a chemical change weigh as much as the substances that went to form them, is a fact which renders the symbol appropriate, but is not all that it is understood to mean. The sign +, connecting the symbols of two molecules, denotes that these molecules disappear, or are formed, simultaneously; it may be expressed in reading by the word 'and.'

Thus the equation

$$K_2 S O_4 + Ba Cl_2 = Ba S O_4 + 2 K Cl$$

should be read thus, 'a molecule of potassium sulphate and a molecule of barium chloride are converted into a molecule of barium sulphate and two molecules of potassium chloride.' To obtain the relations by weight between the different substances we write down,

K₂ — 78	Ba — 137	Ba — 137	K — 39
S — 32	Cl₂ — 71	S — 32	Cl — 35.5
O₄ — 64		O₄ — 64	
K₂SO₄ — 174	Ba Cl₂ — 208	BaSO₄ — 233	K Cl — 74.5

whence we see that 174 weights of potassium sulphate and 208 weights of barium chloride are converted into 233 weights of barium sulphate and 149 weights of potassium chloride.

Since equal volumes of all gases contain, under the same conditions, the same number of molecules, equations representing changes in which gases take part may be read off at once as expressing changes of volume. The common volume occupied by an equal number of molecules of the different kinds of matter in the gaseous state should be called one volume.

Thus, $2 CO + O_2 = 2 CO_2$ may be read 'two volumes of carbon protoxide and one volume of oxygen are converted into two volumes of carbon dioxide.' If in comparing the molecular weights and volumes of different substances we take a gramme for unit instead of the weight of an atom of hydrogen, the common volume occupied by 28 grms. of carbon protoxide, 32 grms. of oxygen, 44 grms. of carbon dioxide, &c. is 22.3 litres.

A. V. H.

APPENDIX E.

I.

TABLES OF WEIGHTS AND MEASURES.

The system of weights and measures used in this book is that which is known as the metric system.

The unit of the system is the **metre**, the length of which is $\frac{1}{40,000,000}$th part of the earth's circumference, as determined in 1796 by Delambre and others.

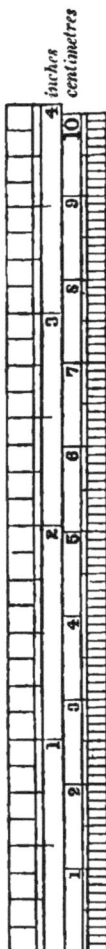

1. Length.

From it the larger and smaller measures of length are all derived on the following simple principle.—

(a) The smaller measures are obtained by taking $\frac{1}{10}$ of the length of the metre for the next smaller measure: $\frac{1}{100}$ of its length for the next smaller one; and $\frac{1}{1000}$ of its length for the smallest. Their names are formed by adding to the word "*metre*" a prefix derived from a **Latin** numeral (*decem, centum, mille*), denoting what fraction of the metre the measure is.

(b) The larger measures of length are obtained by taking 10 times the length of the metre for the next larger measure; 100 times its length for the next larger one; and 1,000 times its length for the largest. Their names are formed by adding to the word "*metre*" a prefix derived from a **Greek** numeral (δέκα, ἑκατὸν, χίλιοι) denoting what multiple of the metre the measure is.

A still larger measure, the myriametre (=10,000 metres) is sometimes, though rarely, used.

The following Table shews the metric measures of length, and their value in English measure.

Kilometre = 1000 metres = 1093.6 yards = 0.6214 of a mile.
Hectometre = 100 ,, = 3937.08 inches = 109.36 yards.
Decametre = 10 ,, = 393.71 ,, = 10.93 ,,
Metre = 1 ,, = 39.37 ,, = 1.09 ,,
Decimetre = $\frac{1}{10}$,, = 3.94 ,,
Centimetre = $\frac{1}{100}$,, = 0.39 ,,
Millimetre = $\frac{1}{1000}$,, = 0.039 ,,

It may be convenient to remember that a decimetre is very nearly four inches, and Fig. 82 shows a decimetre and its subdivisions, compared with a scale of English inches.

2. Volume.

The unit of volume is a cube, each of the sides of which measures one decimetre. It is called a litre, and from it the other measures of volume are derived by taking $\frac{1}{10}$, $\frac{1}{100}$, and $\frac{1}{1000}$ of its size for the smaller ones, and 10 times, 100 times, and 1,000 times its size for the larger ones, precisely as those of length are derived from the metre. Their names also are formed on the same principle by adding to the word "*litre*" prefixes derived from the Latin for the smaller measures, and prefixes derived from the Greek for the larger ones.

Kilolitre = 1000 litres = 220.01 gallons.
Hectolitre = 100 ,, = 22.00 ,,
Decalitre = 10 ,, = 2.20 ,,
Litre = 1 ,, = 1.76 pints.
Decilitre = 100 c.c. = $\frac{1}{10}$,, = 3.5 fluid oz.
Centilitre = 10 c.c. = $\frac{1}{100}$,, = 0.35 ,,
Millilitre = 1 c.c. = $\frac{1}{1000}$,, = 0.035 ,,
(*or* cubic centimetre)

In scientific work, quantities smaller than the litre are almost always expressed in cubic centimetres instead of decilitres, &c. Thus, half a litre would be expressed (not as 5 decilitres but) as 500 c.c.

3. Weight.

The unit of weight is the weight of 1 cubic centimetre of water at the temperature of 4° Centigrade. It is called a gramme, and the other weights are subdivisions and multiples of it, derived in the same way and named on the same principle as above explained.

Kilogramme = 1000 grammes = 2.205 lbs. (avoird.)
Hectogramme = 100 ,, = 3.527 oz. ,,
Decagramme = 10 ,, = 0.35 oz. ,,
Gramme = 1 ,, = 15.43 grains.
Decigramme = $\frac{1}{10}$,, = 1.54 ,,
Centigramme = $\frac{1}{100}$,, = 0.154 ,,
Milligramme = $\frac{1}{1000}$,, = 0.015 ,,

A series of measures of surface ‚or area is sometimes used, of which the unit is the "are," which is a surface 1 decametre square. The hectare is very nearly 2½ acres.

Rules for Reduction.

1. To reduce the smaller measures to metres (litres, or grammes), and metres (litres, or grammes) to the larger measures,—

Divide by the no. expressed in the name of the measure.

2. To reduce the larger measures to metres (litres or grammes) and metres (litres, or grammes) to the smaller measures,—

Multiply by the no. expressed in the name of the measure.

Examples :

Reduce 1881 *milli*metres to metres.

„ 1881 ÷ 1000 = 1·881 metre.

Reduce 24 *deca*grammes to grammes and *centi*grammes—

24 × 10 = 240 grammes.

240 × 100 = 24,000 centigrammes.

Decimal fractions of the measures may always be read off into their equivalents without altering the figures, by simply attending to the value of each figure as fixed by the position of the decimal point. Thus

1·881 metre = 1 metre, $\frac{8}{10}$ metre, $\frac{8}{100}$ metre, $\frac{1}{1000}$ metre.

= 1 metre, 8 decimetres, 8 centimetres, 1 millimetre.

TABLES OF ENGLISH MEASURES.

1. Length.

1	inch		=	2.54 centimetres.
12	inches	= 1 foot	=	30.48 „
3	feet	= 1 yard	=	91.44 „
5½	yards	= 1 pole	=	5.03 metres.
4	poles	= 1 chain	=	20.12 „
40	poles	= 1 furlong	=	201.16 „
8	furlongs	= 1 mile	=	1609.3 „

2. Volume.

1 cubic inch	=	16.38 cubic centimetres.
1728 cubic inches	= 1 cubic foot	= 28.31 litres.
61.027 „	= 1 litre.	

```
1 drachm   = 0.216 cubic inch =    3.55 cubic centimetres.
8 drachms  = 1 ounce          =   28.4    „         „
20 ounces  = 1 pint           =  567.9    „         „
2 pints    = 1 quart          =    1.136 litres.
4 quarts   = 1 gallon         =    4.543  „
                              =  277.274 cubic inches.
```

1 gallon = 70,000 grains (or 10 lbs. avoird.) of distilled water
at the temperature of 60° F (15.5° C).

3. Weight.

(a) *Apothecaries.*

```
1 grain    =                        0.0648 gramme.
20 grains  = 1 scruple =    1.296      „
3 scruples = 1 drachm  =    3.888      „
8 drachms  = 1 ounce   =   31.104      „
12 ounces  = 1 pound   =  373.248      „
```

(b) *Avoirdupois.*

```
1    grain   =                       0.0648 gramme.
27.34 grains = 1 drachm =    1.772      „
16   drachms = 1 ounce  =   28.35       „
16   ounces  = 1 pound  =  453.6        „
112  pounds  = 1 cwt.   =   50.8    kilogrammes.
```

II.

THERMOMETRIC SCALES.

The temperatures mentioned in this book are all expressed on the Centigrade scale.

According to this scale, the space through which the column of mercury in a thermometer moves, when passing from the temperature at which water freezes to the temperature at which water boils, is divided into 100 equal parts or degrees, and the scale is extended below the freezing-point and above the boiling-point of water, in divisions of the same value.

The point at which the mercury column stands when the thermometer is immersed in melting ice is marked 0°, and the degrees below this point are distinguished by a *minus* sign, thus − 1, − 2, &c.

The point at which the mercury column stands when the thermometer is immersed in the steam from boiling water (the barometric column being 760 mm.) is marked 100°.

In the Fahrenheit scale, which is not yet superseded in this country, the space between the freezing-point and the boiling-point of water is divided into 180 degrees, and, moreover, the freezing-point of water is marked 32° instead of 0°, the boiling-point of water being consequently marked 212° instead of 180°.

The points to be borne in mind, then, in converting temperatures expressed on the one scale to the corresponding temperatures on the other scale, are—

1. The same space is divided on the Centigrade scale into 100 parts, on the Fahrenheit scale into 180 parts.

Consequently, 1 degree on the Centigrade scale is equal in length to $(\frac{180}{100}=)$ $\frac{9}{5}$ of a degree on the Fahrenheit scale; and 1 degree on the Fahrenheit scale is equal in length to $(\frac{100}{180}=)$ $\frac{5}{9}$ of a degree on the Centigrade scale.

$$\text{For since} \quad 100°\,C = 180°\,F$$
$$100°\,C : 180°\,F :: 1°\,C : \tfrac{9}{5}°\,F$$
$$\text{and} \quad 180°\,F : 100°\,C :: 1°\,F : \tfrac{5}{9}°\,C.$$

2. The point marked 0° on the Centigrade scale is marked 32° on the Fahrenheit scale.

Consequently, in order to bring the two scales to the same level, we must subtract 32 from the number of Fahrenheit degrees before converting them into Centigrade degrees; and, in the reverse process, after converting the Centigrade degrees into Fahrenheit degrees, we must add 32.

If 0° C expressed the same temperature as 0° F, then, clearly—

$$1°\,C \text{ would } = 1.8°\,F.$$
$$5°\,C \quad \text{,,} \quad 9°\,F.$$
$$10°\,C \quad \text{,,} \quad 18°\,F.$$
$$\&c. \qquad \&c.$$

But 0° C marks a point of temperature $= 32°$ F.

Hence all the above Nos. Fahr. have to be increased by 32 (lifted, as it were, in the row), so that

$$1°\,C = (\tfrac{9}{5} + 32 =) \; 33.8°\,F.$$
$$5°\,C = (\; 9 + 32 =) \; 41°\,F.$$
$$10°\,C = (18 + 32 =) \; 50°\,F.$$

Similarly in reducing Fahr. to Cent. we must take away 32 from the no. of degrees (thus practically lowering it in the scale) *before* apply-ing the rule. Thus

$$59^\circ\,F - 32 = 27^\circ \;\|\; 27^\circ \times \tfrac{5}{9} = 15^\circ C.$$

The rules, therefore, may be thus expressed :—

1. To convert **Fahrenheit** degrees into **Centigrade** degrees,

Subtract 32 from the number of degrees, and multiply the re-mainder by $\tfrac{5}{9}$ (or 0.5)[1].

2. To convert **Centigrade** degrees into **Fahrenheit** degrees,

Multiply the number of degrees by $\tfrac{9}{5}$ (or 1.8) and add 32 to the product[2].

In applying the rule to degrees low in the scale, the character of the sign + or − must be carefully attended to, and considered in its algebraic sense. Thus,—

$$23^\circ\,F - 32 = -\,9^\circ \;\|\; -\,9^\circ \times \tfrac{5}{9} = -\,5^\circ C.$$
$$-40^\circ\,F - 32 = -\,72^\circ \;\|\; -\,72^\circ \times \tfrac{5}{9} = -\,40^\circ C.$$

III.

TABLE OF THE SOLUBILITY OF SALTS.

(See p. 481.)

This Table is intended to indicate the solubility in water and acids of the single salts formed by the union of the different radicles treated of in the foregoing pages, and has been compiled mainly upon the authority of Störer's ' Dictionary of Solubilities.'

The degrees of solubility are, however, so numerous that it is difficult or impossible to draw a sharp line between a soluble and an insoluble substance. The statements in the Table must there-fore be only regarded as approximations; as relatively rather than absolutely true.

It is hardly necessary to remark that in cases where a sub-stance is said to dissolve in an acid, the phenomenon usually consists in the formation, by the action of the acid, of a salt soluble in water.

---

[1] $n^\circ\,F = (n-32)\,\tfrac{5}{9}^\circ\,C.$
[2] $n^\circ\,C = (\tfrac{9}{5}n + 32)^\circ\,F.$

This rule may also stand as follows :—

Double the number of degrees, subtract $\tfrac{1}{10}$ of this product, and add 32 to the remainder.

$$\text{Thus,}\quad 25^\circ C \times 2 = 50$$
$$50 - 5 = 45$$
$$45 + 32 = 77^\circ\,F.$$

INDEX.

www.ingramcontent.com/pod-product-compliance
Lightning Source LLC
Chambersburg PA
CBHW020858210326
41598CB00018B/1708